Digital Tools and Solutions for Inquiry-Based STEM Learning

Ilya Levin
Tel Aviv University, Israel

Dina Tsybulsky
Tel Aviv University, Israel

A volume in the Advances in Educational Technologies and Instructional Design (AETID) Book Series

IGI GLOBAL
DISSEMINATOR OF KNOWLEDGE

www.igi-global.com

Published in the United States of America by
IGI Global
Information Science Reference (an imprint of IGI Global)
701 E. Chocolate Avenue
Hershey PA, USA 17033
Tel: 717-533-8845
Fax: 717-533-8661
E-mail: cust@igi-global.com
Web site: http://www.igi-global.com

Library of Congress Cataloging-in-Publication Data

Names: Levin, Ilya, 1954- editor. | Tsybulsky, Dina, editor.
Title: Digital tools and solutions for inquiry-based STEM learning / Ilya Levin and Dina Tsybulsky, editors.
Description: Hershey, PA : Information Science Reference, [2017]
Identifiers: LCCN 2017005137| ISBN 9781522525257 (h/c) | ISBN 9781522525264 (eISBN)
Subjects: LCSH: Science--Study and teaching--Technological innovations. | Science--Study and teaching--Computer-assisted instruction. | Educational technology.
Classification: LCC Q197 .D54 2017 | DDC 507.8--dc23 LC record available at https://lccn.loc.gov/2017005137

This book is published in the IGI Global book series Advances in Educational Technologies and Instructional Design (AETID) (ISSN: 2326-8905; eISSN: 2326-8913)

British Cataloguing in Publication Data
A Cataloguing in Publication record for this book is available from the British Library.

All work contributed to this book is new, previously-unpublished material.
The views expressed in this book are those of the authors, but not necessarily of the publisher.

For electronic access to this publication, please contact: eresources@igi-global.com.

Advances in Educational Technologies and Instructional Design (AETID) Book Series

ISSN:2326-8905
EISSN:2326-8913

Editor-in-Chief: Lawrence A. Tomei, Robert Morris University, USA

MISSION

Education has undergone, and continues to undergo, immense changes in the way it is enacted and distributed to both child and adult learners. From distance education, Massive-Open-Online-Courses (MOOCs), and electronic tablets in the classroom, technology is now an integral part of the educational experience and is also affecting the way educators communicate information to students.

The **Advances in Educational Technologies & Instructional Design (AETID) Book Series** is a resource where researchers, students, administrators, and educators alike can find the most updated research and theories regarding technology's integration within education and its effect on teaching as a practice.

COVERAGE

- Curriculum Development
- Adaptive Learning
- Game-Based Learning
- E-Learning
- Virtual School Environments
- Bring-Your-Own-Device
- Online Media in Classrooms
- Hybrid Learning
- Higher Education Technologies
- Classroom Response Systems

IGI Global is currently accepting manuscripts for publication within this series. To submit a proposal for a volume in this series, please contact our Acquisition Editors at Acquisitions@igi-global.com or visit: http://www.igi-global.com/publish/.

The Advances in Educational Technologies and Instructional Design (AETID) Book Series (ISSN 2326-8905) is published by IGI Global, 701 E. Chocolate Avenue, Hershey, PA 17033-1240, USA, www.igi-global.com. This series is composed of titles available for purchase individually; each title is edited to be contextually exclusive from any other title within the series. For pricing and ordering information please visit http://www.igi-global.com/book-series/advances-educational-technologies-instructional-design/73678. Postmaster: Send all address changes to above address.

Titles in this Series

For a list of additional titles in this series, please visit:
http://www.igi-global.com/book-series/advances-educational-technologies-instructional-design/73678

Exploration of Textual Interactions in CALL Learning Communities ...
Jonathan R. White (Högskolan Dalarna, Sweden)
Information Science Reference • ©2017 • 195pp • H/C (ISBN: 9781522521426) • US $120.00

Mobile Technologies and Augmented Reality in Open Education
Gulsun Kurubacak (Anadolu University, Turkey) and Hakan Altinpulluk (Anadolu University, Turkey)
Information Science Reference • ©2017 • 366pp • H/C (ISBN: 9781522521105) • US $185.00

Cases on STEAM Education in Practice
Judith Bazler (Monmouth University, USA) and Meta Van Sickle (College of Charleston, USA)
Information Science Reference • ©2017 • 375pp • H/C (ISBN: 9781522523345) • US $195.00

Handbook of Research on Driving STEM Learning With Educational Technologies
María-Soledad Ramírez-Montoya (Tecnologico de Monterrey, Mexico)
Information Science Reference • ©2017 • 590pp • H/C (ISBN: 9781522520269) • US $300.00

Integrating an Awareness of Selfhood and Society into Virtual Learning
Andrew Stricker (The Air University, USA) Cynthia Calongne (Colorado Technical University, USA) Barbara Truman (University of Central Florida, USA) and Fil Arenas (The Air University, USA)
Information Science Reference • ©2017 • 383pp • H/C (ISBN: 9781522521822) • US $200.00

Intercultural Responsiveness in the Second Language Learning Classroom
Kathryn Jones (Lamar University, USA) and Jason R. Mixon (Lamar University, USA)
Information Science Reference • ©2017 • 327pp • H/C (ISBN: 9781522520696) • US $175.00

Handbook of Research on Transformative Digital Content and Learning Technologies
Jared Keengwe (University of North Dakota, USA) and Prince Hycy Bull (North Carolina Central University, USA)
Information Science Reference • ©2017 • 453pp • H/C (ISBN: 9781522520009) • US $230.00

For an enitre list of titles in this series, please visit:
http://www.igi-global.com/book-series/advances-educational-technologies-instructional-design/73678

IGI GLOBAL
DISSEMINATOR OF KNOWLEDGE

www.igi-global.com

701 East Chocolate Avenue, Hershey, PA 17033, USA
Tel: 717-533-8845 x100 • Fax: 717-533-8661
E-Mail: cust@igi-global.com • www.igi-global.com

Editorial Advisory Board

Table of Contents

Detailed Table of Contents

The purpose of this chapter is to explore some conceptual and methodological issues related to computer-based simulation as a teaching and learning method for STEM education. Two major themes will be examined in detail: 1) the barriers to the penetration of simulation into school programs; 2) simulation as a way to enhance scientific imagination in students. It is argued that scientific imagination is linked closely with mental simulation, a fundamental capacity of the human brain which allows us to move from static to dynamic mental representations. The role of mental simulation in understanding scientific concepts is discussed and an explicit statement is provided of the relation between computer simulation and mental simulation. On these grounds, computer simulation is viewed as a tool for extending human cognition by overcoming the limits of mental simulation. Finally, the implications of these findings for designing simulation-based instructional units and conducting lessons are discussed.

This chapter explores use of Design-based learning (DBL) and digital tools to facilitate collaborative learning through design-based projects. Design-based learning (DBL) is an educational approach that incorporates hands-own, authentic, multidisciplinary design tasks to identify problems and design solutions. With DBL, students typically work in teams and are tasked designing solutions to open-ended problems. Teams develop conceptual solutions to problems and then work through the design process to arrive at the creation of an actual artifact. This artifact may be fully functional or simply a model, prototype, or other representation of the complete system. STEM instructors and students should give careful attention to selecting the digital tools for collaboration. Some collaborative tools offer affordances and features that compliment the communication processes in one phase of the design process while another other tool may be better suited for the tasks specific to another phase.

The use of social media in and outside the classroom is increasing in the number of popular applications as well as pervasiveness in our culture. Teachers utilize social media to engage students, connect with experts, and expand their own professional learning. This chapter provides educators with information about the use of social media to support STEM practices. Social media can be used to engage students in active learning and problem-solving through student-posted claims and effective online questioning. Using social media supports the scientific practice of engaging in argument from evidence, as well as emulates how scientists collaborate on their own research and share research findings. Best practices and lessons learned are shared in this chapter, including a case study from a secondary science classroom and suggestions for the use of social media for educator professional learning.

This chapter presents techniques that combine the productive activity aimed at creation of collaborative stories with the research activity based on the analysis of the relationship between the participants of the productive activity. The productive activity is delivered through a wiki site Letopisi.org, where all actions of the participants are recorded in an electronic log. Log records are used as a data source to build sociograms, which constitute the basis for the research activity. The chapter describes different cycles of collaboration based on the usage of social objects. The first cycle is creation of digital story. The second cycle is social inquiry.

The chapter deals with a new research field that has arisen at the intersection of scientific experiment and emerging digital technologies. The classical triad of experimental research ‘subject-instrument-object’ and its implementation in science education are in the focus of the chapter. The triad is studied in its evolution to a so-called digital triad corresponding to the experimental science of digital society. In the digital triad, each of the three components are transformed. The knowing subject - researcher is transformed to the digital scholar; the experimental instrument is transformed on the base of emerging cloud and mobile technologies; the research object comprising hybrid natural-artificial components emerges. The digital transformations of the experimental research triad and educational practices based on the digital triad are manifested in a number of pioneer inquiry-based projects analyzed in the chapter.

This chapter explores digital web-based tools for engaging learners and promoting inquiry-based STEM learning. Specifically the authors analyze a selection of technological supports in STEM education, including remote laboratories and simulations, within the context of inquiry based teaching and learning in physics. Teaching physics through inquiry continue to create high levels of anxiety amongst elementary school teachers, which in turn influences their pedagogical choices and limits the possibility of spontaneous events arising from student exploration in the classroom. The authors maintain that teachers will require professional development opportunities to work within the Technological Pedagogical Content Knowledge (TPACK) framework to ensure that they are able to select from a broad spectrum of technological supports. The authors highlight the potential of web-based digital tools to promote inquiry-based STEM learning, and engage both teachers and students, thus potentially improving attitudes toward teaching and learning STEM content through digital technologies.

The aim of this paper is to examine the use of digital technologies in Science, Technology, Engineering, and Mathematics Education (STEM). We will discuss the effectiveness of the teaching-learning process in terms of the elements that could possibly promote learning with the use of Information and Communication Technologies (ICT) in STEM education. This will be done, first, by taking a learner-centred approach to an activity that students carried out using ICT to performing a task set by the teacher (cognitive engagement in the task, motivation, nature of knowledge built). The aim was to understand how ICT could be a cognitive aid for the student. Second, a teacher-centred perspective to the development of prescribed tasks (form, knowledge carried by the task) was used to identify how ICT can be adapted to aid student learning in STEM education.

This chapter aims to address several limitations of Technological Pedagogical Content Knowledge (TPACK) – a theoretical model used in the application of technology when teaching STEM disciplines. To this end, a supplement to TPACK drawn from the Action on Objects (AO) framework is suggested. To illustrate the value of the proposed enhancement of TPACK, an example integrating science, technology, and mathematics is provided. The Texas College and Career Readiness Standards are used to demonstrate the relationship between the proposed theoretical modification of the leading model and the current teaching practice involving such scientific activities as measuring, record keeping, analyzing, conjecturing and evaluating. Additional suggestions and applications of the TPACK/AO model are provided.

The chapter describes the implementation of collaborative educational technologies in STEM teacher education to support teacher-candidates in acquiring inquiry-based teaching skills and positive attitudes about inquiry learning. The focus is on five different collaborative technology-enhanced pedagogies: (1) Peer Instruction, (2) collaborative design of conceptual questions with PeerWise, (3) data-driven STEM inquiry via using live data collection and analysis, (4) computer modeling-enhanced inquiry, and (5) collaborative reflection on peer teaching. Teacher-candidates experienced these pedagogical approaches first as learners, then reflected on them as future teachers, and lastly incorporated some of them during the practicum. As a result, teacher-candidates gained experience in promoting technology-enhanced inquiry in STEM education and began developing positive attitudes towards technology-enhanced inquiry-based STEM education.

This chapter describes a project-based massive open online course (MOOC) in nanotechnology and nanosensors that was offered in two languages: English and Arabic. A mixed methods research was conducted to examine the role of project-based learning in the process of knowledge construction and motivation to learn a MOOC. The study compared between three groups of science and engineering students: English MOOCers, Arabic MOOCers, and university students. Findings indicated positive attitudes about learning in a project-based MOOC, especially with relation to gaining work experience. Findings also indicated that in a project-based setting, MOOC participants were mostly driven by a desire to establish connections with peers, whereas university students were mostly motivated by their interest in the subject matter. Arabic MOOCers, who were less successful in solving ill-defined problems, narrowed the gap, and at the end of the course received similar grades in the final project.

Foreword

Today, STEM education is at the center of educational systems all over the world. However, two international comparative studies—TIMES and PIZA—demonstrate that only a small segment of the population reach acceptable mastery of the STEM knowledge core. At the third millennium in the midst of the digital age, it is expected that emerging technologies will be able to accelerate scientific literacy and enable the majority of citizens to enjoy the blessing of STEM.

The authors of this book are searching innovative ways to bridge traditional learning strategies, namely the inquiry conceived by Dewy and Bruner, with new digital technologies to overcome the cognitive difficulties in learning complex logical rational concepts. No less important to the authors is to recognize the meaning of those emerging digital technologies in the real world.

Simulations, Virtual worlds, Networking, Design and construction of learning environments are among the topics under study here. We are having a fresh start, and it is the researchers' as well as the educators' role to carry the inquiry about the inquiry as far as we can.

This book sets the significant stage for a very important insight into the STEM universal scene. The talented authors share their attitude and knowledge with the community of modern scientists and skilled educators. The critical involvement of all of us will be a blessing.

David Chen
Tel Aviv University, Israel

David Chen *is the President of the Center for Academic Studies in Or Yehuda, Israel, Professor-Emeritus of Science and Technology Education, School of Education, Tel Aviv University and former Dean of the School. Member of professional societies in the area of educational research both in Israel and worldwide. Held lectureships and visiting positions at a number of universities including Harvard University, MIT, Stanford and Queens College, CUNY. Also, served as National Director of the Amos De Shalit Israel Science Teaching Center. Served as Assistant to the President of Israel for Education and Social Welfare issues. Recipient of the Tel Aviv-Yafo Education Prize and the Kennedy Prize from the Weizmann Institute of Science. Co-authored several books and textbooks as well as many articles. Received a Ph.D. in Bio-Physics from the Weizmann Institute of Science.*

Preface

In the 20th century, science and science education enjoyed unprecedented governmental support in many countries. The great success of scientific research followed by its implementation in technology led governments to the assumption that investing in science has a direct and positive impact on the lives of the citizens. Hence, science came to occupy a central position both in society and in human consciousness.

In recent years, this situation has changed. As a result of the intensive integration of information technology in human life, public interest began to shift away from traditional science. The great achievements in the field of Information and Communication Technologies (ICT) and their transformational impact on human life have affected human consciousness. Technology's role in society, as well as its connection to science, is changing. The traditional perception, according to which science was the main factor affecting the quality of human life, and technology was considered merely an "applied science," is already dated and irrelevant. The newer emerging perception, which is gaining more and more popularity, is that technology (and not science) is the predominant factor in human life, and that science plays an ancillary role in the new technological society. To a great extent, these shifts in social consciousness reflect our global transformation into a digital society. Our fundamental perceptions of the world around us are changing. These changes are most noticeable in the shift from scarcity to abundance of information, and from the primacy of entities to the primacy of interactions; as well as in the blurring of distinctions between reality and virtuality and between people, machines, and nature (Floridi, 2014). Moreover, the distinction between science and technology has faded (Levin, 2016). Today, scientific research is unimaginable without the emerging technologies, which in turn are becoming an inseparable part of the scientific endeavor. The interpenetration of scientific and technological research methodologies is another instance in which prior distinctions are becoming less significant. Thus, an increasing number of scientific analytical research methods are being implemented in engineering research, and vice-versa: many technological, design-oriented methods are gaining popularity in scientific research.

These substantial transformations in science, technology and society have significant implications for education, in general, and for STEM education, in particular; these are the focus of our book. Obviously, if the contemporary changes in education could be addressed by merely equipping schools with ICT facilities, the problem would be much more modest in scope and could be managed by promoting technological literacy. In effect, the problem is far more complex. Our awareness of the fact that the changes in STEM education correspond to general transformations in society provides another perspective on the problem. Our underlying assumption is that the new fundamental role of technology in science and society ought to be reflected in each and every component of STEM education: learning, teaching, environment, curriculum, and assessment. Additionally, along with the traditional ways of integrating ICT in education, STEM education can learn from the ways in which scientific research was transformed through digitalization.

Our main intent in this book is to take an in-depth look at the integration of ICT in STEM. Our exploration considers: a) inquiry-based learning to be the key method in STEM learning; b) digitalization to be a means to enhance inquiry-based learning; c) a conscious change in teachers' and students' perceptions to be a crucial factor for transforming STEM education in digital society.

The main motivation for writing this book was our realization that the digital era has transformed STEM education. We asked ourselves: "what are the changes and the corresponding directions that STEM education can pursue in light of the digital turn?" We tried to answer this question by curating a selection of recent and interesting research studies in the corresponding fields.

The chapters included in the book paint a multifaceted picture of these changes, representing a range of different approaches, from theoretical aspects of technology integration to practical implications of the digitalization process. Each approach is uniquely distinct and significant.

The book comprises two sections. The five chapters in Section 1, "Inquiry-Based Learning of STEM in the Digital Era", examine both new models of learning and existing models that have been significantly revised and upgraded to conform to today's pedagogical standards. The review of these models addresses questions, dilemmas, trends and solutions related to learning STEM in the Digital era. Section 2, "Inquiry-Based Teaching of STEM in the Digital Era", complements Section 1 by discussing various aspects of the inquiry-based teaching process. The six chapters in this section deal with the efficacy of innovative teaching strategies, methods, and approaches developed to integrate digital technologies in STEM education. The implications of these developments and the challenges they present in terms of teacher training and professional development constitute a focal interest in the second section.

A brief description of each of the chapters follows.

SECTION 1: INQUIRY-BASED LEARNING OF STEM IN THE DIGITAL ERA

Chapter 1 is devoted to the discussion about the opportunities and challenges of integrating technology into preschool science inquiry-based learning. This study was intended to address the widely-recognized concern regarding young children's protracted screen time, on the one hand, coupled with the apparent necessity and possible advantages of digital literacy acquisition, on the other hand. Should technology be introduced to children at a young age, or are they liable to lose valuable capabilities as we rush to discard habitual analog educational tools in favor of recent new gadgets? How can educators create stimulating yet safe digital environments that enhance scientific curiosity and exploration among young children? These questions motivated the chapter of Ornit Spektor-Levy and her coauthors.

The remarkable role of modeling and computer simulation in the learning of STEM is analyzed in Chapter 2. Modeling and simulation rely on great learning theories, from classical constructivism to epistemological constructionism, and thus can be introduced in education successfully. Moreover, simulation technology represents the so-called computer scientific research paradigm, which considers the inquiry-based study of computer models of scientific phenomena a main method of research. The author of the chapter, Franco Landriscina, discusses inquiry-based learning of computer models and the problems and difficulties of implementing pedagogical simulations. Specifically, the author considers the relationship between student ability to conduct mental simulations, on the one hand, and the computer-based simulation, on the other hand.

As mentioned, the emergence of new learning methods, in general, and innovative design-oriented methods, in particular, is the main tendency in STEM education nowadays. Chapter 3 deals with design-based learning (DBL) developed and used to facilitate collaborative learning of STEM. The author of the chapter, Joycelyn Streator, shows that integrating ICT made it possible to utilize the potential of the DBL in inquiry-based STEM learning. Today, a number of efficient Web-based tools are available to support this type of learning. The author demonstrates that design has to be considered an inquiry-based activity in STEM learning.

One of the important phenomena that emerged in the digital era is social media, which is already considered a significant part of the native habitat of today's students. Chapter 4 of the book deals with social media intended to support STEM inquiry-based practices, particularly scientific argumentation. The authors, Jana Craig-Hare and her colleagues, consider social media to provide an educational environment that supports the type of collaborative scientific inquiry in which students are involved. Scientific activity has a social nature and as such it requires viewing students as "digital citizens" living in a digital habitat. A study of the behavior of digital citizens as scientists in a digital society is the focus of the chapter.

The fact that both social media and networking are deeply rooted in pedagogy is demonstrated in Chapter 5. Its author, Evgeny D. Patarakin, proposes a new form of educational activity, which has arisen at the intersection of social media, network analytics and computer modeling. According to the proposed approach to learning, students' reflection regarding their place and functioning within the network is utilized as a valuable learning activity. Actually, social media is combined with an agent-based educational environment where students can carry out innovative scientific inquiry.

SECTION 2: INQUIRY-BASED STEM TEACHING IN THE DIGITAL ERA

Chapter 6 deals with experimental, inquiry-based learning in the digital era. Authors of the Chapter, Dina Tsybulsky and Ilya Levin, traced the sequential changes that occurred in the experimental triad (subject-instrument-object) during the history of science education. The authors discuss the evolutionary process, beginning with the classic experimental triad and following it through to the digital era triad. They show that each of the triad's components has changed in the digital era. The digital triad represents a new scientific entity, which, in turn, has begun to appear in STEM educational practices.

Dealing with ICT in STEM education calls for an examination of the pedagogical potential of available online educational resources. After all, the accessibility of the resources on the Net determines to a great extent the success of ICT in STEM learning. Chapter 7 explores the potential of Web-based tools for engaging learners and promoting inquiry-based STEM learning. The authors, Isha DeCoito and Tasha Richardson, analyze the most promising ICT tools: simulations and remote laboratories in physics education.

Chapter 8, written by Jean-François Hérold and Jacques Ginestié, deals with the problem of learning efficiency when using digital tools in STEM education. The authors study how ICT could be considered a cognitive aid for the student. Both the learner-centered and the teacher-centered views are applied. The authors demonstrate the necessity for teachers to understand the ability (or inability) of certain students to learn in an ICT-enhanced class. They demonstrate the importance of emotion and motivation and their effects on learning and on the teacher-learner dynamics in the classroom.

Technology-enhanced inquiry is the focus of Chapter 9. The authors, Michael L. Connell and Sergei Abramovich, emphasize the value of the constructivist approach to ICT integration in STEM education, when “learners enter into an intellectual partnership with the computer” (Jonassen, 2000). This remarkable approach, rooted in the seminal works of Seymour Papert and his followers, views human-computer interaction through the lenses of the so-called “glass boxes”, rather than relying on the “black boxes” with their traditional “right or wrong” answer. As an illustration, the approach is mapped onto a collaborative STEM-related activity.

In the transition to the digital era, teacher training obviously attracts special attention. Chapter 10 discusses the implementation of collaborative educational technologies in the education of STEM teachers. The main aim of teacher education is defined as providing teachers with inquiry-based teaching skills and positive attitudes regarding inquiry-based learning. This can be achieved by studying a wide spectrum of technological means used in the contemporary classroom. The author, Marina-Milner Bolotin, provides a comprehensive study of some collaborative educational technologies and demonstrates how these technologies can be implemented in STEM teacher education.

The classical educational *constructionism–instructionism* dilemma (Papert, 1980) corresponds to two alternative approaches to ICT-enhanced education. Traditionally, MOOC is considered an instance of ICT-based instrumentalism, which is a far cry from inquiry-based education. The authors of Chapter 11, Miri Barak and Abeer Watted, discuss using the MOOC platform for conducting project-based learning activities. They present a project-based MOOC course in nanotechnology. Such a combination provides an alternative view, showing the potential of MOOCs in STEM learning.

We believe that this book provides a much-needed reference source in the field of Inquiry-based STEM education. We hope that educators, researchers and students will find it useful for gaining insight into the regularities and features of STEM education in the Digital Age, as well for envisioning new ways to transform and enhance STEM education.

Ilya Levin
Tel Aviv University, Israel

Dina Tsybulsky
Tel Aviv University, Israel

REFERENCES

Floridi, L. (2014). *The fourth revolution: How the infosphere is reshaping human reality*. OUP Oxford.

Jonassen, D. (2000). *Mindtools for schools: Engaging critical thinking with technology*. Columbus, OH: Merrill.

Levin, I. (2016). Cyber-physical systems as a cultural phenomenon. *International Journal of Design Sciences & Technology, 22*(1).

Papert, S. (1980). *Mindstorms: Children, computers, and powerful ideas*. Basic Books, Inc.

Section 1

Inquiry-Based Learning of STEM in Digital Era

Chapter 1
Integrating Technology in Preschool Science and Inquiry

Ornit Spektor-Levy
Bar Ilan University, Israel

Naama Israeli
Bar Ilan University, Israel

Inna Plutov
Bar Ilan University, Israel

Netta Perry
Bar Ilan University, Israel

ABSTRACT

Technology has brought about considerable changes in our private, social and professional lives, as well as in our culture and values. Therefore, educational frameworks should make an effort to become more relevant for young students and prepare them for the future in all aspects of career and life, with a focus on Science, Technology, Engineering and Math (STEM). This chapter will discuss the opportunities and challenges of integrating technology into preschool classrooms (3-6 years of age). It attempts to determine the essence of judicious, proportionate, and beneficial integration of technology in preschool, with a particular focus on science and inquiry. Consideration is given to maintaining the children's creativity, their joy of play, their concrete and sensory exploration, their unmediated observation of their environment, their social interactions, and their safety. Examples of actual practices from preschool classrooms are presented followed by recommendations for successful technology integration in preschool curriculum.

INTRODUCTION

In July of 1859, a British philosopher named Herbert Spencer published an essay titled "What Knowledge is of Most Worth?" Spencer wished to define the biggest thing that ought to be taught in the educational system at that time. This same question is still asked today (NEA, 2010; Luna Scott, 2015): What should be studied in educational frameworks to prepare our children for life in a global digital economy?

DOI: 10.4018/978-1-5225-2525-7.ch001

Technology has brought about considerable changes in our private, social and professional lives, as well as in our culture and values. The technologies that exist today are just the tip of the iceberg; some are only in their initial phase of development, and others will be replaced by newer and more sophisticated technologies. Consequently, certain knowledge, skills, and competencies have become redundant, while other knowledge, skills, and competencies have become essential (Zhao, 2012). The Next Generation Science Standards (NGSS, 2013) addressed the need for new Science and Technology standards, noting the competitive global economy, the need to prepare for the demands of new careers in the modern workforce, and the need for an educated society, literate in science and technology (NAE & NRC, 2014). Educational frameworks, therefore, should not only make an effort to become more relevant for young students, but should also prepare them for the future in all aspects of career and life, with a focus on Science, Technology, Engineering and Math (STEM) (Scott, 2012; Zhang, et al. 2010).

This chapter will discuss the opportunities and challenges of integrating technology into preschool classrooms (3- to 6- years of age) in general, with a particular focus on science education and inquiry. The term *technology* has a broad meaning. In this chapter, *technology* refers to digital technology (Instance & Kools, 2013); i.e., all kinds of modern computers, information and communication technology (ICT), and digital media.

BACKGROUND

The Dilemmas

With the ever-growing ubiquity and accessibility of technology and digital devices, today's children spend increasingly more time in front of all types of screens (television, computer, tablet computer, smart phone, hand-held mobile devices, game consoles, and other digital means). One of the dilemmas related to the use of computers, information and communication technology (ICT) and digital media in the preschool classroom concerns children's screen time. A child who sits in front of a screen for protracted periods of time is physically inert. This leads to dwindling physical and motor activity, which result in additional health problems, all of which can be traced to prolonged sitting and staring at a flickering screen (APP, 2016). To avoid these health concerns authorities recommended that screen time limits be set for young children (De Decker, et al., 2012; Lauricella, Wartella, & Rideout, 2015), both at school and at home.

The American Academy of Pediatrics (AAP, 2016), spoke out against allowing screen time of any kind for children under the age of eighteen months. Parents of children aged 18 months to 2 years who choose to introduce digital media into their children's lives should opt for high-quality programs. The parents should watch these programs together with their children, to help the children understand what they are seeing and how it is related to their own life and environment. The AAP recommended that screen time not exceed one hour a day for children between the ages of 2 to 5 years. Other authorities working to prevent obesity in children (Reynolds, et al., 2013), and those who work for the benefit of public health (Hobbs Vinluan & Hofman, 2014) are even more stringent. These authorities have also called upon educators to convince parents to limit their children's screen time at home. These recommendations underscore the depth of concern regarding the ubiquity and accessibility of digital means of all kinds (Lerner & Barr, 2014; Tremblay, et al., 2012).

Another dilemma is related to the benefits of computers and digital media in early childhood. For more than two decades, the professional literature has been investigating whether computer use inhibits (Cordes & Miller, 2000) or enhances (Christakis, 2014; Clements & Sarama, 2003) young children's learning and development. Although the academic discourse has not yet resolved the matter (Radetsky, Schumacher, & Zuckerman, 2015; Strasburger, 2015), research persists to examine how to improve learning and development in technology-rich environment (Billington, 2016; Stephen & Plowman, 2008). Educators (Blackwell, et al. 2013; McCarrick & Li, 2007; Siraj-Blatchford & Siraj-Blatchford, 2006) claim that computers can support young children's learning if appropriate pedagogy and mediation are employed, if the technology is thoughtfully integrated into the natural learning environment, and the teachers are aware to each child's needs and experiences (Billington, 2016; Nikolopoulou & Gialamas, 2009).

Beneficial Use of Technology

Digital Literacy is a term that has emerged in the context of the pedagogical reforms taking place over the last two decades. This involves the ability to aggregate, analyze, synthesize, and communicate data using technology tools and sources (Bair & Stafford, 2016). Digital literacy is not only a matter of the technical skills needed to operate digital tools and computing devices, but also the cognitive, psychological, and social processes that are beneficial for working in a technology-rich environment.

The social aspect of digital literacy was further conceptualized to Digital Citizenship (2016). Digital citizenship concerns appropriate behavior when using the Internet and digital technologies; responsible and ethical behavioral norms that respect the rights of others on the Internet; safeguarding privacy; safety and security.

Adults and children need to show responsibility when using digital technology, to know the meaning of inappropriate use of digital resources, and to avoid harming others (Digital Citizenship, 2016).

Recent years have seen a rise in the presence of digital mobile devices operating on innovative wireless and cellular technologies. Mobile learning was defined by O'Malley et al. (2005) as "any sort of learning that happens when the learner is not at a fixed predetermined location or learning that happens when the learner takes advantage of the learning opportunities offered by mobile technologies".

Mobile technologies include devices such as laptops, tablet computers, smart phones, and digital cameras. Mobile technology enables embeddedness - the device becomes part of the user's thinking as a tool to solve problems, interact with the world, and perform common functions (Puentedura, 2011). One of the most ubiquitous uses of mobile devices is photography. Photography is the most common and widespread way to document events, processes, and situations in daily life, workplaces, research, etc. Photography can promote higher-order thinking processes, and provide children with an alternative that may complement verbal and written language (Burnett & Myers, 2002; Clark & Moss, 2001). Thus, mobile technology can support the development of children's capabilities in various domains (Wu & Zhang, 2010).

Given the above, early childhood educators are entrusted not only with the responsibility of mediating the technology-rich environment for young toddlers, but also making beneficial use of technology, examining the impact of technology on young children, and using technology in ways that are developmentally appropriate for young children 3- to 8-years-old (NAEYC, 2012). This leads to questions such as: How young children should be prepared for life as future digital citizens? Should "preparations" begin as young as preschool? Should computers and digital media be integrated as early as preschool?

In literature, there is a growing body of research that supports digital technology integration as early as preschool. Billington (2016) claimed that technology can play an important role in supporting early communication, language, and literacy by offering new opportunities, such as interactive and intuitive story telling e-books and apps, as well as other services, such as online video calling. Others indicated enhanced sense of self-efficacy (Segers & Verhoeven, 2002), social and cognitive efficacy (Hsin, Li, & Tsai, 2014). Nuemann and Neumann (2014) found that the use of tablet computers can support increased alphabetical knowledge and emergent writing in preschool children. Children who use computers have been found to show greater structural knowledge, problem solving, and language skills compared with those who do not use technology in their learning (Clements & Samara, 2003; Haugland, 1999; Linebarger & Vaala, 2010; Swaminathan & Wright, 2003; Vernadakis et al., 2005). Mouza (2005) reported that when children in kindergarten through second grade used digital tools, the result was increased motivation in math, reading, and writing. Copple and Bredekamp (2009) elaborated on how learning activities in digital environments enhance children's eye-hand coordination, motor and cognitive development, paving the way for later academic learning. Children who use computers have also been found to show greater gains in science, math and technology (Judge, Floyd, & Jeffs, 2015; Orr, Kamdar, Lewis Presser, & Vahey, 2015; Tang, Li, Hsin, & Tsai, 2016; Weiss, Kramarski, & Tallis, 2006).

In light of these evidences, this chapter attempts to determine the essence of judicious, proportionate, and beneficial integration of technology in preschool, with a particular focus on Science, Technology, Engineering and Math (STEM) education. Consideration is given to maintaining the children's creativity, their joy of play, their concrete and sensory exploration, their unmediated observation of their environment, their social interactions, and their safety.

Developmentally Appropriate Use of Technology

Before discussing the integration of technology in preschool life, some basic principles regarding the fundamental nature of preschool frameworks shall be highlighted:

- Despite all the shifts and changes, preschool must remain a fun and safe place, based on physical activity, on playing in the yard and in the preschool classroom, on unmediated sensory exploration, on inquiry and discovery (Ceppi & Zini, 1998; Roskos & Neuman, 2011).
- Technology should be used in preschool classrooms only when it offers added value (Blackwell, Lauricella, & Wartella, 2014).
- Technological devices cannot replace mediation by the teacher and by the educational staff at the preschool (NAEYC, 2012), nor discussions and interactions amongst the children and their peers (NAEYC, 2012).

Nevertheless, ever since the 1980s, when computer technologies became pervasive in schools around the world, it has been commonly believed that a technology environment is agreeable for young learners (Drigas & Kokkalia, 2014). Benefits include:

- Individualized learning: tailored to the unique needs of every learner (Blackwell, Lauricella, & Wartella, 2014; Instance & Kools, 2013).
- Multimodal representations, mainly of a visual nature, facilitate understanding and memory (Shams & Seitz, 2008; Tang, Delgado, & Moje, 2014).
- Interactivity, interactions and sharing (NAEYC, 2012).

- Self-initiated learning - increased intrinsic motivation encourages the children to share with others, to ask questions, to locate sources of information, to apply elements of research (Stephen, 2006).

Software programs and mobile applications that are developmentally inappropriate for young children can be detrimental to their creativity (Clements, 2002), especially those based mainly on drill and practice (e.g., practicing simple arithmetic exercise, memorizing letters and words). Haugland (1999) developed a scale to identify developmentally appropriate software by means of ten criteria: (1) Age appropriate; (2) Child in control of the computer environment; (3) Clear instructions; (4) Expanding complexity - a gradual increase in the level of difficulty of the tasks; (5) The child's ability to work independently; (6) Non-violent content; (7) Learning and thinking process oriented - processes preferred over content; (8) The software presents real world models; (9) Technical features; (10) Transformations - options to modify commands and icons in the environment based on what the user wants. Each criterion consists of its own essential characteristics, above which hangs an additional criterion, which Haugland (1999) called an "antibias": The software must reflect the diversity of human society. Based on the above scale, Haugland (1999) evaluated different software programs. Scores ranged from 1.0 to 9.0. Software that received a score of more than 7 were considered developmentally appropriate for children.

Findings from Haugland's (1992) study showed that when preschoolers work with software that is developmentally appropriate, they have a high sense of efficacy, and their capacity for creativity is high.

As educators, we are aware that the process of young children's gradual development begins with direct and active use of tools and materials, after which there is a typical transition from exploration (inquiry and discovery) to expertise, and later, to the functional use of the tools to fulfill various tasks that are more complex (transfer) (Deák, 2014). A similar process occurs in the way that young children work with technology tools. Children need time to discover the functionality of the technology before they use it for communication and to complete complex tasks (Geist, 2012).

The Role of the Teacher and the Educational Staff

Computers and digital technology in the preschool learning environment are tools that offer the children additional opportunities for inquiry, discovery, enrichment, and representation. The extent to which these versatile tools can make a contribution depends on the preschool teacher's perceptions and attitudes toward the integration of technology (Nikolopoulou & Gialamas, 2015a,b) and how the teacher chooses to integrate technology in the preschool curriculum. Based on Shulman's (1986) Pedagogical Content Knowledge (PCK) framework, Mishra and Koehler (2006) called this group of specifications the Technological Pedagogical Content Knowledge (TPACK), and described the knowledge required to effectively teach with technology. Specifically, teachers need to know how technology can: 1) enhance the representations of specific content and make subject matter easier or harder to learn; 2) be matched with specific pedagogical practices that enhance the teaching of specific subject matter; and 3) build on students' prior knowledge and instigate new learning (Mishra & Koehler, 2006; Koehler, Mishra, & Cain, 2013). Factors such as technology self-efficacy, attitudes toward technology, and general pedagogical beliefs all can influence TPACK.

Blackwell, Lauricella, and Wartella (2014) examined the relationship between TPACK contextual factors (effective technology integration practices) and the specific ways in which early childhood educators integrate tablet computers into the learning environment. Blackwell, Lauricella, and Wartella (2014) concluded that although NAEYC (2012) suggested that teachers use technology in developmentally appropriate ways, and although the TPACK literature suggested that effective technology integration should apply student-centered pedagogies (Mishra & Koehler, 2006; Koehler, Mishra, & Cain, 2013), their study revealed that many early childhood educators are not using tablet computers to support a student-centered learning environment. Therefore, more work is needed to develop policies that ensure teachers undergo sufficient professional development relating to educational technology to appropriately deliver on these recommendations (Blackwell, Lauricella, & Wartella, 2014, p. 24).

Preschool teacher training and in-service professional development should include acquainting the teachers with the TPACK conditions, along with suggestions for judicious integration of computers and digital media in preschool. For example, (Israeli Ministry of Education, 2016; NAEYC, 2012):

- **Authentic Work:** Learning technology tools within a particular context and skills needed in everyday life enables better acquaintance of the tools' benefits and limitations.
- **Appropriate Selection of Software:** Teachers should devote time to evaluate and select the appropriate technology and to carefully observe the children who use it, identify opportunities or problems, and then make appropriate adaptations.
- **Individual Support:** Educators must adapt the technology to the individual needs, learning style, and preferences of each and every child.

Another vital role of the preschool teacher is supporting and mediating the children who work in technology-rich environments. A study by Nir-Gal and Klein (1999) indicated that children who were exposed to adult-mediated computer activities showed improvement in measures of cognitive performance (i.e. measures of abstract thinking, planning competencies, vocabulary, visuo-motor coordination, and measures of response styles, including measures of reflective thinking), compared to children in groups that worked at computers without adult mediation, or with very little mediation (accompaniment).

It must be remembered that the child's interaction does not have to be with the preschool teacher alone; interaction among the children is valuable as well, because they can provide assistance for each other and learn from their peers, which establishes very important social competencies (Hsin, Li, & Tsai, 2014).

The preschool teacher plays an important role in monitoring computer use (Spektor-Levy, 2012):

- Ensuring that all preschoolers have opportunities to use the computers and experience working on them.
- Supervising to ensure that computers are only used for specific purposes known to the teacher.
- Controlling the amount of time each child sits by the computer.
- Locating the computer in the learning environment. There are different approaches to the physical location of the computer in the preschool classroom:
    - Computer is relegated to a peripheral spot in the classroom, so as not to distract the children who are not working on the computer. Moreover, children who have less mastery of the computer may feel embarrassed when other children stand behind them and watch what they are doing - they may prefer to work in an isolated and quiet environment.
    - Computer is located where it is accessible and exposed, just like all the other tools and materials in the classroom, and just like any source of information from which the children learn about the world.
    - Computer is placed inside a learning center, or in a space that concentrates on one particular topic. The computer serves as one of the sources of information and activities for that topic. Thus, the computer can be placed in different locations in the preschool classroom at different times.

There appears to be an advantage to combining all three approaches. If computers are tools, then they must be moved around based on needs. Laptop computers, tablets and other mobile devices can provide this mobile approach.

Scientific Literacy, Inquiry, and Technology

Scientific literacy is a well-recognized educational term, which has received different interpretations over the years. Overall, scientific literacy can be defined by a scheme that includes three different categories (Laugksch, 2000): Science content and intellectual abilities; Competency (the extent of the ability to conduct science-related tasks); Function in society (the ability to use scientific knowledge and implement it in a wide variety of contexts in order to function effectively in society and achieve well-being, as consumers and citizens).

Children are born with a natural curiosity and an intrinsic motivation to inquire into the world around them (Jirout & Klahr, 2012). Making use of this natural curiosity to develop scientific literacy among young children has been acceptable for a long time; however, only recently has it become obligatory in many countries. Over the past two decades, studies have shown evidence that even infants possess cognitive capabilities enabling them to employ scientific thinking and inquiry skills (Eshach, 2006; Gopnik, 2012). It has been revealed that young children who participated in different scientific programs showed an improvement in cognitive and linguistic achievements, a better understanding of scientific inquiry processes and of important life science concepts (Samarapungavan, Mantzicopoulos, & Patrick, 2008), an increased motivation for science learning (Patrick et al., 2009), as well as enhanced abilities to participate in a scientific discourse and to use scientific vocabulary (French, 2004; Samarapungavan et al., 2008).

Learning science through inquiry is a genuine learning experience (Kuhn et al., 2000). Teachers facilitate inquiry when they encourage children who ask questions about the natural world to seek answers by making their own predictions, designing the proper experiences, collecting, recording, and analyzing the information, and drawing their own conclusions from the evidence.

The term “inquiry-based science teaching” conveys a way of teaching science in the manner actually practiced by scientists; i.e., by problem solving through formulating and testing hypotheses. This may provide students with the opportunities to learn how scientific knowledge is generated and to participate in the practices of science (Kuhn et al., 2000).

Inquiry involves making predictions, gathering new ideas (from investigations, visualizations, or observations), using evidence to distinguish among new and existing ideas, and developing explanations (National Research Council, 2012; Schroeder, Scott, Tolson, Huang, & Lee, 2007). Inquiry demands higher-order thinking skills (i.e., questioning, designing an experimental array, critical and logical thinking, metacognition and reflection, Zion et al. 2004), ownership, and responsibility for determining the purpose of the investigation (Reid &Yang, 2002). These and other competencies are known as 21st century skills; namely, knowledge and skills that are relevant for a changing reality and for the ability to function optimally in the 21st century (Israeli Ministry of Education, 2009; Partnership for 21st Century Learning, 2008). Recent national and international policies have underscored the need for 21st century skills, which are seen as instrumental for coping with the fast-paced changes of the world, as well as being able to integrate into the workforce (Johnson, 2009). The development of these competencies can be supported through inquiry-based teaching (Maass & Doorman, 2013) and the integration of technology.

In literature, one may find various studies that have explored the contribution of digital technology to the implementation and performance of inquiry capabilities among very young children. Digital cameras, smartphones, tablet computers, and other digital mobile devices have opened new opportunities for young children to engage easily in auto, multimodal documentation. Most mobile devices include a built-in digital camera, making photography accessible and easy to share. Photography enables students to capture "just-in-time" photos, audio, and video recording, and to upload these files to the network for instant sharing. It can also allow students to edit captured photos for specific learning needs without the constraints of time and space (Song, 2014).

Rikala and Kankaanranta, (2014) explored whether the concept of the mobile application 'Nature Tour' meets the needs of early childhood education on field trips. The idea was for the children to take photos using the mobile application to record and compare their observations about nature, as well as to arouse their interest in nature. The teachers who participated in the study noted that the children had never paid so much attention to nature as during the experiment. The mobile application, therefore, opened up a whole new world for the children; they began to construct knowledge and awareness of nature. Thus, mobile learning is characterized with "any time any place learning" and considered as authentic, collaborative and personal learning (Kearney, Schuck, Burden, & Aubusson, 2012; Wong & Looi, 2011). Mobile learning is contextual (Huang, 2011; Sharples, Taylor & Vavoula, 2005(. When children are using mobile devices, they can express their own perspectives more freely. Mobile systems can improve creativity and ability to explore and absorb new knowledge (Shih et al., 2011; Song, 2014).

Aronin and Floyd (2013) reported on intervention in inclusive preschool classrooms were preschoolers were provided with iPads and apps focused on pre-engineering, math, and science content to interact with the technology and gain understanding of STEM concepts and careers. Four important principles were considered when the apps were selected for the preschoolers: the student should be the source of the action to make the outcome more scientific; the students should be able to see cause and effect relationships by changing the beginning action and seeing how it reflects the outcome; the outcome of changing the variable must be observable to the preschooler; the action and reaction must happen immediately for the child to see and make connections between the cause and effect.

Aronin and Floyd (2013) reported that while interacting with the play-based apps, preschoolers gained confidence and independence with the iPad and this outcome provided the opportunity for the teachers to lead discussions and expand the learning from the apps to the classroom environment and to careers in various STEM fields.

FROM THEORY TO PRACTICE

Digital technology provides interactive multi-channel, multi-sensory digital environments designed to support knowledge integration, data analysis, creative thinking, problem solving, communication, social relationships, and cooperation with other children as well as with adults. All of these are capabilities needed for the scientific inquiry process (National Research Council, 2012).

This section describes practical examples from actual preschools of how different technologies can be used to promote inquiry in preschool classrooms. The first example was followed by a short pilot study to examine the actual contribution of a digital mobile device as will be described next.

Documenting an Experiment With a Mobile Device

This section describes a scientific activity with preschoolers, followed by a pilot study. The aim was to examine how performing documentation via scientific drawing (with markers and crayons) combined with a digital mobile device develops inquiry capabilities, compared to documenting by drawing only.

The authors followed 10 young children (7 boys and 3 girls) aged 4-5, from a medium-high socioeconomic urban neighborhood. Three qualitative data collection methods were employed:

- **Participant Observation:** The observation took place during scientific experimentation that demonstrated the reaction of yeast to sugar and water (anaerobic respiration). The observation documented the children's responses. Observation was also implemented when the children created a final artifact - the retrospective presentation of the scientific experiment. The observation was recorded on video.
- **Semi-Structured Interview:** The interview consisted of open questions (Table 1) based on the *"Scientific Demonstration with Documented Response Questionnaire" (SDDRQ)* (Kesner Baruch, Spektor-Levy, & Mashal, 2014).

Table 1. Documenting Scientific Demonstration of foamy reaction: The study stages, inquiry capabilities and the procedures

Stage	Inquiry capabilities	Procedure
Pre-reaction	Questioning, Hypothesizing	• Demonstrated experiment: Two cups, two spoons, jar of water, small bowl with sugar (about 100g), small bowl with dried active yeast (about 100g). Each child was given the opportunity to engage in sensorimotor interactions with the materials and tools (taste, smell, touch). • Interview: Please describe what you see. Suggest what you think we can make with these materials and tools. What questions do you want to ask? Are you curious to see what is going to happen? What do you think is going to happen? • Documentation: All children were asked to draw what they saw. The intervention - group children were also asked to take photos with an iPad.
During reaction-fermentation	Observing, describing, documenting the results	• Demonstrated experiment: The demonstrator placed two cups on the table and put yeast, sugar, and warm water in one (cup 1), but only yeast and warm water in the other (cup2). Each child was given the opportunity to engage in sensorimotor interactions with the materials. The experimenter did not explain or ask questions. The children observed the foamy reaction in cup 1. • Documentation: All children were asked to draw what they saw. The intervention group - children were also asked to take photos with the iPad.
Post-reaction	Concluding, preparing presentation, sharing knowledge.	• Interview: Please describe what you saw? What do you think happened? Please explain. Do you remember what you answered when I asked: What do you think is going to happen - before we began? Would you like to know why this happened? What questions would you like to ask? Are you curious to observe more scientific experiments like this? • Documentation: Demonstrator asked each child to think about what he/she saw and thought during the scientific demonstration, and to prepare for story telling - to tell a friend who had not seen the experiment about it. Intervention group: Used Adobe Voice, an iPad application, to combine their drawings (scanned), their photos and to record themselves explaining the experiment, thus creating a short movie with voiceover. Comparison group: Placed their two drawings on the table like an exhibition and explained - told the 'story' of the experiment.

The interview lasted 5-10 minutes, and was conducted in two parts. One before the scientific demonstration and the second immediately after the final artifact.

- **Analysis of Final Artifacts - Retrospective Presentation of the Scientific Experiment:** Final artifacts were created by all children after the scientific experiment. They were asked to prepare a retrospective presentation of the scientific experiment. Participants were randomly divided into two groups. Five children who were part of the comparison group were asked to place their drawings (with markers and crayons) on the table and explain what they had seen, what had happened during the experiment, and what their conclusions were. The intervention group included five children that were asked to document the experiment with drawings (with markers and crayons), in addition to photographing with the digital camera of the iPad. Then, using the Adobe Voice application, they were asked to explain what they saw, what happened during the experiment and what their conclusions were. They were asked to prepare a short movie (about 2 minutes) that would include scans of their drawings and the pictures they had photographed.

The study procedure is presented in Table 1.

Figure 1 presents the two methods of documentation that N. performed as a participant in the intervention group. She was asked to carefully observe the demonstration and the reaction, and to draw (with markers and crayons), what she saw (Method A). When she said, she had finished drawing, she was asked to hold the iPad and to take a picture of the same phenomena she had drawn (Method B). In Method A, the drawing shows imaginary elements: N. added a straw in each cup.

Figure 1. Documentation: Second stage of the experiment - During reaction - fermentation. N.'s drawing and the photo she took with the iPad.

Method A

N.'s drawing with markers

Method B

N.'s photo with the iPad's digital camera

She also added a cover or lid for each cup. The left cup was drawn with horizontal lines and one vertical line. The cup on the right was drawn with a scribble. These additions did not exist in reality. In Method B, N. took a photo of the same two cups. The photo illustrates exactly what the two cups looked like, with the cup on the left full of foam as a consequence of the anaerobic fermentation of the yeast (in the presence of the sugar and water). The second cup, on the right, held only yeast and water and no foamy reaction was observed. This picture presents the reality more accurately. It took N. only a few seconds to take the picture. While taking the picture, she did not observe the phenomena as carefully as she did when she drew the same phenomena. This part of the study exemplifies the advantages and disadvantages of each documentation method.

In the third stage of the study, all the children were asked to prepare the final artifact and to describe the scientific demonstration as though they were telling a story. The intervention group prepared a short video clip. The comparison group exhibited their drawings. All qualitative data was transcribed verbatim. The qualitative analysis revealed differences between the two groups:

Children were asked: Do you remember what your hypothesis was regarding what would happen during the demonstration - before we began? All children in the intervention group remembered what they had hypothesized: "It will turn into lemonade", "It will become dough", "It will turn into clay", "I thought there would be bubbles", "I thought it could become juice or become a different color". In the comparison group, only one child remembered his hypothesis: "There will be bubbles in both cups".

When the children were asked to relate what had happened during the scientific demonstration, 4 of 5 children in the intervention group were able to describe in detail what was mixed in each of the two cups, in addition to the fact that the sugar contributed to the foaming reaction. In the comparison group, 3 of 5 children found it difficult, noting that: "I don't remember… One swelled up", " I don't remember", "Nothing happened… but I told you already".

Children were also asked: What questions would you like to ask?

Three children from the intervention group suggested questions such as: "What will happen if we put the experiment in the refrigerator?", "Why is there juice at the bottom of the cup?". However, no child in the comparison group was able to ask a question.

When asked, "Would you like to know what happened?", all children in the intervention group said, "yes. One child even said "yes, very much." In the comparison group, 2 children answered "no".

The qualitative analysis of the final artifacts revealed that children in the intervention group could better describe and explain the experiment. The digital documentation combined with the drawings better supported children's comprehension and better restored the process. Similar to other studies that examined digital documentation and assistance to the learning process (Song, 2014), the authentic photos assisted the children to restore and understand what happened because they documented the demonstrated experiment precisely as it was. The drawings were not precise and usually included various interpretations of reality; therefore, drawings may not present the process in an accurate way (Figure 1). The photos helped the children to remember, and to present the demonstrated experiment through the video in the correct chronological order. Using the iPad and its applications increased the children's motivation to prepare the final artifact. It encouraged them to ask more questions; they were more curious to know what had happened, why it had happened; and they were eager to conduct more experiments. These findings are in line with what Rikala and Kankaanranta (2014) found. The combination of detailed observation while drawing the phenomena and the usage of digital tools (taking pictures that documented the phenomena in a precise and realistic way, and preparing a short video), contributed the most to understanding the scientific demonstration and its results.

Outdoors Documentation with a Mobile Device

Preschool teacher D. always goes outside for walks with her preschoolers. They carry an Explorer's Bag that contains a magnifying glass, a plant identification chart that has been adapted for the preschool setting, binoculars, and small containers to inspect insects (and set them free after observing them). D. also takes along a digital camera, and lets the children take pictures of phenomenon and document things they experience on the way.

After transferring the photos to the computer in the preschool classroom, the preschoolers can watch them in Adobe Photoshop, Microsoft Paint program, Microsoft PowerPoint or in a Microsoft Word document. They can write about the photos (Figure 2) they took (after acquiring keyboard typing skills or with the help of the teacher), and share the experience with their parents or friends.

They can continue by editing an album that includes the photos taken by the children and their descriptions, and email it to the parents and other family members, friends, etc. They can upload the album to the preschool website.

One day when the children went out for a walk in the overgrown field near the preschool, the teacher brought along a flower identification chart. The preschoolers decided that the chart was too difficult for them, so they prepared their own chart with descriptions that made it easier for them to identify the plants. During the hike to the overgrown field the children photographed the flowers in the field with a digital camera and back in the preschool classroom, they created a "flower identification chart" on the computers. Every time they went out for a walk, a

Figure 2. Documenting natural phenomena outdoors. H. took a picture of the same tree in the winter after she had noticed that since the spring the leaves were gone, and the tree was bare.

flower was added to the "identification chart". The chart was prepared using a word processor - Microsoft Word.

During the documentation activity, the children learn to use digital tools, to transfer photos from the camera to the computer, to use a graphics program to edit the photo, to open files, to manage folders, and to build a portfolio. This kind of activity also promotes literacy skills. The documentation can also preserve rare and precious moments, continuous processes, or momentary events. The documentation evokes discussion about phenomena and about the situations that were photographed, thus develops a scientific discourse and the usage of scientific terms. This process engages the children in learning to use the software through mutual help and peer learning.

Googale: A Safe Digital Environment to Support Inquiry

Googale (2009 LTD) is a digital web based environment aimed at providing young children with tools to manage knowledge acquisition, information retrieval, gathering and analyzing quantitative data, visual representations and communication. It was developed to assist young children whose reading and writing abilities are limited.

The Googale environment includes tools like a simple survey, designing different kinds of graphs, weather chart, e-mail, etc. All tools are part of a closed and safe environment, available only for registered members.

Information Literacy: The database contains information in topics that are relevant for young children. Information is retrieved using symbols. Information is presented in audio readings, pictures, and sounds.

Electronic Mail: The mailing list is a closed environment where children can communicate by writing, inserting photos, or drawing pictures. This develops their verbal and non-verbal communication. Children send e-mail to their peers, parents, teacher, etc.

Special Features: The "How Much? How Many" tool can log data and track various phenomena from the children's daily lives. This tool assists the children in organizing data, building representations, and drawing conclusions. It is a simple graphical tool that enables the children to practice scientific thinking and scientific inquiry, as well as mathematical thinking - discussing quantities, numbers, digits, units, measurements, cardinal and ordinal numbers, etc.

All the tools in the Googale environment let the children write, paint, decorate, and design. Table 2 presents examples of Googale activities and the graphical representations that the children can design and discuss.

The Whole World in My Hand: Virtual Travels

In her preschool classroom, teacher A. was hearing the children constantly utter the word "boring": "This is boring..." "That's boring..." The teacher told them that "Boring" is the name of a place in the United States. The preschoolers were fascinated, and wanted to know more. At first, the children and the teacher located the United States on a big globe that stands in the preschool classroom (Figure 3). This prompted a search on the Internet, and the search results were projected onto the smart board in the preschool classroom (Figure 4). The preschoolers were able to ask questions and look for answers and photos about Boring, to learn about how people live in a small farming town, and to take a virtual walk along the streets of the town using Google Street View.

This is an example of how arousing the children's curiosity triggers a search for various kinds of information in the inexhaustible knowledge database that comprises the Internet, using only the assortment of learning tools available in the preschool classroom.

In another classroom, Teacher R. and her preschool students held a discussion about the transformation of day and night: How is day "created" and what is day? How is night "created" and what is night? R. explained the phenomenon to the preschoolers, using models and digital animations. "And then," she recalled, "I

Table 2. Examples of graphical representations children can design and discuss as part of inquiry and project-based learning processes conducted in the Googale environment

Activity	Procedure	Graphical representations
Scientific inquiry: Tracking plants growth	• Choosing the plants • Retrieving information in database about the plant and the best season (month) for planting. • Setting goals and inquiry questions like: Compare growth in natural light and in the dark. • Wording hypotheses. • Planning and setting up the experiment. • Data collection (measuring the length of plants) • Organizing and representing findings in tables and graphs. • Explaining, reasoning, discussing, drawing conclusions, reporting to peers.	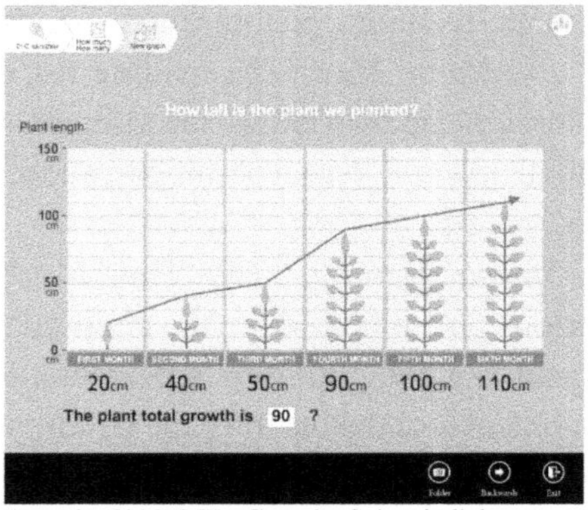 Googale (2009) LTD. Growth of plant in light.
Survey	• Choosing the topic • Retrieving information in database. • Setting goals and inquiry questions like: How many children are in your family? • Wording hypotheses. • Planning and setting up the questions for the survey. • Organizing and representing findings in tables and graphs. • Explaining, reasoning, discussing, drawing conclusions, reporting to peers.	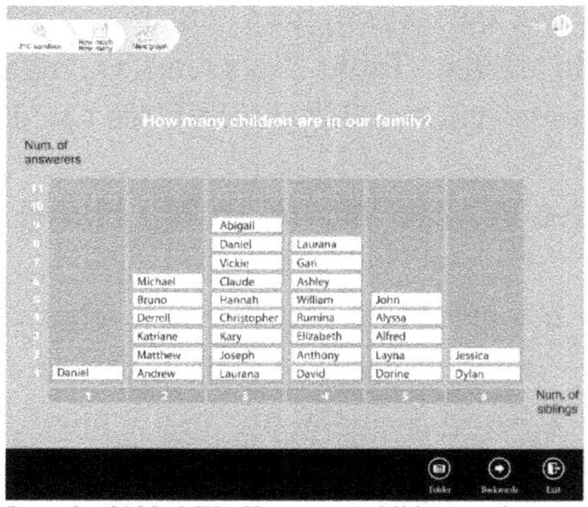 Googale (2009) LTD. How many children are in our family?

told the children that my brother lives in New York. They marked the location on the globe. At 10 AM local time (Israel) I asked the children what time they thought it was in New York, and whether it was day or night over there. We took another look at where the New York is located compared to Israel. At that moment, I called my brother using the Skype application with a webcam. My brother appeared on the screen in the preschool classroom and said: 'Good night'. We asked my brother what time it was over there, and we also asked him to turn the camera around so that we could see what was going on outside the window of his apartment. We saw that it was dark. That's how we illustrated the difference between day and night in different places around the world."

Figure 3. Children and the teacher located the United States on a big globe

Figure 4. This prompted a search on the Internet, and the search results were projected onto the smart board in the preschool classroom

RECOMMENDATIONS

Judicious and beneficial integration of technology can occur only when policy makers, teacher trainers, in-service teacher professional development programs ensure that preschool teachers and the educational support staff at the preschools receive proper and worthwhile training to be able to work in technology-rich environments with young children.

The educational staff must become acquainted with the various tools that are constantly changing as they are updated; the opportunities to use digital media; the tools for assessment, data analysis, presentation of knowledge, and so on. Support must be provided all year round, every year. Besides pedagogical and emotional support for the educational staff, policy makers must allocate an adequate budget of financial and human resources for immediate technical support when something malfunctions, and for maintenance of the equipment. However, it must be emphasized that it is not necessary to have a lot of knowledge about computers and software in order to get started. *The computer is only one of the tools in the preschool, and it can be adapted for the optimal comfort level of all the users.*

Beyond these, additional conditions may contribute to the success of technology integration in preschools:

- Educational policy makers in collaboration with teachers must draft a clear and detailed policy regarding the integration of technology and digital media in preschools (e.g. NAEYC, 2012).
- Ongoing evaluation should be taken, and examination of the aims, objectives, and achievements of all stakeholders: children, preschool teachers, supervisors, and policy makers.
- Allocating national, local government resources and school specific resources not only for the purchase of equipment, but also for ongoing and immediate pedagogical and technical support, training and supervision for the teachers.
- To abate the parents' concerns about prolonged sitting at the computer and exposure to undesirable content, the parents must be involved in the educational approach and in ways to integrate computers and technology at preschool.
- A computer is a tool for inquiry and discovery, and is not a reward. Not only should those children who behave well be allowed to work at the computer.
- Computers should not be used for drill and practice. This kind of use fails to make judicious use of a computer environment at a young age, where creative thinking, inquiry and social cooperation are more important (NAEYC, 2012).

- Young children need to develop knowledge and experience with technology tools and media: They must differentiate between appropriate and inappropriate use, and begin to understand the consensus that has grown on the matter of inappropriate use. Issues of safety on the Internet - protecting private information, and knowing that there is an adult they can count on and turn to - these are the elements of a budding digital citizenship that starts at an early age. Children must be safeguarded by their teachers and parents against commercialization and exploitation.
- In preschools with an Internet connection, ensure that filters are used to block access to harmful websites.

FUTURE RESEARCH DIRECTIONS

In this chapter, we described a pilot study and practical examples from actual preschools of how different technologies can be used to promote inquiry in preschool classrooms. These scientific activities should be followed by structured studies to examine the actual contribution of the digital mobile devices to young children's scientific understanding, scientific curiosity, high-order thinking and inquiry skills. These probable gains and children's abilities should be compared to children who will experience equivalent scientific activities with no digital devices.

Moreover, further research should examine larger samples of children in order to gain better reliability and generalization of the results.

CONCLUSION

Computerized digital devices are already playing a part in the lives of our young children. The appropriate way to utilize technology in early childhood education settings requires a balance, where technology use is matched with the curriculum, the children's needs, and social interactions. Technology use in early childhood settings also requires sufficient support, in particular because the new digital media and information technology in early childhood settings is a quite seldom used (Rikala & Kankaanranta, 2014).

Thus, we must investigate how to use technology judiciously and beneficially in the preschool classroom, and only when technology offers a clear advantage over other means. Technology should be used in a reasonable measure, as one element in the range of tools and aids that are available in the preschool classroom. It makes no sense to use technology to teach the same old materials in the same ways. Efforts should be made to use technology in innovative and worthwhile ways to empower each child, based on the child's cognitive and emotional development, and to meet his/her capabilities and needs.

As we demonstrated in this chapter, digital technology can support the development of young children's inquiry capabilities and scientific literacy. By using digital technology in ways that are developmentally appropriate, employing mediation, judicious selection of software and devices, educators can create stimulating yet safe digital environment that enhance scientific curiosity and exploration among young children.

The extent of the success of technology integration in preschool depends on various factors such as the quality of the classroom learning environment, the socio-economic status of the children, teacher training, the teachers' attitudes, and teacher's mediation (Blackwell, Lauricella, & Wartella, 2014; Cviko, McKenney, & Voogt, 2014). A computer is not a magic wand, but it can certainly contribute to preschool students' development. Technology can assist children, empower, enrich, and advance them - but always alongside the many other tools that exist in the preschool.

As previously indicated, preschool must remain an experiential and safe framework. Activities should be based on physical movement, on playing in the yard and in the preschool classroom, on unmediated sensory exploration, inquiry, and discovery of the natural and artificial environment, and on learning in an interesting setting that welcomes independent learning in diverse ways. Preschool should be a safe place where every child wants to learn, and where every child wants to be.

REFERENCES

American Academy of Pediatrics (AAP), Council on Communications and Media. (2016). Media and Young Minds. *Pediatrics*, *138*(5), 2591. PMID:27940793

Aronin, S., & Floyd, K. K. (2013). Using an iPad in inclusive preschool classrooms to introduce STEM concepts. *Teaching Exceptional Children*, *45*(4), 34–39. doi:10.1177/004005991304500404

Bair, R., & Stafford, T. (2016). Digital literacy. *Training & Development*, *43*(3), 10–11.

Billington, C. (2016). *How digital technology can support early language and literacy outcomes in early years settings: A review of the literature*. London: National Literacy Trust.

Blackwell, C. K., Lauricella, A. R., & Wartella, E. (2014). Factors influencing digital technology use in early childhood education. *Computers & Education, 77*, 82–90. doi:10.1016/j.compedu.2014.04.013

Blackwell, C. K., Lauricella, A. R., Wartella, E., Robb, M., & Schomburg, R. (2013). Adoption and use of technology in early education: The interplay of extrinsic barriers and teacher attitudes. *Computers & Education, 69*, 310–319. doi:10.1016/j.compedu.2013.07.024

Burnett, C., & Myers, J. (2002). "Beyond the frame": Exploring children's literacy practices. *Reading Literacy and Language, 36*(2), 56-62.

Ceppi, G., & Zini, M. (Eds.). (1998). *Children, spaces, relations: Meta-project for an environment for young children.* Reggio Emilia, Italy: Reggio Children.

Christakis, D. A. (2014). Interactive media use at younger than the age of 2 years: Time to rethink the American Academy of Pediatrics guideline? *JAMA Pediatrics, 168*(5), 399–400. doi:10.1001/jamapediatrics.2013.5081 PMID:24615347

Clark, A., & Moss, P. (2001). *Listening to young children: The Mosaic approach.* London: National Children's Bureau for the Joseph Rowntree Foundation.

Clements, D., & Sarama, J. (2003). Strip mining for gold: Research and policy in educational technology – A response to 'Fool's Gold'. *Educational Technology Review, 11*(1), 7–69.

Clements, D. H. (2002). Computers in early childhood mathematics. *Contemporary Issues in Early Childhood, 3*(2), 160–181. doi:10.2304/ciec.2002.3.2.2

Copple, C., & Bredekamp, S. (Eds.). (2009). *Developmentally appropriate practice in early childhood programs serving children from birth through age 8* (3rd ed.). Washington, DC: National Association for Education of Young Children.

Cordes, C., & Miller, E. (Eds.). (2000). *Fool's gold: A critical look at computers in childhood.* Retrieved July 19, 2007, from http://www.allianceforchildhood.net/projects/computers/computers_reports.htm

Cviko, A., McKenney, S., & Voogt, J. (2014). Teacher roles in designing technology-rich learning activities for early literacy: A cross-case analysis. *Computers & Education, 72*, 68–79. doi:10.1016/j.compedu.2013.10.014

De Decker, E., De Craemer, M., De Bourdeaudhuij, I., Wijndaele, K., Duvinage, K., Koletzko, B., & Cardon, G. et al. (2012). Influencing factors of screen time in preschool children: An exploration of parents perceptions through focus groups in six European countries. *Obesity Reviews*, *13*(1), 75–84. doi:10.1111/j.1467-789X.2011.00961.x PMID:22309066

Deák, G. O. (2014). Development of adaptive tool-use in early childhood: Sensorimotor, social, and conceptual factors. *Advances in Child Development and Behavior*, *46*, 149–181. doi:10.1016/B978-0-12-800285-8.00006-6 PMID:24851349

Digital Citizenship. (2016). Retrieved August, 13, 2016, from: www.digitalcitizenship.net

Drigas, A., & Kokkalia, G. (2014). ICTs in Kindergarten. *International Journal of Emerging Technologies in Learning*, *9*(2), 52–58. doi:10.3991/ijet.v9i2.3278

Eshach, H. (2006). *Science literacy in primary schools and pre-schools*. Dordrecht: Springer Verlag. doi:10.1007/1-4020-4674-X

French, L. (2004). Science as the center of a coherent, integrated early childhood curriculum. *Early Childhood Research Quarterly*, *19*(1), 138–149. doi:10.1016/j.ecresq.2004.01.004

Geist, E. A. (2012). A qualitative examination of two year-olds interaction with tablet based interactive technology. *Journal of Instructional Psychology*, *39*(1), 26–35.

Gopnik, A. (2012). Scientific thinking in young children: Theoretical advances, empirical research, and policy implications. *Science*, *337*(6102), 1623–1627. doi:10.1126/science.1223416 PMID:23019643

Haugland, S. (1999). The newest software that meets the developmental needs of young children. *Early Childhood Education Journal*, *26*(4), 245–254. doi:10.1023/A:1022915706904

Haugland, S. W. (1992). The effect of computer software on preschool children's developmental gains. *Journal of Computing in Early Childhood Education*, *3*(1), 15–30.

Hobbs Vinluan, M., & Jennifer Hofman, J. (2014). Transforming Out-of-School Time Environments: The Ys Commitment to Healthy Eating and Physical Activity Standards. *American Journal of Health Promotion*, *28*(3), 116–118. doi:10.4278/ajhp.28.3s.S116 PMID:24380456

Hsin, C. T., Li, M. C., & Tsai, C. C. (2014). The Influence of Young Children's Use of Technology on Their Learning: A. *Journal of Educational Technology & Society*, *17*(4), 85–99.

Huang, D. (2011). Mobile cloud computing. *IEEE COMSOC Multimedia Communications Technical Committee (MMTC) E-Letter*, *6*(10), 27–31.

Israeli Ministry of Education. (2009). *Tochnit limudim be'mada ve'technologia be'gan hayeladim hamamlachty ve'hamamlachty-dati* [Curriculum of science and technology for state and state-religious kindergartens]. Department for Curriculum Planning and Development. Retrieved July 19, 2010, from http://cms.education.gov.il/educationcms/units/tochniyot_limudim/kdamyesodi/tochniyotlimudim/madatechnologia.htm (in Hebrew)

Israeli Ministry of Education. (2016). *Digital learning environments in preschool.* Retrieved August 5, 2016 from: http://edu.gov.il/minhalpedagogy/preschool/subjects/science-and-technology/digital-media-lib/Pages/digital-environments.aspx (in Hebrew)

Istance, D., & Kools, M. (2013). OECD Work on Technology and Education: Innovative learning environments as an integrating framework. *European Journal of Education*, *48*(1), 43–57. doi:10.1111/ejed.12017

Jirout, J., & Klahr, D. (2012). Childrens scientific curiosity: In search of an operational definition of an elusive concept. *Developmental Review*, *32*(2), 125–160. doi:10.1016/j.dr.2012.04.002

Johnson, L., Adams Becker, S., Cummins, M., Estrada, V., Freeman, A., & Ludgate, H. (2013). *Technology outlook for community, technical, and junior colleges 2013-2018: An NMC horizon project sector analysis*. Austin, TX: The New Media Consortium. Houston Community College ISBN.

Johnson, P. (2009). The 21st Century Skills Movement. *Educational Leadership*, *67*(1), 11.

Judge, S., Floyd, K., & Jeffs, T. (2015). *Using mobile media devices and apps to promote young children's learning. In Young Children and Families in the Information Age* (pp. 117–131). Springer Netherlands.

Kearney, M., Schuck, S., Burden, K., & Aubusson, P. (2012). Viewing mobile learning from a pedagogical perspective. *Research in Learning Technology*, *20*(1), 14406. doi:10.3402/rlt.v20i0.14406

Kesner Baruch, Y., Spektor-Levy, O., & Mashal, N. (2014). Pre-schoolers verbal and behavioral responses as indicators of attitudes and scientific curiosity. *International Journal of Science and Mathematics Education, 14*(1), 125–148. doi:10.1007/s10763-014-9573-6

Koehler, M. J., Mishra, P., & Cain, W. (2013). What is technological pedagogical content knowledge (TPACK)? *Journal of Education, 193*(3), 13–19.

Kuhn, D., Black, J., Keselman, A., & Kaplan, D. (2000). The development of cognitive skills to support inquiry learning. *Cognition and Instruction, 18*(4), 495–523. doi:10.1207/S1532690XCI1804_3

Laugksch, R. C. (2000). Scientific literacy: A conceptual overview. *Science Education, 84*(1), 71–94. doi:10.1002/(SICI)1098-237X(200001)84:1<71::AID-SCE6>3.0.CO;2-C

Lauricella, A. R., Wartella, E., & Rideout, V. J. (2015). Young childrens screen time: The complex role of parent and child factors. *Journal of Applied Developmental Psychology, 36*, 11–17. doi:10.1016/j.appdev.2014.12.001

Lerner, C., & Barr, R. (2014). Screen sense: Setting the record straight – Research-based guidelines for screen use for children under 3 years old. *Zero to Three.*

Linebarger, D. L., & Vaala, S. E. (2010). Screen media and language development in infants and toddlers: An ecological perspective. *Developmental Review, 30*(2), 176–202. doi:10.1016/j.dr.2010.03.006

Lipponen, L., Rajala, A., Hilppö, J., & Paananen, M. (2015). Exploring the foundations of visual methods used in research with children. *European Early Childhood Education Research Journal*, 1–11.

Luna Scott, C. (2015). *The Futures of Learning 3: what kind of pedagogies for the 21st century?* Paris: UNESCO Education Research and Foresight.

Maass, K., & Doorman, M. (2013). A model for a widespread implementation of inquiry-based learning. *ZDM, 45*(6), 887–899. doi:10.1007/s11858-013-0505-7

McCarrick, K., & Li, X. (2007). Buried treasure: The impact of computer use on young children's social, cognitive, language development and motivation. *AACE Journal, 15*(1), 73–95.

Mishra, P., & Koehler, M. J. (2006). Technological pedagogical content knowledge: A framework for teacher knowledge. *Teachers College Record, 108*(6), 1017–1054. doi:10.1111/j.1467-9620.2006.00684.x

Mouza, C. (2005). Using technology to enhance early childhood learning: The 100 days of school project. *Educational Research and Evaluation, 11*(6), 513–528. doi:10.1080/13803610500254808

National Academy of Engineering (NAE) & National Research Council (NRC). (2014). STEM integration in K-12 education: Status, prospects, and an agenda for research. Washington, DC: The National Academies Press.

National Association for the Education of Young Children (NAEYC). (2012). *Technology and Interactive Media as Tools in Early Childhood Programs Serving Children from Birth through Age 8*. Position Statement. NAEYC and the Fred Rogers Center. Retrieved August 10 2015, from http://www.naeyc.org/files/naeyc/file/positions/PS_technology_WEB2.pdf

National Education Association (NEA). (2010). *Preparing 21st Century students for a global society*. National Education Association.Retrieved November 12, 2016 from http://www.nea.org/assets/docs/A-Guide-to-Four-Cs.pdf

National Research Council (NRC). (2012). *A Framework for K-12 Science Education: Practices, Crosscutting Concepts, and Core Ideas*. Committee on a Conceptual Framework for New K-12 Science Education Standards. Board on Science Education, Division of Behavioral and Social Sciences and Education. Washington, DC: The National Academies Press.

NGSS Lead States. (2013). *Next generation science standards: For states, by states. The Need for Standards*. Retrieved November 12, 2016 from http://www.nextgenscience.org/need-standards

Nikolopoulou, K., & Gialamas, V. (2009). Investigating pre-service early childhood teachers views and intentions about integrating and using computers in early childhood settings: Compilation of an instrument. *Technology, Pedagogy and Education, 18*(2), 201–219. doi:10.1080/14759390903003837

Nikolopoulou, K., & Gialamas, V. (2015a). Barriers to the integration of computers in early childhood settings: Teachers perceptions. *Education and Information Technologies, 20*(2), 285–301. doi:10.1007/s10639-013-9281-9

Nikolopoulou, K., & Gialamas, V. (2015b). ICT and play in preschool: Early childhood teachers beliefs and confidence. *International Journal of Early Years Education, 23*(4), 409–425. doi:10.1080/09669760.2015.1078727

Nir-Gal, O., & Klein, P. (1999). The Use of the Computer in Early Childhood with or without Adult Mediation–the role of the teacher in the computerized environment. [Hebrew]. *Dapim, 29*, 76–99.

Nuemann, M. M., & Nuemann, D. L. (2014). Touch screen tablets and emergent literacy. *Early Childhood Education Journal, 42*(4), 231–239. doi:10.1007/s10643-013-0608-3

O'Malley, C., Vavoula, G., Glew, J., Taylor, J., Sharples, M., Lefrere, P., Lonsdale, P., Naismith, L., & Waycott, J. (2005). *Guidelines for learning/teaching/tutoring in a mobile environment.* Public deliverable from the MOBILearn project (D.4.1).

Orr, J., Kamdar, D., Lewis Presser, A., & Vahey, P. (2015). Next Generation Preschool Math: Researchers, public media, teachers, and children team up to identify the best practices for tablet technology in the classroom. In D. Rutledge & D. Slykhuis (Eds.), *Proceedings of Society for Information Technology & Teacher Education International Conference 2015* (pp. 2022-2027). Chesapeake, VA: Association for the Advancement of Computing in Education (AACE). Retrieved November 12, 2016 from https://www.learntechlib.org/p/150279

Partnership for 21st Century Skills. (2008). *A report and mile guide for 21st Century skills.* Retrieved October 15, 2010, from Http://www.21stCenturyskills.org/downloads/P21_Report.pdf

Patrick, H., Mantzicopoulos, P., & Samarapungavan, A. (2009). Motivation for learning science in kindergarten: Is there a gender gap and does integrated inquiry and literacy instruction make a difference. *Journal of Research in Science Teaching, 46*(2), 166–191. doi:10.1002/tea.20276

Puentedura, R. R. (2011). The iPad and Research Design (Data / Research Sessions I, II, III). In Leveraging Learning: The iPad in Primary Grades. Auburn, ME: Academic Press.

Radetsky, J. S., Schumacher, J., & Zuckerman, B. (2015). Mobile and interactive media use by young children: The good, the bad, and the unknown. *Pediatrics, 135*(1), 1–3. doi:10.1542/peds.2014-2251 PMID:25548323

Reid, N., & Yang, M. J. (2002). Open-ended problem solving in school chemistry: A preliminary investigation. *International Journal of Science Education, 24*(12), 1313–1332. doi:10.1080/09500690210163189

Reynolds, M. A., Jackson Cotwright, C., Polhamus, B., Gertel-Rosenberg, A., & Chang, D. (2013). Obesity prevention in the early care and education setting: Successful initiatives across a spectrum of opportunities. *The Journal of Law, Medicine & Ethics, 41*(2), 8–18. doi:10.1111/jlme.12104 PMID:24446993

Rikala, J., & Kankaanranta, M. (2014). The Nature Tour mobile learning application. Implementing the mobile application in Finnish early childhood education settings. In *CSEDU 2014: Proceedings of the 6th International Conference on Computer Supported Education* (vol. 3). SCITEPRESS-Science and Technology Publications.

Roskos, K., & Neuman, S. B. (2011). The classroom environment. *The Reading Teacher, 65*(2), 110–114. doi:10.1002/TRTR.01021

Samarapungavan, A., Mantzicopoulos, P., & Patrick, H. (2008). Learning science through inquiry in kindergarten. *Science Education, 92*(5), 868–908. doi:10.1002/sce.20275

Schroeder, C. M., Scott, T. P., Tolson, H., Huang, T. Y., & Lee, Y. H. (2007). A meta-analysis of national research: Effects of teaching strategies on student achievement in science in the United States. *Journal of Research in Science Teaching, 44*(10), 1436–1460. doi:10.1002/tea.20212

Scott, C. (2012). An investigation of science, technology, engineering and mathematics (STEM) focused high schools in the US. *Journal of STEM Education Innovations and Research, 13*(5), 30–39.

Segers, E., & Verhoeven, L. (2002). Multimedia support of early literacy learning. *Computers & Education, 39*(3), 207–221. doi:10.1016/S0360-1315(02)00034-9

Shams, L., & Seitz, A. R. (2008). Benefits of multisensory learning. *Trends in Cognitive Sciences, 12*(11), 411–417. doi:10.1016/j.tics.2008.07.006 PMID:18805039

Sharples, M., Taylor, J., & Vavoula, G. (2005). Towards a theory of mobile learning. *Proceedings of mLearn*, 1(1), 1-9.

Shih, J. L., Chu, H. C., Hwang, G. J., & Kinshuk, . (2011). An investigation of attitudes of students and teachers about participating in a context-aware ubiquitous learning activity. *British Journal of Educational Technology, 42*(3), 373–394. doi:10.1111/j.1467-8535.2009.01020.x

Shulman, L. S. (1986). Those who understand: Knowledge growth in teaching. *Educational Researcher, 15*(2), 4–14. doi:10.3102/0013189X015002004

Siraj-Blatchford, I., & Siraj-Blatchford, J. (2006). *A guide to developing the ICT curriculum for early childhood education*. Stoke-on-Trent, UK: Trentham Books.

Song, Y. (2014). Bring Your Own Device (BYOD) for seamless science inquiry in a primary school. *Computers & Education, 74*, 50–60. doi:10.1016/j.compedu.2014.01.005

Spektor-Levy, O. (2012). Chashiva al machshev bagan [Thoughts about ICT in preschool classes]. *Alon Da-Ga, 5*, 56–69.

Stephen, C. (2006). *Early Years Education: Perspectives from a Review of the International Literature*. Academic Press.

Stephen, C., & Plowman, L. (2008). Enhancing learning with information and communication technologies in pre-school. *Early Child Development and Care, 178*(6), 637–654. doi:10.1080/03004430600869571

Strasburger, V. (2015). Should babies be watching and using screens? The answer is surprisingly complicated. *Acta Paediatrica (Oslo, Norway), 104*(10), 967–968. doi:10.1111/apa.13105 PMID:26378634

Swaminathan, S., & Wright, J. L. (2003). Education technology in the early and primary years. In J. P. Isenberg & M. R. Jalongo (Eds.), *Major trends and issues in early childhood education: Challenges, controversies, and insights* (2nd ed.; pp. 136–149). New York: Teachers College Press.

Tang, K. S., Delgado, C., & Moje, E. B. (2014). An integrative framework for the analysis of multiple and multimodal representations for meaning-making in science education. *Science Education, 98*(2), 305–326. doi:10.1002/sce.21099

Tang, K. Y., Li, M. C., Hsin, C. T., & Tsai, C. C. (2016). A Co-Citation Network of Young Children's Learning with Technology. *Journal of Educational Technology & Society, 19*(3), 294–305.

Tremblay, M. S., LeBlanc, A. G., Carson, V., Choquette, L., Connor Gorber, S., Dillman, C., & Kho, M. E. et al. (2012). Canadian sedentary behaviour guidelines for the early years (aged 0–4 years). *Applied Physiology, Nutrition, and Metabolism, 37*(2), 370–380. doi:10.1139/h2012-019 PMID:22448609

Vernadakis, N., Avgerinos, A., Tsitskari, E., & Zachopoulou, E. (2005). The use of computer assisted instruction in preschool education: Making teaching meaningful. *Early Childhood Education Journal, 33*(2), 99–104. doi:10.1007/s10643-005-0026-2

Weiss, I., Kramarski, B., & Tallis, S. (2006). Effects of multimedia environments on kindergarten childrens mathematical achievements and style of learning. *Educational Media International, 43*(1), 3–17. doi:10.1080/09523980500490513

Wong, L.-H., & Looi, C.-K. (2011). What seams do we remove in mobile-assisted seamless learning? A critical review of the literature. *Computers & Education, 57*(4), 2364–2381. doi:10.1016/j.compedu.2011.06.007

Wu, J., & Zhang, Y. (2010). Examining potentialities of handheld technology in students academic attainments. *Educational Media International*, *47*(1), 57–67. doi:10.1080/09523981003654977

Zhang, B. H., Looi, C.-K., Seow, P., Chia, G., Wong, L.-H., Chen, W., & Norris, C. et al. (2010). Deconstructing and reconstructing: Transforming primary science via a mobilized curriculum. *Computers & Education*, *55*(4), 1504–1520. doi:10.1016/j.compedu.2010.06.016

Zhao, Y. (2012). *World class learners: Educating creative and entrepreneurial students*. Corwin Press.

Zion, M., Slezak, M., Shapira, D., Link, E., Bashan, N., Brumer, M., & Valanides, N. et al. (2004). Dynamic, open inquiry in biology learning. *Science Education*, *88*(5), 728–753. doi:10.1002/sce.10145

KEY TERMS AND DEFINITIONS

21st Century Skills: Skills, knowledge, and expertise to be mastered by the time students around the world graduate from education frameworks to ensure successful careers, as well as personal, social, and economic self-sufficiency. 21st Century Skills include skills that are necessary for today's technology-rich world, such as critical thinking, problem solving, communication, and collaboration; the skills required for digital literacy.

Application: Software program to use on a computer or on a mobile device such as a smart phone or tablet.

Digital Citizenship: Understanding of the use, abuse, and misuse of technology as well as the norms of appropriate, responsible, and ethical behaviors related to online rights, roles, identity, safety, security, and communication.

Digital Literacy: Digital Literacy involves the ability to aggregate, analyze, synthesize and communicate data using technology tools and sources. It is not only a matter of the technical skills needed to operate digital tools and computing devices, but also the cognitive, psychological and social processes that are beneficial for working in a technology-rich environment.

Digital Media: Digital materials, including software programs, applications (apps), TV programs, broadcast and streaming media, e-books, the Internet, and other forms of content designed to facilitate active and creative use and to encourage social interactions.

Mobile Learning: Learning that takes place when the learner moves from one place to the other while using mobile technological device such as smart phone, hand-held computers, and laptop.

Screen Time: The time individual spends watching screens such as TV, computer, Laptops, smart phones.

Chapter 2
Computer-Supported Imagination:
The Interplay Between Computer and Mental Simulation in Understanding Scientific Concepts

Franco Landriscina
University of Trieste, Italy

ABSTRACT

The purpose of this chapter is to explore some conceptual and methodological issues related to computer-based simulation as a teaching and learning method for STEM education. Two major themes will be examined in detail: 1) the barriers to the penetration of simulation into school programs; 2) simulation as a way to enhance scientific imagination in students. It is argued that scientific imagination is linked closely with mental simulation, a fundamental capacity of the human brain which allows us to move from static to dynamic mental representations. The role of mental simulation in understanding scientific concepts is discussed and an explicit statement is provided of the relation between computer simulation and mental simulation. On these grounds, computer simulation is viewed as a tool for extending human cognition by overcoming the limits of mental simulation. Finally, the implications of these findings for designing simulation-based instructional units and conducting lessons are discussed.

DOI: 10.4018/978-1-5225-2525-7.ch002

INTRODUCTION

Will polar bears survive global warming? What happened in the first moments after the Big Bang? How can we stop epidemics from spreading? What are the effects of increasing atmospheric carbon dioxide levels on agriculture in various world regions? How can we design car interiors to improve passenger crash safety? Every day scientists and engineers all over the world use computer simulation to try to answer questions like these, and the more accurate their simulations are, the more reliable their answers will be.

To this end, universities, national laboratories, and corporate research centers are investing in supercomputing facilities where the technological limits of simulation push the envelope on our current simulation capacity. To illustrate current progress in the field, consider, for example, that biologists recently managed to create a complete computer model of an entire single-celled organism and can now simulate the basic processes of cell physiology, from DNA replication to metabolic reactions (Covert, 2013). In industry, many new products are carefully simulated before manufacturing, and simulation also helps streamline the production process. Simulation as a research method has been extended to the social sciences: economists, sociologists, and anthropologists are using simulations to understand phenomena such as the onset of financial crises, the emergence of cooperation among groups, and the fall of ancient civilizations. Furthermore, simulation is a recurrent theme in many science fiction novels, movies, and video games, to the extent that it influences our conceptions of reality itself. According to Winsberg (2010), *the last part of the twentieth century has been, and the twenty-first century is likely to continue to be, the age of computer simulation* (p. 9).

The ubiquity of simulation is due to the many advantages it offers to scientists and engineers alike. Simulation allows them to view otherwise invisible phenomena, to practice without space or time constraints, to proceed by trial and error safely and cost-effectively, to verify alternative hypotheses, and to carry out actions that are unfeasible in the real world.

These same advantages suggest that simulation can be a valuable adjunct to STEM education. Think of the possibility to visualize the air flow around the wing of an airplane or oceanic currents, to be able to see the ways in which a substance's molecules react to changes in pressure or temperature, to control the flow of electrons in an electronic circuit, to study photosynthesis in a virtual laboratory, or to change the reflectance of ice and water at the North Pole. Furthermore, innovative user interfaces give students new ways to interact with simulations, such that they can actually touch intermolecular forces with haptic devices which simulate tactile feedback (Han & Black, 2011), use sensor-augmented technologies to conduct experiments which merge the virtual and the physical world (Xie, 2012a), and control a simulation using their bodies and gestures (Lindgren et al., 2016).

Thus, many researchers and policymakers consider simulation to be an effective teaching method and are encouraging its implementation in school settings. For example, a National Academy of Education white paper states that *carefully designed simulations and computer models can help students visualize and understand complex concepts about objects and processes* (Kilpatrick & Quinn, 2009, p. 8). Smetana and Bell (2012) reviewed the literature on the effectiveness of simulation for science teaching and learning, and concluded that *simulations can be as effective, and in many ways more effective, than traditional (i.e., lecture-based, textbook-based and/or physical hands-on) instructional practices in promoting science content knowledge, developing process skills, and facilitating conceptual change* (p. 1337). Similarly, a recent comprehensive meta-analysis supports the view that that *using computer-based simulations in science, in many different configurations and contexts within the classroom, does improve student science achievement compared to not using simulations* (D'Angelo et al., 2014 p. 40).

One might therefore imagine that simulation is already a widely used instructional method in schools and universities, in addition to more traditional methods. Yet, little headway has actually been made in integrating simulation into school curricula. In fact, according to the report entitled *Learning Science Through Computer Games and Simulations* (National Research Council, 2011) *several barriers slow large-scale development and use of games and simulations for science learning in K-12 and higher education* (p. 175). Some of the barriers highlighted in the report are: need for a closer alignment with school curricula, teachers' professional support needs, and school policies that allocate funds for hardware and software purchases.

The relative absence of simulation in science education represents a notable paradox when compared to its widespread use in science and engineering. Furthermore, the limited diffusion of simulation in school curricula is paralleled by a lack of in-depth research on the instructional effectiveness of simulation. As evidenced by the above-cited report: *much research has been exploratory, making it difficult to generalize, because researchers and developers have not always clearly defined the desired learning outcomes or the mechanisms by which the simulation or game is expected to advance these outcomes* (p. 120). As a consequence, principles and guidelines on ways to maximize the instructional effectiveness of simulation for different educational contexts and learning goals are actually quite scarce.

This chapter's aim is to present an interpretation of simulation-based learning in terms of the interplay between the cognitive ability of mental simulation and computer-based simulation. It proposes that mental simulation provides a deeper level of processing for students than do formal reasoning and exclusively verbal explanations. It suggests that students can tap this ability to better understand and apply novel scientific concepts with an emphasis on interdisciplinary concepts that are common to science, technology and engineering.

BACKGROUND

Type of Simulation

There has at times been confusion in the field of simulation-based learning about what an instance of simulation is. It is important, for example, to distinguish simulation from other computer-based technologies that can play a key role in instruction, but which differ nonetheless from simulation, such as visualizations of scientific data, computer algebra programs, or molecular graphics tools. The distinguishing feature of simulation is the reproduction of selected aspects of a system's behavior, in function of a user's decisions, which allows a student or scientist to modify the system's time evolution by altering the simulation's initial conditions or other parameters.

Another point to consider concerns the distinctive features of different types of simulation, which may be subdivided into 2 broad types:

- **Experiential Simulations:** Based on the creation of a virtual situation or event to be experienced by one or more participants;
- **Model-Based Simulations:** Based on the construction of the theoretical model of a real or imagined system (also known as "theoretical" or "conceptual" simulations).

Experiential simulations are typically used to teach a given procedure or operation by providing learners with scenarios resembling reality. In industry training, they can be virtual reality environments for users to practice real-world skills - such as piloting an airplane or a train. In STEM education, they are often virtual laboratories in which the learning goal is to learn how to perform an activity which would otherwise require a physical lab. An example is that of simulated electrical circuit construction kits, in which students can create their own electrical circuits and do measurements on it.

The instructional goals and learning processes involved in experiential and model-based simulations are, in general, different and combining data from these two types of simulation to draw general conclusions about the effectiveness of simulation as a teaching method can be misleading. The rest of the present chapter will focus on model-based simulations, because these are the most frequently used by scientists to learn new aspects of natural and artificial systems, and by engineers to design new products and processes; thus, from here on in the term "simulation" will refer exclusively to this type of simulation.

Difference Between Simulation and Games

The terms "simulation" and "game" are frequently used interchangeably, but the two activities are diverse in nature:

- Whereas a simulation is based on the imitation of a system or a situation, a game is not bound by this constraint;
- The element of competition is not necessarily present in simulations, but conversely is a typical component of games.[1]

Consequently, different cognitive processes are involved, and not distinguishing between simulations and digital games can lead to confusion about the conclusions of a study or classroom intervention.[2]

Simulation-Using vs. Simulation-Building

To understand the ways in which simulation can foster learning in so many different contexts, it is important to note that simulation-based learning occurs in two main ways: *simulation-building* and *simulation-using*. In the first case, students must use either a programming language or modeling and simulation software to create a simulation model on their own. The activities involved are:

- Analyzing a specific system;
- Developing a conceptual model of it;
- Creating a computational model;
- Implementing the computational model as a simulation program; and
- Conducting numerical experiments on it.

Conceptual and methodological contributions from Modeling and Simulation (M&S) can be very helpful to students and teachers wanting to build a simulation. This field developed in the context of systems science and engineering, and M&S experts are able to describe in detail the activities and outputs of a typical simulation project, independently from the software or type of simulation utilized (see: Birta & Arbez, 2007; Robinson, 2004; Sokolowsky & Banks, 2010 for an overview of the field).

On their part, instructional scientists have emphasized the instructional value of having students build, evaluate, revise, and elaborate their own models (e.g., Gobert

& Buckley, 2000; Jonassen, 2000). Yet, simulation building is not always feasible in the classroom, due to the difficulty, time, and costs involved. Therefore, students more frequently use a previously developed simulation, typically an instructional resource available on the Web. The relevant student activities in this case are:

- Using the simulation program as required by the learning task at hand;
- Constructing their own explanations of the system or problem;
- Comparing their own explanations with the target conceptual model; and,
- Using the knowledge, they acquire about the conceptual model to gain a better understanding of the system represented by the model.

Degree of Opacity

Students nearly always do not interact directly with the underlying model, but exclusively through the mediation of the simulation program's user interface, and/or through teacher-designed learning activities. One factor that can significantly increase or diminish the instructional effectiveness of a simulation is the underlying model's degree of opacity. It is possible to distinguish between *black-box model* simulations, in which students can explore a system's behavior - but the underlying conceptual and/or computational models remain hidden and can only be inferred by what is viewed on the screen - and *glass-box model* simulation, in which the model is made explicit to students, e.g., by showing a causal diagram or the equations/rules that underlie the model's behavior.

Design Issues

Simulation programs are typically characterized by a high number of interacting elements (such as texts, pictures, animations, levers, and graphs), which require simultaneous processing in working memory and produce a high intrinsic cognitive load thereby (Plass, Homer, & Hayward, 2009). If the user interface is not well-designed, it can cause high extraneous cognitive load. The key learning factors in situations such as these will be students' prior knowledge and attentional abilities. It is therefore important to understand the degree of cognitive load caused by a simulation, as it significantly interacts with other factors to produce different outcomes.

Type of Learning Goal

Simulation can be used for many different instructional purposes; explicitly defining a simulation's learning goal is therefore a crucial first step in designing the simulation's features and relative student activities. The main types of learning goals for simulation are:

- Increasing students' interest in science (e.g., the motivation to pursue scientific studies);
- Acquiring more specific subject matter knowledge (e.g., learning the concepts and principles of a scientific discipline);
- Developing core scientific concepts and practices (e.g., unifying concepts and practices of an interdisciplinary nature);
- Developing scientific-inquiry skills (e.g., asking questions, conducting investigations, analyzing data, constructing explanations, communicating results);
- Developing thinking skills (e.g., critical-thinking skills, problem-solving skills, systems thinking).

These differences in learning goals require different learning activities and types of assessment, making it difficult to compare results in the literature. The present chapter is focused on conceptual understanding and scientific-inquiry skills, positing that simulations can be particularly suited to advance these types of learning goals.

Support Provided to Students

A factor that notably influences the effectiveness of simulation is the type of cognitive support provided to students by teachers, or by the simulation program itself. Simulation-based learning environments frequently feature instructional supports and resources aimed at facilitating and enriching the students' learning experience, such as questions, hints, explanations, exploration guides, exercises, graphing tools, and planning tools (de Jong, 2006).

As occurs with the variable of type of learning goal, the teaching strategy selected can significantly impact a simulation's instructional efficacy.

Instructional Design Issues

Several decisions that affect learning outcomes pertain to the sequence in which simulation is presented to students and to its potential for integration with other teaching methods and techniques. With respect to sequencing, it is important to note that simulation can be used to present new content or as a kind of practice - similarly to laboratory activities - to help a student reinforce and apply knowledge acquired by other means. For instance, Brant, Hooper, and Sugrue (1991) investigated whether it is more effective to use simulation before or after a lesson. They found that student participants receiving simulation as a framework for understanding prior to formal classroom instruction scored significantly higher on an applications post-test than did students using simulation as an integrating activity following formal instruction.

Epistemic Beliefs About Simulation

The beliefs with which students and teachers view simulation as a learning method can significantly influence on their attitudes toward the approach. As shown by research on personal epistemologies, these types of beliefs significantly affect different aspects of the learning process. One key obstacle to broader acceptance by teachers is the idea that simulation is somehow an imperfect imitation or reproduction of an objective reality, and therefore, is an only partially reliable source of knowledge. Consequently, some teachers may view the method with suspicion, believing that it risks showing students a poor or distorted representation of reality.[3] The issue warrants further investigation because, on a broader scale, it can negatively impact the integration of simulation into school curricula.

SIMULATION IN THE CONTEXT OF MODEL-BASED LEARNING

Model-Based Learning and Teaching (Gobert & Buckley, 2000; Clement, 2000), also known as Model-Centered Learning and Instruction (Seel, 2003) is an approach that focuses on students' mental models, which are assumed to be knowledge structures that students construct to process new tasks or to solve problems. It also highlights the ways in which these internal models are expressed through external models, and the changes that occur in them as students learn new concepts. Specifically, the learning process is viewed as a progression of mental models, which leads from an initial mental model based on student's preconceptions and intuitions, to a target model, which students should possess after instruction, through a succession of intermediate models.[4]

Many parallels can be made between simulation-based learning and model-based learning. For example, when building a simulation, students express their initial mental models in the form of hypotheses about the system under study. These hypotheses are then specified as conceptual models, which include the model's objectives, inputs, outputs, content, assumptions and simplifications. Based on the experiments conducted with a given simulation, changes can be made in its conceptual and/or computational models (see the two-way arrows in Figure 1). A great deal of iteration may be required before the computational model can be considered to satisfactorily represent the conceptual one.

Figure 1. Simulation-based learning

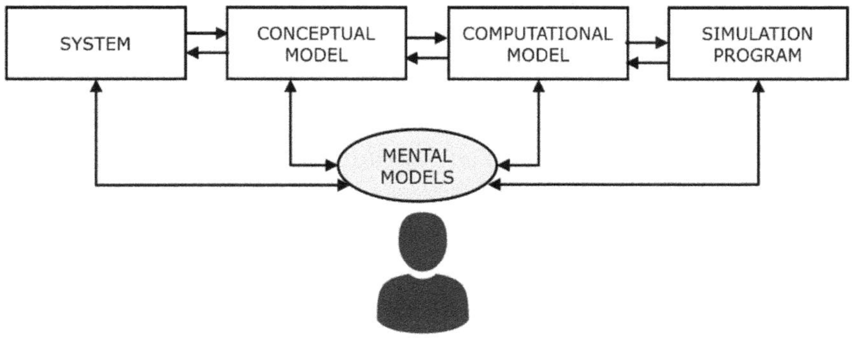

Similarly, when using an existing simulation, students also construct and revise their mental models to explain the phenomena and the data they see on the screen. To summarize the similarities between simulation-based learning and model-based learning, both approaches involve (1) an interplay of internal and external models, (2) the use of different and coordinated modes of representation, and (3) learning processes based on a progression of mental models.

The main difference between simulation-based learning and model-based learning is in the fact that computational models are a special kind of *dynamic model*, which are able to reproduce some aspects of a system's time evolution and also allow students to interact with the model by changing the simulation parameters and exploring the consequences of different decisions thereby. These aspects affect students' cognitive processes and, in particular, changes in their mental models. As stated by Seel (2012), dynamic modeling provides a new perspective, called *learning by system modeling*. In this perspective, *Learning occurs by comparing the expected results of operations on a system with the observed consequences of transformations. In the case of gaps between expectations and observations, the outcomes are used to update or revise the mental model.* (Seel, 2012, p. 1053).

In other words, learning by system modeling entails: (1) using a mental model to generate a prediction about the system's future behavior, (2) observing the actual behavior by running a computer simulation, and (3) updating or revising the mental model based on the simulation's outcomes.

Yet, how can a mental model be used to make predictions? A possible answer, is that it allows students to run a kind of "mental simulation", that is, a cognitive operation that consists in mentally manipulating the objects involved in a situation to perform transformations on it that may occur in the real world. This cognitive operation, and its consequences for scientific reasoning and learning, will be examined in the next section.

SIMULATION AS A COGNITIVE OPERATION

In earlier studies on mental models, many of which date back to the 1980s, it was assumed that mental models are in some way analogous to simulation models. Kahneman and Tversky (1982) first noted that *there appear to be many situations in which questions about events are answered by an operation that resembles the running of a simulation model* (p. 201). The analogy was put forth by researchers in the field of Artificial Intelligence, who underscored that *it should be possible for people to 'run' their models mentally* (Norman, 1983, p. 12), and that *mental models often permit mental simulation: the sense of being able to run a mental model internally, so that one can observe how it will behave and what the outcome of the process will be* (Gentner, 2002, p. 9684). It was argued that *we could 'run a mental simulation' and imagine the events that would take place in the world when we performed a particular action. This mental model would allow us to perform actions entirely internally and to judge the consequences of our actions, interpret them, and draw conclusions based on them* (Rumelhart, Smolensky, McClelland, & Hinton, 1986, p. 42). Overall, mental simulation is conceived as a relatively common type of reasoning that occurs in many everyday situations - for example, when figuring out the steps required to assemble a piece of furniture, deciding how to handle a difficult conversation, making the decision to buy a pizza the way home from work, or reading a novel and experiencing the characters' thoughts and emotions.

The idea that mental simulation is a mechanism underlying many different cognitive abilities has garnered particular attention in the "embodied cognition" and "grounded cognition" approaches to the study of the mind.[5] A currently influential mental simulation definition has been proposed by Barsalou (2008): *Simulation is the re-enactment of perceptual, motor, and introspective states acquired during experience with the world, body, and mind* (p. 618). Furthermore, Barsalou offered an account of how mental simulation works as a key process in in categorization and concept formation: *As an experience occurs (e.g., easing into a chair), the brain captures states across the modalities and integrates them with a multimodal representation stored in memory (e.g., how a chair looks and feels, the action of sitting, introspections of comfort and relaxation). Later, when knowledge is needed to represent a category (e.g., chair), multimodal representations captured during experiences with its instances are reactivated to simulate how the brain represented perception, action, and introspection associated with it* (Barsaloi, 2008, p.618). In this view, a "simulator" forms after several experiences with individual instances of a category, and corresponds to a concept, as intended in traditional cognitive theories. Importantly, once a simulator is formed, it can reactivate a subset of its perceptual components as specific simulations and can thereby create an infinite number of simulations, in function of a given context. For instance, as described above, a simulator for the concept of "chair" forms after several experiences with this type of object; it can then create simulations of events, such as those of standing on a chair, sitting in the armchair of a cinema, lifting a chair and taking it to another room, etc. It is the context that determines which simulation will be activated (for this reason, the terms *situated simulations* and *situated conceptualizations* are also used, see: Barsalou, 2015, Connell and Lynott, 2014).

It is important, however, to note two important points: mental simulations never completely recreate the original experience, and they can be either unconscious - as is the case with the simulations associated with perception or motor movements - or conscious, as in the case of mental imagery or memory of specific episodes. (For the rest of the present chapter the term 'mental simulation' will be used exclusively in reference to conscious simulation, which entails an intentionally constructed representation in working memory).

"Mechanical reasoning" is a line of mental simulation research that has particular relevance for STEM education. It is the study of mental representations people form to understand the functioning of simple mechanical systems, starting from descriptions of these systems in the form of texts and diagrams. Hegarty (2004) reviewed the evidence that conscious mental simulation is sometimes used in this kind of reasoning. She also underscored a key difference between visual imagery and mental simulation, by stating that visual imagery is based on the holistic inspection of a mental image of the moving system, whereas mental simulation (a) is based on the step-by-step simulation of the events, (b) incorporates non-visible properties (e.g., force or density), and (c) it often includes the representation of associated motor actions.

A further difference between visual imagery and mental simulation is that the latter can be used with "if-then" rules. In fact, Schwartz and Black (1996) found that participants knowing verbal rules to infer a movement rely on these, rather than on simulation, to solve problems more quickly. The two researchers proposed that people use mental simulation in novel situations for which they have no rules available or for which they have inadequate rules, and that, they otherwise rely on the application of verbal rules (e.g., the "parity rule" for determining the motions of linked gears in a mechanical system, which states that "if there are an odd number of gears connected, then the first will go in the same direction as the last").

The strategy of relying on verbal rules can be related to the fact that mental simulation involves the construction of temporary representations in working memory and consequently imposes a high load on the processing capacity of working memory's visual-spatial components. Consequently, mental simulation has a high cognitive cost as compared to other, more automatic mental processes, such as applying known verbal rules.

Based on the fact that mental simulation makes it possible to generate inferences by creating mental models and simulating them, one might wonder whether scientists use mental simulation as a complement to logic-based reasoning. Notably, the answer is "yes", given that recent studies on practical scientific inquiry methods have shown that scientists use cognitive processes akin to mental simulation to generate hypotheses (Clement, 2008, 2009), create novel concepts (Nersessian, 2008), and to interpret data in complex knowledge domains (Trickett & Trafton, 2007). Moreover, it appears that scientists and experts most frequently use the tool of mental simulation when evaluating hypotheses and in situations involving uncertain information - e.g. when the data available are unclear or anomalous.

Yet, do students faced with the task of learning scientific concepts also use this cognitive operation? The answer is affirmative once again, as explained in the next section, which will also explore the conditions under which mental simulation can lead to enhanced learning.

SIMULATION AND LEARNING: A GROUNDED COGNITION PERSPECTIVE

When studying simulation as an educational method, a simulation program's features (e.g., technology, visualization methods, level of interactivity) may garner more attention, but the mental processes involved in students' interaction with them must also be considered.

Imagine a group of students attempting to learn about the concept of heat and the related notion of heat transfer. Typically, they would be given verbal definitions in the form of written texts or teacher explanations, complemented by images (e.g., photos or videos of natural phenomena involving heat, drawings of experimental apparatus, technical schemas, diagrams), lab experience (e.g., a heat conduction experiment comparing the thermal properties of different materials), and classroom discussions. What then, would be the most appropriate mental counterparts for learning these two concepts?

A growing number of studies on embodied and grounded cognition and on semantic memory have been focusing on the mental processes that underlie the understanding of concepts and can now provide some indications (e.g., see Barsalou, 2015; Barsalou, Santos, Simmons, & Wilson, 2008; Binder & Desai, 2011; Casasanto & Lupyan, 2015; Yee & Thompson-Schill, 2016). In brief:

- The meaning of a given concept is grounded in sensorimotor, introspective, and affective experience;

- The brain does not store static representations of concepts but produces dynamic representations in response to the demands of a specific task and context (i.e., situated conceptualizations);
- Understanding a concept entails the ability to mentally simulate some of the properties and events related to the concept;
- Language can guide the construction of complex and time-extended mental simulations as well as the formation of more abstract schemas fleshed out from sensory-motor representations.

The above-listed points support the view that mental simulation plays a key role in conceptual processing, and that it should therefore receive careful consideration when designing learning activities and instructional materials. With respect to the previous example of heat transfer, imagine now that the students also have the opportunity to use a simulation showing the difference between natural convection and conduction, such as the one showed in Figure 2. This simulation would allow

Figure 2. Image from a simulation conducted with the Energy2D simulation program, courtesy of Charles Xie, The Concord Consortium. Web site: http://energy.concord.org/energy2d.

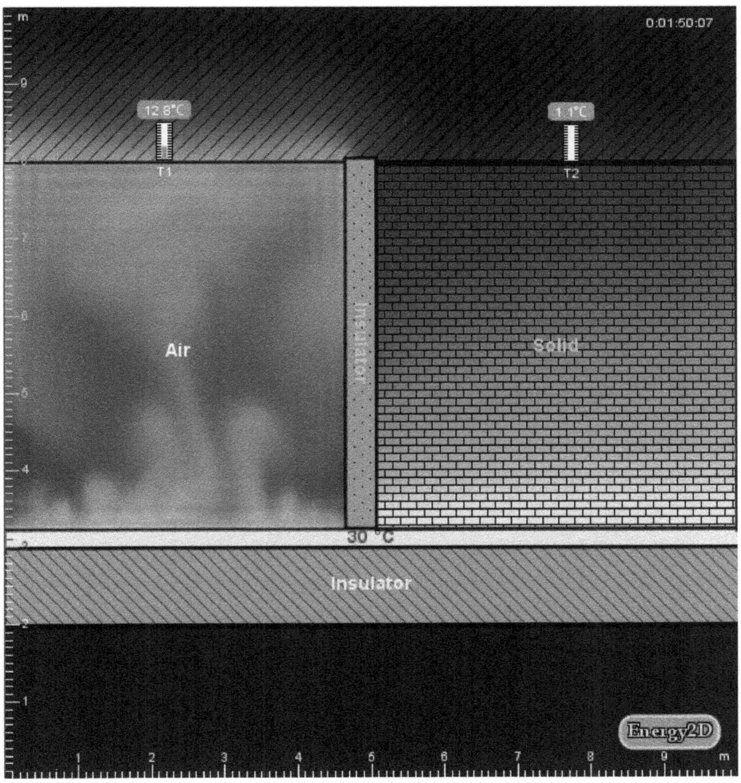

them to set the temperature of a heater which warms a chamber of air and a solid of equal size, separated from each other by a thermal insulator, to observe how the system changes with time, and to use virtual sensors, such as thermometers and air velocity anemometers, to measure properties at different locations.

Students could be prompted with targeted questions, such as: "In which chamber will temperature increase more rapidly?", "What causes the difference in temperature distribution?", and "What would happen if you suddenly lowered the heater's temperature?" Based on their prior knowledge about heat-related phenomena, they would generate hypotheses about what might happen in the two chambers, and these hypotheses could lead to mental simulation for predicting the outcomes of their "computational experiments". During computer simulation, heat propagation through the two materials would be vividly represented by turbulent and hot currents in the air chamber and by slow changes in the solid's colors, illustrating the property of heat thereby, which is an otherwise invisible physical quantity. Moreover, the students could interact with the computer simulation in real time, and the images, animations, and numerical information shown on the screen would change immediately, in function of their own actions and decisions. Students could therefore run a series of simulations on the computer and compare their predictions with the phenomena and data visualized.

In this way, students can ground their learning in a rich sensory-motor experience and are helped to move from a static concept of heat to a dynamic one. Furthermore, they can develop the ability to mentally simulate what will happen in comparable situations, for example, in a room heated by different types of radiators. It is important to note, however, that these processes do not always occur automatically, due to the high cognitive cost of mental simulation. Indeed, students will most likely not engage their mental resources to construct complex simulations if the effort is not actually required by the task at hand.

Generally speaking, students are more likely to use mental simulation: (1) in novel situations for which they have no rule available, or when their rules are inadequate (a strategy also used by scientists); or, (2) when the learning task requires that a specific analogy or metaphor be used for inferences.

In recent years, a series of studies have also illustrated the interplay between mental simulation and language. This is particularly evident in experiments (reviewed in Barsalou, 2008) showing that *language can activate mental simulations* - e.g. to represent the meaning of sentences, readers can simulate the situation described in texts - and that, reciprocally, *simulations can activate language* - e.g., individuals involved in problem-solving tasks frequently verbalize the solution process, or use inner speech to plan their actions. Importantly, emergent "hybrid" approaches to cognition (e.g., Barsalou, Santos, Simmons, & Wilson, 2008; Evans, 2015; Louwerse & Jeuniaux, 2010; Vigliocco, Meteyard, Andrews, & Kousta, 2009) posit that conceptual understanding requires both mental simulation and linguistic processing, with the relative contribution of the two systems, upon requirements of the task at hand.

It follows that teachers and designers should give careful consideration to the *experiential information* the simulation provides and to the associated *verbal information* given to students in the form of narrations, instructions, questions, explanations, and discussions, so as to ensure that the two sources of information are coherent and that they complement each other.

Narration can also greatly enhance the learning process, because stories are a way to transmit event-related knowledge in a more dynamic and stimulating format than are simple descriptions.

In conclusion, this section has described how understanding a concept entails the ability to:

- Construct a sufficiently accurate mental model of the concept and run the mental simulations corresponding to it;
- Linguistically express the content of these mental simulations;
- Compare the outcomes of the mental simulations with empirical evidence.

Correspondingly, any difficulties in concept comprehension and conceptual change frequently pertain to:

- Lack of domain-specific knowledge (essential for constructing and simulating an adequate mental model);
- Difficulty in grounding the new knowledge in an embodied sensorimotor experience;
- Difficulty in comparing the outcomes of mental simulation with contrary empirical evidence;
- High extraneous cognitive load (which exceeds students' available working memory capacity);
- Individual differences in mental simulation capacity (which might be linked to different cognitive styles and cognitive capacities).

The next section will investigate how the interplay of mental- and computer simulation can increase the effectiveness of concept learning and can obviate, or at least mitigate, the difficulties listed above.

COMPUTER-SIMULATION AS EXTENDED COGNITION TOOL

A closer examination of the two different types of simulation that have been examined herein - digital-external and mental-internal - reveals some intriguing correspondences between the two. Indeed, both represent a facet of an actual or potential reality - e.g., a natural phenomenon, a product to be developed, a past or future state of affairs - and the resulting representations are based, respectively, on computational and mental dynamic models. From a psychological perspective, both kinds of simulation serve identical functions: to envision the future, imagine alternative perspectives, understand the causes of a phenomenon, and to plan a course of action. These functions are linked to higher-order forms of learning, such as concept learning and problem solving.

It was similarly proposed in Landriscina (2013) that computer simulation can extend the cognitive capacity of mental simulation, such that the relation between the two might be considered a relation of complementarity between internal and external resources: The computer becomes a "partner" in a larger, coordinated cognitive system, in which each component plays a distinct role. In this instance, the enhancement of mental processes does not coincide merely with a reduction in the amount of cognitive effort required to carry out a mental task (as in the case, for example, of using Google to search for information more rapidly that by relying on our own memory, see Sparrow, Liu & Wegner, 2011). It pertains, rather, to more general limits of the human mind in effectively simulating the behavior of dynamic systems.

The limits of mental simulation have been highlighted by researchers in the field of System Dynamics, a branch of simulation dealing with problems arising in social, managerial, economic, or ecological systems. In the words of Forrester: *The human mind is well adapted to building and using models that relate objects in space. Also, the mind is excellent at manipulating models that associate words and ideas. But the unaided human mind, when confronted with modern social and technological systems, is not adequate for constructing and interpreting dynamic models that represent changes through time in complex systems.* (1968, p. 3). Specifically, our brain seems to be "hard-wired" to effectively simulate situations presenting an effect and a cause that are spatially or temporarily adjacent by linking them in a linear relation (A→B). It encounters difficulties, however, when a cause manifests its effect only after a longer time interval (e.g., as occurs with the effects of carbon dioxide emissions on the Earth climate or the effects of smoking on health) or when causality is circular (A→B→A) - conditions that frequently occur in the real world.

Another branch of simulation highlighting these limits is Agent-Based Modeling (Wilensky & Rand, 2015), in which a system is represented as being made up of many individual elements interacting with each other through local rules. The phenomena yielded through the agent-based simulations can give the appearance of behavior regulated by complex laws, when only simple local rules conversely are at work. It is, in fact, quite difficult or even impossible to predict the behavior of any computational model-based on knowledge of its equations or rules only. Even physical or biological systems describable with few variables - such as the movement of a pendulum or ecosystem models with populations of only two species - can generate complex and counter-intuitive behavior, when nonlinear effects are considered. On the other hand, mental simulation has a strictly qualitative nature, which does not naturally lead to quantitatively accurate results, nor does it allow for the exploration of behavior determined by incremental parameter changes.

The qualitative-analogical aspect of mental simulation and the quantitative-digital aspect of computer simulation, however, can complement each other very well. For example, the graphical user interface elements - "widgets" - users can manipulate to input their decisions to a simulation program are frequently of an analog nature - e.g., horizontal or vertical sliders and rotating knobs, which mimic controls on the front panels of analog electronic devices. The same holds for computer simulations in which the user can directly manipulate the elements that represent the system by "grabbing" a virtual object and dragging it to a different location or onto another virtual object.

The interplay of computer- and mental simulation is still a relatively unexplored field. From a science education perspective, Monaghan and Clement (1999) conducted an innovative study by investigating the use of computer simulation to help high school physics students learn relative motion concepts. Students first interacted with a simulation presented in a "predict-observe-explain" format and were then asked to solve post-test problems. The method of protocol analysis was used to examine the students' ability to run mental relative motion simulations on their own, both during and after computer simulation use. The researchers found that computer simulation interaction also facilitated mental simulation off-line, by helping students learn to solve related target problems when away from the computer.

In summary simulation models, can extend our biological capacity to carry out mental simulations. A form of "cognitive partnering" can be set up between student and simulation, resulting in a circular interaction that renders the computer a very helpful thinking tool. In Landriscina (2009; 2013), this circular interaction as being represented by an Epistemic Cycle model, is described. The model is focuses on the ways in which students can acquire knowledge through simulation in an instructional context (see the diagram in Figure 3).

Figure 3. The Epistemic Cycle

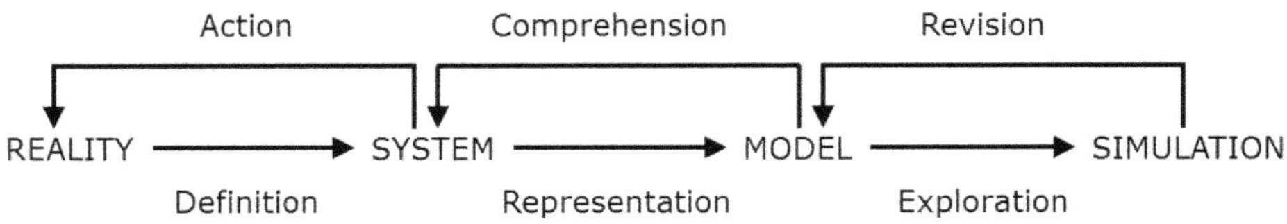

FUTURE RESEARCH DIRECTIONS

When focusing on the characteristics and functions of mental simulation, one may note that this cognitive ability is not that different from the human faculty commonly termed "imagination". In psychological terms, imagination generally refers to the consciousness of objects that are not present to the senses - a state of mind that is typical of conscious mental simulation. In truth, research in the field of cognitive neuroscience has shown a close link between mental simulation and imagination.

According to this line of research, memory and imagination consist respectively in the simulation of past and future events, and are functions that are strongly interlinked with each other. Many studies support the hypothesis that the ability to remember past experiences as well as imagine ourselves in future scenarios relies on a common network of brain regions, in which a large-scale brain system - the "default network" - plays a key role in internally directed or self-generated thought (for recent reviews, see Andrews-Hanna, Smallwood, & Spreng, 2014; Buckner, Andrews-Hanna, & Schacter, 2008). Researchers have hypothesized that this network is involved in *constructing dynamic mental simulations based on personal past experiences such as used during remembering, thinking about the future, and generally when imagining alternative perspectives and scenarios to the present* (Buckner et al., 2008, p. 18). At the same time, however, the brain's mental simulation capacity is not restricted to representing specific past- or future-oriented episodes, because it can also represent general states of the world - i.e. semantic knowledge (Szpunar, Spreng and Schacter, 2014). Thus, mental simulation can support more abstract forms of imagination, such as those involved in scientific thinking.

It is important to note that imagination is a more complex faculty than suggested by ordinary use of the term. Some features of imagination can be drawn from philosophy, where the notion of "imagination" has been the subject of in-depth investigation from antiquity to the present day (for a review, see Kind, 2016). Firstly, it is important to distinguish between r*eproductive imagination* - i.e., the partial reenactment of perceptions which have previously appeared in our consciousness

- and *productive* (or *creative*) *imagination* - i.e., the presence in consciousness of objects and situations never encountered before. It is also possible to distinguish between *sensory imagination* - the forming of visual images (or other perceptual-like representations) - and *propositional* (or *cognitive*) *imagination* - i.e., the ability to construct new sentences which can serve as objects of reflection. According to this distinction, imagination is not limited to mental imagery. In other words, it is not simply like "watching a movie in our head", but is also like reasoning about the movie plot, changing the course of the story and exploring alternatives scenarios, and comparing different movie endings. As elegantly expressed by the Italian novelist Italo Calvino: *my procedure aims at uniting the spontaneous generation of images and the intentionality of discursive thought. Even when the opening gambit is played by the visual imagination, putting its own intrinsic logic to work, it finds itself sooner or later caught in a web where reasoning and verbal expression also impose their logic* (1988, p. 90). It is precisely the interplay between sensory experience and reason that fuels imagination with the power to construct imaginary worlds.

The history of science is rich in studies that underscore the constructive role of imagination in scientific endeavor. Holton (1978) pioneered the study of scientific imagination by investigating its role in the formation of new theories, drawing on case studies from the lives of famous scientists. Along the same lines, Miller (1986) and Rocke (2010) analyzed the role of mental imagery in scientific thought. In these and in similar studies, imagination has been frequently linked to *creativity*, which is a quality deemed fundamental in the process of scientific discovery, due to its role in generating hypotheses worthy of investigation and testing.

In an effort to delineate the relation between imagination and mental simulation in scientific contexts is presented in Landriscina (2015). It delineates the relation between imagination and mental simulation in scientific contexts by characterizing scientific imagination as *the informed and disciplined use of mental simulation for envisioning a system's behaviors and drawing testable inferences*. The terms *informed* and *disciplined*, respectively, mean that the usefulness of a mental simulation depends on the amount and type of knowledge an individual has about the system under study, and on the degree to which the mental simulation is directed toward a specific goal.

In other words, the above-cited classifications from the philosophical literature propose that scientific imagination is a form of productive imagination that involves both sensory and propositional aspects. In psychological terms, it corresponds to a class of mental simulations that requires higher-level cognitive skills to focus on relevant information, visualize and manipulate mental objects, plan action sequences, track changes, and to generate and test predictions.

Teachers may be concerned about students with "too much" imagination, believing they are at risk of inability to think logically and concretely - two qualities considered to be essential for scientific thinking - or even of a tendency to daydream. Yet, Bransford and Donovan state that: *One of the most important aspects of science - yet perhaps one of the least emphasized in instruction - is that science involves processes of imagination. If students are not helped to experience this for themselves, science can seem dry and highly mechanical.* (2005, p. 406). Importantly, cognitive studies of science have emphasized that scientists' imaginative practices do not differ substantially from those that individuals use in everyday reasoning and students activate to achieve a deep understanding of novel scientific concepts (e.g., see Clement, 2008; Trickett & Trafton, 2007)[6].

Thus, STEM education can draw on the same type of practices to render instruction more effective and interesting. Computer simulation can support student scientific imagination capacity, by extending and reorganizing the mental processes through which students approach the task of learning new and difficult concepts.

An approach that can serve in the effort to fully develop and implement the imagination-enhancing potential of computer simulation was introduced by the National Research Council's *A Framework for K-12 Science Education: Practices, Crosscutting Concepts, and Core Ideas* (2012; hereafter referred to as "the *Framework*"), whose principles guided the writing of the *Next Generation Science Standards* (NGSS Lead States, 2013). According to the vision described in the Framework, three dimensions of science learning must be brought together: (1) scientific and engineering practices, (2) crosscutting concepts, and (3) disciplinary core ideas. The report highlighted the role this set of concepts can have in connecting knowledge from various disciplines into a coherent scientific view of the world. The interdisciplinary approach to science, which informed the Framework, is also a distinctive feature of simulation, in which a relatively small number of computational models can be applied to a wide range of phenomena, on different scales, and in seemingly unrelated disciplines. For example, the differential equation governing the harmonic oscillator originated in physics and has applications in molecular and cell biology, as it can also describe the rate at which mRNA produces proteins in a cell, and in neuroscience, given that harmonic oscillators can model rhythmic neural activity in the central nervous system. In fact, many scientific simulations can be mapped quite precisely onto the above-listed crosscutting concepts (for more examples of using simulation to connect crosscutting concepts, see Landriscina (2013)).

One way simulations can effectively support scientific imagination is by facilitating the construction of conceptual links among different levels of natural or artificial systems. For example, students can use simulations to conceptually interpret macroprocesses, such as chemical reactions, state changes of matter, and electricity in terms of emergent properties that link macroscopic- and molecular- or atomic-level processes (Sengupta & Wilensky, 2009; Stieff & Wilensky, 2003; White & Frederiksen, 1998).

CONCLUSION

This chapter has illustrated how the use of simulation technologies can help students and teachers overcome the limitations of mental simulation, and - given the right conditions - how a cognitive partnership can be set up between mind and computer.

Evidence has been cited showing that mental simulation and imagination are two interrelated cognitive abilities involved in remembering the past and thinking about possible futures. In particular, scientific imagination has been hypothesized to be linked to a class of mental simulations directed toward the goal of envisioning a system's behaviors, drawing testable inferences, and involving both imaginative thought and cognitive control. Computer simulations can provide special support for these mental processes by simplifying complex phenomena, visualizing abstract entities, moving among different scales and levels of representation, and rapidly exploring the consequences of different hypothesis.

Many avenues for future research remain open. Cognitive neuroscience can contribute a great deal to our knowledge of the processes involved in concept learning. Innovative simulation-based learning environments using hand gestures as input to control the simulation can help students build better mental models of phenomena usually concealed from everyday experience, such as the interactions of atoms and molecules or the Earth's rotation.

A promising area of research is the use of virtual reality technologies to achieve direct immersion in a simulation. Imagine not only seeing a visualization on a screen, but being inside the simulation itself and directly interacting with the system under study! Students will be able to enter a disk of gas and dust in orbit around a star and see the formation of a planetary system from there; they will be able to assume the dimension and point of view of an atom inside a protein, or visit the interior of a running engine turbine. On a broader scale, this type of progress in the culture of simulation may contribute not only to science education, but also to developing a new and hopefully more human and learning-oriented digital society.

REFERENCES

Andrews-Hanna, J. R., Smallwood, J., & Spreng, R. N. (2014). The default network and self-generated thought: Component processes, dynamic control, and clinical relevance. *Annals of the New York Academy of Sciences*, *1316*, 29–52.

Barsalou, L. W. (2008). Grounded cognition. *Annual Review of Psychology*, *59*(1), 617–645. doi:10.1146/annurev.psych.59.103006.093639 PMID:17705682

Barsalou, L. W. (2015). Situated conceptualization: Theory and applications. In Y. Coello & M. H. Fischer (Eds.), Foundations of embodied cognition: Vol. 1. Perceptual and emotional embodiment (pp. 11-37). East Sussex, UK: Psychology Press.

Barsalou, L. W., Santos, A., Simmons, W. K., & Wilson, C. D. (2008). Language and simulation in conceptual processing. In M. De Vega, A. M. Glenberg, & A. C. A. Graesser (Eds.), *Symbols, Embodiment, and Meaning* (pp. 245–283). Oxford, UK: Oxford University Press. doi:10.1093/acprof:oso/9780199217274.003.0013

Binder, J. R., & Desai, R. H. (2011). The neurobiology of semantic memory. *Trends in Cognitive Sciences, 15*(11), 527–536. doi:10.1016/j.tics.2011.10.001 PMID:22001867

Black, J. B. (2010). An embodied/grounded cognition perspective on educational technology. In M. S. Khine & I. M. Saleh (Eds.), *New Science of Learning. Cognition, Computers and Collaboration in Education* (pp. 45–52). New York, NY: Springer. doi:10.1007/978-1-4419-5716-0_3

Bransford, J. D., & Donovan, M. S. (2005). Scientific Inquiry and How People Learn. In *How Students Learn: History, Mathematics, and Science in the Classroom* (pp. 397–419). Washington, DC: The National Academies.

Brant, G., Hooper, E., & Sugrue, B. (1991). Which comes first the simulation or the lecture? *Journal of Educational Computing Research, 7*(4), 469–481. doi:10.2190/PWDP-45L8-LHL5-2VX7

Buckner, R. L., Andrews-Hanna, J. R., & Schacter, D. L. (2008). The Brains Default Network Anatomy, Function, and Relevance to Disease. *Annals of the New York Academy of Sciences, 1124*(1), 1–38. doi:10.1196/annals.1440.011 PMID:18400922

Calvino, I. (1988). *Six Memos for the Next Millennium.* Cambridge, MA: Harvard University Press.

Casasanto, D., & Lupyan, G. (2015). All Concepts are Ad Hoc Concepts. In E. Margolis & S. Laurence (Eds.), *The Conceptual Mind: New Directions in the Study of Concepts* (pp. 543–566). Cambridge, MA: The MIT Press.

Clark, D., Tanner-Smith, E., Killingsworth, S., & Bellamy, S. (2013). *Digital Games for Learning: A Systematic Review and Meta-Analysis. Executive Summary.* Menlo Park, CA: SRI International.

Clement, J. J. (2000). Model based learning as a key research area for science education. *International Journal of Science Education, 22*(9), 1041–1053. doi:10.1080/095006900416901

Clement, J. J. (2008). *Creative Model Construction in Scientists and Students. The Role of Imagery, Analogy, and Mental Simulation*. New York, NY: Springer. doi:10.1007/978-1-4020-6712-9

Clement, J. J. (2009). The Role of Imagistic Simulation in Scientific Thought Experiments. *Topics in Cognitive Science*, *1*(4), 686–710. doi:10.1111/j.1756-8765.2009.01031.x PMID:25163453

Connell, L., & Lynott, D. (2014). Principles of representation: Why you can't represent the same concept twice. *Topics in Cognitive Science*, 1–17. PMID:24895329

Covert, M. W. (2013). Simulating a Living Cell. *Scientific American*, *310*(1), 44–51. doi:10.1038/scientificamerican0114-44 PMID:24616970

D'Angelo, C., Rutstein, D., Harris, C., Bernard, R., Borokhovski, E., & Haertel, G. (2014). *Simulations for STEM Learning: Systematic Review and Meta-Analysis*. Menlo Park, CA: SRI International.

de Jong, T. (2006). Technological advances in inquiry learning. *Science*, *312*(5773), 532–533. doi:10.1126/science.1127750 PMID:16645080

Evans, V. (2015). What's in a concept? Analog versus parametric concepts in LCCM Theory. In E. Margolis & S. Laurence (Eds.), *The Conceptual Mind: New Directions in the Study of Concepts* (pp. 251–290). Cambridge, MA: The MIT Press.

Forrester, J. W. (1968). *Principles of Systems*. Cambridge, MA: Productivity Press.

Gendler, T. (2013). Imagination. In E. N. Zalta (Ed.), *The Stanford Encyclopedia of Philosophy (Fall 2013 Edition)*. Retrieved from http://plato.stanford.edu/archives/fall2013/entries/imagination/

Gentner, D. (2002). Psychology of Mental Models. In N. J. Smelser & P. B. Bates (Eds.), *International Encyclopedia of the Social and Behavioral Sciences* (pp. 9683–9687). Amsterdam: Elsevier Science.

Gibbs, R. W. Jr. (2006). *Embodiment and Cognitive Science*. New York, NY: Cambridge University Press.

Gobert, J. D., & Buckley, B. C. (2000). Introduction to model-based teaching and learning in science education. *International Journal of Science Education*, *22*(9), 891–894. doi:10.1080/095006900416839

Han, I., & Black, J. B. (2011). Incorporating haptic feedback in simulation for learning physics. *Computers & Education*, *57*(4), 2281–2290. doi:10.1016/j.compedu.2011.06.012

Hegarty, M. (2004). Mechanical reasoning as mental simulation. *Trends in Cognitive Sciences*, *8*(6), 280–285. doi:10.1016/j.tics.2004.04.001 PMID:15165554

Holton, G. (1978). *The Scientific Imagination: Case Studies*. Cambridge, UK: Cambridge University Press.

Kahneman, D., & Tversky, A. (1982). The simulation heuristic. In Judgment Under Uncertainty: Heuristics and Biases (pp. 201-208). Cambridge University Press. doi:10.1017/CBO9780511809477.015

Kilpatrick, J., & Quinn, H. (Eds.). (2009). *Science and Mathematics Education*. Washington, DC: National Academy of Education.

Kind, A. (Ed.). (2016). *The Routledge Handbook of Philosophy of Imagination*. Routledge.

Landriscina, F. (2009). Simulation and Learning: The Role of Mental Models. *Journal of E-Learning and Knowledge Society*, *5*(2), 23–32.

Landriscina, F. (2012). Simulation and Learning: The Role of Mental Models. In N. M. Seel (Ed.), *Encyclopedia of the Sciences of Learning* (pp. 3072–3075). New York, NY: Springer.

Landriscina, F. (2013). *Simulation and Learning. A Model-Centered Approach*. New York, NY: Springer.

Landriscina, F. (2015). The role of mental simulation in understanding and in creating scientific concepts. In F. Corni (Ed.), *Proceedings from the 3d Conference on Innovation in Science Education in Primary School and Kindergarten*. University of Modena and Reggio Emilia.

Lindgren, R., Tscholl, M., Wang, S., & Johnson, E. (2016). Enhancing learning and engagement through embodied interaction within a mixed reality simulation. *Computers & Education*, *95*, 174–187. doi:10.1016/j.compedu.2016.01.001

Louwerse, M. M., & Jeuniaux, P. (2010). The linguistic and embodied nature of conceptual processing. *Cognition*, *114*(1), 96–104. doi:10.1016/j.cognition.2009.09.002 PMID:19818435

Lynott, D., & Connell, L. (2010). Embodied conceptual combination. *Frontiers in Psychology*, *1*, 1–14. doi:10.3389/fpsyg.2010.00212 PMID:21833267

Mason, R. A., & Just, M. A. (2016). Neural Representations of Physics Concepts. *Psychological Science*, *27*(6), 904–913. doi:10.1177/0956797616641941 PMID:27113732

McComas, W. (1998). The Principal Elements of the Nature of Science: Dispelling the Myths. In W. McComas (Ed.), *The Nature of Science in Science Education: Rationales and Strategies* (pp. 53–72). Dordrecht: Kluwer Academic Publishers.

Miller, A. I. (1986). *Imagery in Scientific Thought*. Cambridge, MA: The MIT Press.

Monaghan, J. M., & Clement, J. (1999). Use of a computer simulation to develop mental simulations for understanding relative motion concepts. *International Journal of Science Education, 21*(9), 921–944. doi:10.1080/095006999290237

National Research Council. (2011). Learning Science Through Computer Games and Simulations. Washington, DC: The National Academies Press.

National Research Council. (2012). *A Framework for K-12 Science Education: Practices, Crosscutting Concepts, and Core Ideas*. Washington, DC: The National Academies Press.

Nersessian, N. J. (2008). *Creating Scientific Concepts*. Cambridge, MA: The MIT Press.

NGSS Lead States. (2013). *Next Generation Science Standards: For States, By States*. Washington, DC: The National Academies Press.

Norman, D. (1983). Some observations on mental models. In D. Gentner & A. Stevens (Eds.), *Mental Models* (pp. 7–14). Hillsdale, NJ: Lawrence Erlbaum Associates.

Plass, J. L., Homer, B. D., & Hayward, E. (2009). Design factors for educationally effective animations and simulations. *Journal of Computing in Higher Education, 21*(1), 31–61. doi:10.1007/s12528-009-9011-x

Remmer, A. M., & Jernstedt, C. (1982). Comparative effectiveness of simulation games in secondary and college level instruction: A meta-analysis. *Psychological Reports, 51*(3), 742. doi:10.2466/pr0.1982.51.3.742

Rocke, A. J. (2010). *Image and Reality. Kekulé, Kopp, and the Scientific Imagination*. Chicago, London: The University of Chicago Press. doi:10.7208/chicago/9780226723358.001.0001

Rumelhart, D. E., Smolensky, P., McClelland, J. L., & Hinton, G. E. (1986). Schemata and Sequential Thought Processes in PDP models. In J.L. McClelland, D.E. Rumelhart, & The PDP Research Group (Eds.), *Parallel Distributed Processing. Explorations in the Microstructure of Cognition: Vol. 2. Psychological and Biological Models* (pp. 7–57). Cambridge, MA: The MIT Press.

Schwartz, D. L., & Black, J. B. (1996). Shuttling between depictive models and abstract rules: Induction and fall-back. *Cognitive Science, 20*(4), 457–497. doi:10.1207/s15516709cog2004_1

Seel, N. (2003). Model-Centered Learning and Instruction. *Tech., Inst. Cognition and Learning, 1*, 59–85.

Seel, N. (2012). Dynamic Modeling and Analogies. In N. Seel (Ed.), *Encyclopedia of the Sciences of Learning* (pp. 1051–1054). New York, NY: Springer.

Sengupta, P., & Wilensky, U. (2009). Learning electricity with NIELS: Thinking with electrons and thinking in levels. *International Journal of Computers for Mathematical Learning, 14*(1), 21–50. doi:10.1007/s10758-009-9144-z

Smetana, L. K., & Bell, R. L. (2012). Computer simulations to support science instruction and learning: A critical review of the literature. *International Journal of Science Education, 34*(9), 1337–1370. doi:10.1080/09500693.2011.605182

Sokolowski, J. A., & Banks, C. M. (Eds.). (2010). *Modeling and Simulation Fundamentals. Theoretical Underpinnings and Practical Domains*. Hoboken, NJ: John Wiley & Sons. doi:10.1002/9780470590621

Sparrow, B., Liu, J., & Wegner, D. M. (2011). Google effects on memory: Cognitive consequences of having information at our fingertips. *Science, 333*(6043), 776–778. doi:10.1126/science.1207745 PMID:21764755

Stieff, M., & Wilensky, U. (2003). Connected Chemistry - incorporating interactive simulations into the chemistry classroom. *Journal of Science Education and Technology, 12*(3), 285–302. doi:10.1023/A:1025085023936

Szpunar, K. K., Spreng, R. N., & Schacter, D. L. (2014). A taxonomy of prospection: Introducing an organizational framework for future-oriented cognition. *Proceedings of the National Academy of Sciences of the United States of America, 111*(52), 18414–18421. doi:10.1073/pnas.1417144111 PMID:25416592

Thomas, R. A., & Hooper, E. (1991). Simulations: An opportunity we are missing. *Journal of Research on Computing in Education, 23*(4), 497–513. doi:10.1080/08886504.1991.10781978

Trickett, S. B., & Trafton, J. G. (2007). What if. . .: The use of conceptual simulations in scientific reasoning. *Cognitive Science, 31*(5), 843–875. doi:10.1080/03640210701530771 PMID:21635319

Vigliocco, G., Meteyard, L., Andrews, M., & Kousta, S. (2009). Toward a theory of semantic representation. *Language and Cognition*, *1*(2), 219–247. doi:10.1515/LANGCOG.2009.011

White, B. Y., & Fredericksen, J. R. (1998). Inquiry, Modeling and Metacognition: Making Science Accessible to All Students. *Cognition and Instruction*, *16*(1), 3–118. doi:10.1207/s1532690xci1601_2

Wilensky, U., & Rand, W. (2015). *An Introduction to Agent-Based Modeling: Modeling Natural, Social, and Engineered Complex Systems with NetLogo*. Cambridge, MA: The MIT Press.

Wilensky, U., & Resnick, M. (1999). Thinking in levels: A dynamic systems perspective to making sense of the world. *Journal of Science Education and Technology*, *8*(1), 3–19. doi:10.1023/A:1009421303064

Winsberg, E. (2010). *Science in the Age of Computer Simulation*. Chicago, IL: The University of Chicago Press. doi:10.7208/chicago/9780226902050.001.0001

Xie, C. (2012a). Framing Mixed-Reality Labs. *@Concord*, *16*(1), 8-9.

Xie, C. (2012b). Interactive heat transfer simulations for everyone. *The Physics Teacher*, *50*(4), 237–240. doi:10.1119/1.3694080

Yee, E., & Thompson-Schill, S. L. (2016). Putting concepts into context. *Psychonomic Bulletin & Review*. doi:10.3758/s13423-015-0948-7

KEY TERMS AND DEFINITIONS

Experiential Simulations: Simulations based on the creation of a virtual situation or event to be experienced by one or more participants.

Imagination: The informed and disciplined use of mental simulation for envisioning a system's behaviors and drawing testable inferences.

Mental Models: Knowledge structures that students construct to process new tasks or to solve problems.

Mental Simulation: The mental re-enactment of perceptual, motor, and introspective states acquired During experience with the world, body, and mind.

Model-Based Simulations: Simulations based on the construction of the theoretical model of a real or imagined system (also known as "theoretical" or "conceptual" simulations).

Model-Based Learning and Teaching: An instructional approach that focuses on students' mental models, and highlights the ways in which these internal models are expressed through external models, and the changes that occur in them as students learn new concepts.

Simulation: The reproduction of selected aspects of a system's behavior, in function of a user's decisions, which allows a student or scientist to modify the system's time evolution by altering the simulation's initial conditions or other parameters.

ENDNOTES

1. D'Angelo et al. (2014) also discussed the game/simulation distinction. They defined a game as having clear goal states and a built-in reward system tied to those goal states and requiring the user to complete levels or achievements in order to progress.
2. See Clark, Tanner-Smith, Killingsworth and Bellamy (2013) for a review and meta-analysis on digital games for learning.
3. Some of the epistemic features of simulation in relation to teaching and learning are examined in Landriscina (2013).
4. A similar concept of a learning-dependent progression of mental models was introduced by White and Fredericksen (1998) in the context of studies examining the effects of inquiry-based instructional methods on students' learning of physics concepts.
5. These approaches have been proposed as alternatives to classical cognitive science. The core idea of the embodied view is that cognition arises from the interaction of the brain with the body and the world (Gibbs, 2006). Specifically, mental representations are not viewed as fixed entities to be manipulated from formal rules, but neural activity patterns that are coupled with perceptual and motor states. Barsalou (2008) calls his approach "grounded cognition", as he maintains that cognition can be grounded in many ways, including through simulation, language, and situated actions, not only through body states.
6. This point is also supported by a recent neuroimaging study on the neural representation of physics concepts (Mason & Just, 2016), that demonstrates how acquiring abstract scientific concepts in an educational setting evoke activation patterns that rely on repurposing neural structures that were originally evolved for other or general purposes.

Chapter 3
Collaborative Systems for Design-Based Learning

Joycelyn Streator
Georgia Institute of Technology, USA

ABSTRACT

This chapter explores use of Design-based learning (DBL) and digital tools to facilitate collaborative learning through design-based projects. Design-based learning (DBL) is an educational approach that incorporates hands-own, authentic, multidisciplinary design tasks to identify problems and design solutions. With DBL, students typically work in teams and are tasked designing solutions to open-ended problems. Teams develop conceptual solutions to problems and then work through the design process to arrive at the creation of an actual artifact. This artifact may be fully functional or simply a model, prototype, or other representation of the complete system. STEM instructors and students should give careful attention to selecting the digital tools for collaboration. Some collaborative tools offer affordances and features that compliment the communication processes in one phase of the design process while another other tool may be better suited for the tasks specific to another phase.

INTRODUCTION

This chapter explores use of design-based learning (DBL) and digital tools to facilitate collaborative learning through design-based projects. Design-based learning is an educational approach that incorporates hands-on, authentic, multidisciplinary design tasks to identify problems and design solutions. As STEM educators seek new ways of incorporating inquiry-based methods, DBL provides a valuable method of bringing student-led learning to the forefront.

DOI: 10.4018/978-1-5225-2525-7.ch003

With DBL, students typically work in teams and are tasked with designing solutions to open-ended problems. Teams develop conceptual solutions to problems and then work through the design process to arrive at the creation of an actual artifact. There are five basic phases to the design process. In the first phase, students identify the key issue or problem to be addressed and then brainstorm potential solutions. In the early phases of the design process, students become proficient in identifying the criteria for selecting an optimal solution. This initial phase requires a high degree of collaboration and one-to-many communication streams. Next, students gain experience in determining requirements for the chosen solution and documenting these requirements. These requirements serve as the blueprint for what will be accomplished during the project and what features are determined to be within its scope. In the final phases of the project, students will implement and evaluate the artifact created as a result of the process. This artifact may be fully functional or simply a model, prototype, or other representation of the complete system.

STEM instructors and students should give careful attention to selecting the digital tools for collaboration. Certain collaborative tools offer affordances and features that complement the communication processes in one phase of the design process while another other tool may be better suited for the tasks specific to another phase. The tools used during the first phase of the design process must be suitable not only for communicating, but also capturing and refining free-flowing ideas. As students endeavor to select a design project choice among competing ideas, students need collaboration tools that support a systematic and logical approach to selecting the optimal choice. The mix of collaboration tools used in the final phases should allow teams to communicate the status of the project to a larger audience and receive feedback on the resulting artifact.

Misappropriation of digital communication tools can undermine team efficiency and focus efforts on technical struggles with software applications rather the design project itself. However, when used effectively, the tools outlined in this chapter can support all phases of design-based learning in a manner that is effective yet unobtrusive to the learning process.

The design-based learning pedagogical approach has the potential to increase students' understanding of the design process through engaging in active and authentic learning projects. STEM students must have expertise in applying a systematic process to translating technological innovation to create value-added solutions.

This chapter explores electronic resources for use in design-based STEM education. Using design-based learning can improve knowledge of design processes and increase competence in problem-solving and student motivation. Specifically, this chapter discusses the use of collaborative tools for project creation, collaboration, peer assessment, and knowledge management. The discussion includes a number of technologies (e.g., email, text messaging, social media, wikis, project web sites, and blogging, among others). While specific applications may be referenced, this chapter emphasizes more salient features that are prevalent across a particular category of software applications. Given the rapid evolution of existing technologies and the disruptive nature of emerging technologies, instructors must continually reassess the value of new collaborative tools to support learning within the community of STEM students.

BACKGROUND

It has long been held that the most effective pedagogy incorporates deep (vs. surface) learning (Cope, Staehr, & Horan, 2002). With deep learning, students endeavor to understand the process by which effective solutions are identified and designed. This is in contrast to surface learning, in which students simply seek "the right answer" as expeditiously as possible. One of the biggest challenges facing STEM educators is incorporating learning experiences within an academic context that prepares students for complex, open-ended challenges found in professional work. For example, information technology (IT) education has been criticized for the disconnect between the IT curriculum and professional practice (Cohen, 2002a; Lee, Koh, Yen, & Tang, 2002; Trauth, Farwell, & Lee, 1993). Research indicates that IT practitioners place a greater emphasis on creative and critical thinking than educators (Lee et al., 2002), and that the ability to plan and design strategic IT solutions is a crucial skill for IT managers (Wu, Chen, & Chang, 2007).

STEM educators must seek news ways of facilitating inquiry-based methods and bring student-led learning to the forefront. In recent years, the importance of design thinking has been emphasized in both practitioner and scholarly circles. Design thinking is characterized as the cognitive strategies and frameworks used to solve what has been termed "wicked problems" (Buchanan, 1992). Buchanan conceptualized wicked problems as that are difficult to solve due to incomplete, dynamic, and often have unknown requirements. In a practitioner space, graduates of STEM programs encounter wicked problems as a matter of course. Developing cognitive frameworks to support design thinking during the formal educational process engenders greater confidence and an organized approach when encountering complex problems.

From a pedagogical perspective, design thinking is reflected in the tangible artifacts produced by student project teams. These artifacts address various levels of cognition and awareness of problem spaces and their complexities (Leinonen & Durall-Gazulla, 2014). For example, on the most basic level, a student may develop a deeper understanding of design based on physical attributes of an artifact. With a deeper initiation into design thinking the student would consider the behavior of the artifact in differing scenarios. Going further, the student would begin to understand design from a systems perspective. That is, how does the artifact and its behavior fit within a larger system of interconnected elements. In the higher levels of design thinking the student develops a greater awareness of the user and the social implications surrounding the design. For instance, the designer of app may consider requirements related to accessibility or cultural matters. Perhaps the designer of a new gadget may consider green manufacturing techniques and re-usability of materials.

In the practitioner realm, many corporations and research firms are positioning themselves to be design leaders and seek to develop an organizational culture centered on innovation (Dunne & Martin, 2006). Hiring managers recognize the value of recruiting individuals with a capacity for design thinking. The importance of developing a pipeline of talent with who can apply creative problem solving skills is highlighted increased level of corporate sponsorship of design activities and competitions in higher education (Collins & Han, 2004).

In the past, STEM faculty typically designed curricula around applying scientific principles to defined, technological problems. However, from an industry approach, the graduates are ill-prepared for unstructured real world problems. More recently faculty are seeking pedagogical approaches to foster design thinking. This shift is seen in both STEM (Razzouk & Shute, 2012) and non-STEM disciplines (Johansson-Sköldberg, Woodilla, & Çetinkaya, 2013). In academia, there is a growing recognition of the importance of developing a strong field of graduates prepared to face wicked, real world problems.

Among the many pedagogical approaches best suited to address this gap between the academic and practitioner worlds is design-based learning. This chapter explores electronic resources for use in design-based STEM education. Using design-based learning can improve knowledge of design processes, foster greater competence in problem-solving, and enhance student motivation. Specifically, this chapter discusses the use of collaborative tools for project creation, collaboration, peer assessment, and knowledge management. While the context of the discussion is in the discipline of information technology, the same approach and tools can be used across STEM fields.

DESIGN-BASED LEARNING

Inquiry-based STEM pedagogy encourages students to pose questions and arrive at solutions, thus creating student-centric learning paths. Design-based learning is an educational approach that incorporates hands-on, authentic, multidisciplinary design tasks (Doppelt, Mehalik, Schunn, Silk, & Krysinski, 2008; Wijnen, 2000) to identify problems and design solutions. Design-based learning is rooted in a constructivist pedagogical approach in that it relies on learning that emerges through experimentation in authentic contexts (Duffy & Jonassen, 1992; Fosnot, 1996).

With the DBL approach, students play a larger role in defining the design problem. Therefore, the resulting projects remain topical and relevant to the current class of students. In addition, DBL has the common stages of the design process to serve as a guide for students through all stages, from conceptualization to development to evaluation.

While similar in origins to problem-based learning (PBL), DBL differs from PBL approaches. Problem-based learning centers on predetermined complex problems that students are challenged with solving ((Duch, Groh, & Allen, 2001; Savery & Duffy, 1995). By contrast, design-based learning places emphasis on the integration of scientific knowledge with emergent inquiries. Learning is facilitated as interdisciplinary teams endeavor to create authentic artifacts (Gómez Puente, Van Eijck, & Jochems, 2011). By contrast, PBL has received criticism as instructor-defined problem contexts tend to become outdated and less relevant to students (Wells, Warelow, & Jackson, 2009). In addition, research has shown that students often experience anxiety and confusion over the minimal guidance inherent in the PBL process (Kirschner, Sweller, & Clark, 2006). A design-based learning approach addresses these issues. Students play a larger role in defining the design problem and, therefore, the resulting projects remain topical and relevant to the current class of students. DBL also has the general tenets of the design process to serve as a guide for students through all stages, from conceptualization to development to evaluation.

Design-based learning engages students in design activities that reflect authentic contexts and guide students through generating specifications, outcome prediction, artifact creation, evaluation, and communication. As students step through iterations of each stage, they develop reasoning and critical thinking skills needed to address open-ended design challenges in the real world. DBL incorporates open-ended design assignments, hands-on experience, real-life scenarios, and multidisciplinary content (Puente, van Eijck, & Jochems, n.d.).

Importance of Design Process Knowledge

Graduates of STEM programs, especially IT, must have expertise in applying a systematic approach to translating technological innovation into valuable, real-world solutions. In order to meet this requirement, IT curriculum guidelines suggest that IT students must develop skills in the process of designing new solutions (Topi et al., 2010). These design skills include designing and implementing solutions that provide a high-quality user experience. Indeed, students who are competent in the design process should possess a greater ability to provide solutions for dynamic, nonlinear problems and anticipate unintended consequences from interrelated system components (Moazzen et al., 2013). While these guidelines are specific to IT, problem-solving and the ability to design effective solutions in complex environments is an important aspect of all STEM fields.

Key outcomes for design-based learning include increased student motivation, greater student emphasis on deep (versus surface) learning, and the implicit instruction provided by feedback on design solutions.

Design and Peer Assessment

Traditional methods for assessing a designed solution include observation of design activities, interviews, and analysis of design documents (Moazzen et al., 2013). However, the advent of the Internet and web-based tools for providing feedback has contributed to a rise in the use of peer assessment as a tool for evaluating student work. Recently, the role of peer assessment in the work of IT students has received considerable attention (Sitthiworachart & Joy, 2003; Tsai, Lin, & Yuan, 2002; Tseng & Tsai, 2007a). Peer assessment in IT projects not only benefits the recipient of the feedback, but also creates a learning mechanism for the student providing feedback. Students develop judgments about what constitutes effective, quality design through reviewing and providing feedback to peers (van Zundert, Sluijsmans, & van Merriënboer, 2010). In addition, research indicates that the quality of peer feedback is often highly correlated with critique provided by experts (Tseng & Tsai, 2007a).

Student Motivation and Deep Learning

When students are engaged in projects that involve real-world scenarios and authentic learning, research suggests that student motivation is greatly increased (Cohen, 2002a). In addition, actively involving students in peer project evaluation has also been recognized as a factor in increasing student motivation (Tseng & Tsai, 2007a). As the level of student motivation increases, students begin to take ownership of learning, develop an intrinsic interest, and shift from surface to deep learning. When STEM students adopt a deep-learning approach, they seek a deeper understanding by developing a holistic view of the relationship between the design tasks, technology skills, and content-related concepts (Cohen, 2002a).

SOLUTIONS AND RECOMMENDATIONS

With the rapid increase in web-based tools available to students and instructors, the challenge in applying technology to STEM learning is not in finding tools, but in identifying the best tools for the learning task. There are numerous solutions available to support the design-based approach to STEM learning. With technology constantly evolving, this discussion focuses on broad categories of collaborative tools, but also includes specific software examples. Most of the basic tools are available as a standard part of learning management software or as stand-alone options.

As students apply the design process to explore unstructured problems and complex outcomes, it's beneficial to have communication structures in place to organize and gather discussions, knowledge, solutions, and results. Below is an overview of each phase of the design process and examples of supporting technologies (see Figure 1).

Phase 1 Collaborative Tools: Idea Generation

The first phase of the design process involves idea generation and brainstorming potential solutions or project topics. Brainstorming invites team members to pitch ideas loosely related to a given subject. The goals for brainstorming sessions include

Figure 1. Phases of the design process and examples of collaborative tools

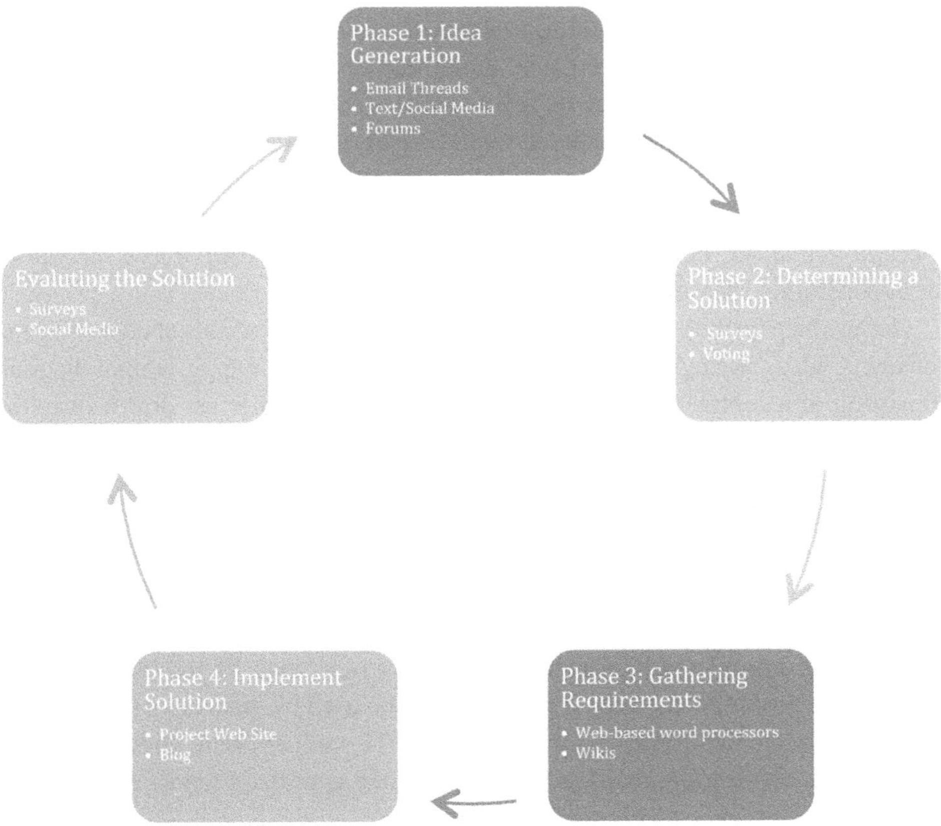

increasing the number ideas, building on existing ideas, and freeing participants to contribute with criticism from teammates and without regard to scope or feasibility. In this stage, students are engaging in highly unstructured communications.

The benefits of using collaborative tools to facilitate brainstorming has been well documented (Dennis, Wixom, & Vandenberg, 2001). However, initiating and organizing ideas is a challenge for many learners facing ambiguous and ill-defined inquiries. The most common collaborative tool used during this phase of the design process is email. However, email threads among project team members can be difficult to track and identify when competing for the user's attention in crowded inboxes. In addition, students must be diligent in their use of email features such as "reply all" to ensure that every group member remains included in the email discussion.

Beyond email, many students use texting and social media to communicate. While these platforms are rapidly emerging, there are key disadvantages to these communication tools, particularly texting. Though texting provides instant distribution and access to communication, there can be issues with group members being inadvertently omitted from communication. Moreover, there is no central repository for archiving of student ideas.

Unlike email and text messaging, use of discussion groups or forums provides a more structured means of facilitating collaboration during the idea generation phase. Students can join the forum or discussion group or be enrolled by the instructor. After initial member enrollment, there's no longer a need to manually check to ensure that all group members are included in all conversations. Since conversations are organized by topic or thread within most forum and discussion group tools, students have the ability to categorize ideas and contribute thoughts in a more efficient and intelligible manner. Forums and discussion groups can also serve as a repository for documenting ideas in the brainstorming stage.

Going a step further, mind mapping tools provide another means of capturing and conveying ideas during the idea generation phase. A mind map is a radial diagram consisting of a central node that represents an idea. Branches are drawn from the central note to show subtopics or concepts related to the central idea. The branches can contain words or pictures and utilize color for categorization.

Mapping cognitive processes creates knowledge structures in the early phases of many endeavors. As students seek to refine ideas, mind mapping tools provide a means of exploring relationships among ideas and identifying conceptual gaps. Just as maps are used in literature to aid in pre-writing, natural sciences maps can be used to group related biological processes or outcomes. In the brainstorming phase of design-based learning, mind maps can be used by teams to generate ideas and cluster related concepts. As an example, if IT students were challenged with designing an application, maps aid in capturing the genesis of a potential solution and possible features (see Figure 2).

Figure 2. Mind map used to design an IT application using Mindmup (mindmup.com)

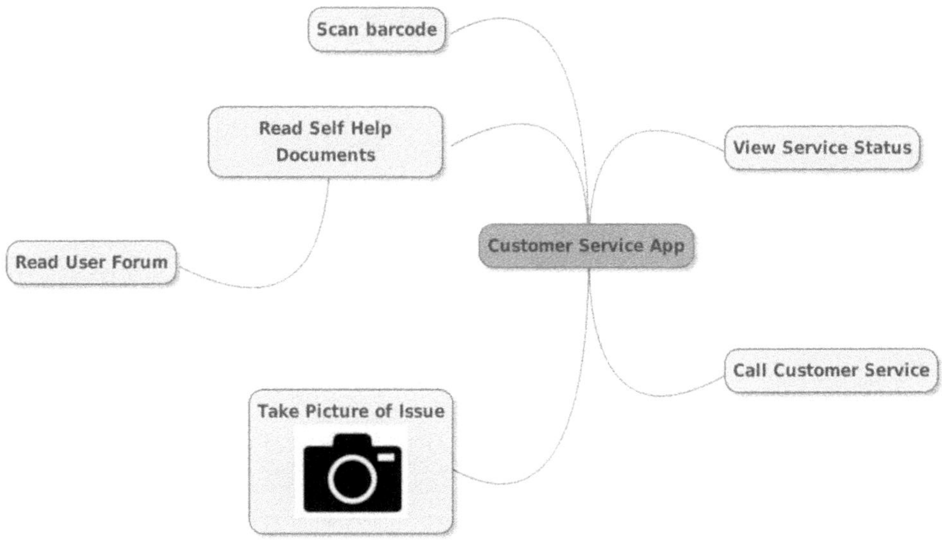

Though mind mapping techniques existed prior to the onset of the digital age, collaborative tools allow easier creation and collaboration. Many mapping tools are available via the web and allow teams to develop and share maps in real time or asynchronously. Changes made by individuals can be viewed or revised by other team members. Thus, the map can be updated as learners refine ideas. The evolution of the project can be documented via updates to the map. In most cases, web-based mapping applications store results in the cloud or can be downloaded and embedded in common formats. With mind mapping tools, teams can explore and collaborate on the potential features of several competing solutions.

The list of features and affordances associated with mind mapping software is constantly evolving. For example, some mind mapping tools allow for a richer object set. Beyond text, some systems support images, external data sources, links, and other forms of media. A valuable feature for team-based collaborative projects is real-time collaboration, which allows users to modify and respond to the document from their own devices in real time. Other features that are of interest for team-based mind mapping include the ability to share documents, cloud storage, and the ability to track document history. In addition, tools that have a free version available provide a practical option for students. Table 1 contains a sampling of mind mapping tools and a summary of features.

Table 1. Examples of mind mapping tools

Tools	Features					
	Web-based	Real-time Collaboration	Online Storage	Document History	Free Version	Additional Features
Mindmup	X		X	X	X	• Project planning • Post to social media
Coggle	X	X		X	X	• Email change notifications • Presentation mode
Sketchboard	X	X	X		X	• Freehand drawing • Unified Modeling Language
VUE			X		X	• Import academic citations • Merge multiple maps • Supports external data sources • Presentation support
xMind	X		X	X	X	• Gantt view • Presentation mode • Create PowerPoint slides

Phase 2 Collaborative Tools: Determining the Optimal Solution

In the next stage of design-based learning, teams will need to identify the optimal solution. The method by which students assess what defines an optimal solution is specific to the discipline and topic. However, the method with which students arrive at their decision may ultimately consist of the judgment of individual group members. Some potential criteria for consideration are:

- Fit between the proposed solution and the original problem statement
- Completeness of the proposed solution
- The feasibility of the proposed solution by the skills possessed
- The ease of the proposed solution
- The overall appeal of the proposed to the design team and/or potential users.

In order to obtain the most favorable solution, students can create criteria for assessing their project. Each group member then rates potential project ideas, or the team can conduct a simple vote. Either way, surveys and questionnaires provide an ideal vehicle for rating, ranking, and voting. While there are a number of free web-based survey tools available, survey and assessment features are often built into learning management systems and web-based office productivity applications.

Approaches to evaluating potential solutions via web-based survey may include open-ended questions, closed-format questions, or a combination of both. If students poll team members or potential users with open-ended questions, the data obtained will be every rich, but highly unstructured and will take additional time to parse. Another means of evaluating ideas via web-based survey would be structuring the survey as an evaluation matrix. (See Figure 3) The survey questions would consist of an array of criteria for assessing the merits of an idea. The evaluator rates each idea (e.g., on a scale of 0-5) based on how well it meets each criterion. The results of the survey provide a criterion-by-criterion score as well as an overall score for each idea. The scores can be compared and the highest scoring idea selected.

Phase 3 Collaborative Tools: Requirements Gathering

In the third phase of design-based learning, students gather requirements for implementing the chosen solution. Successful requirements gathering requires the collaborative development of a document that specifies the resources needed to complete the project. Students may work individually or in teams at this stage.

Figure 3. Evaluation matrix created using SurveyMonkey (surveymonkey.com)

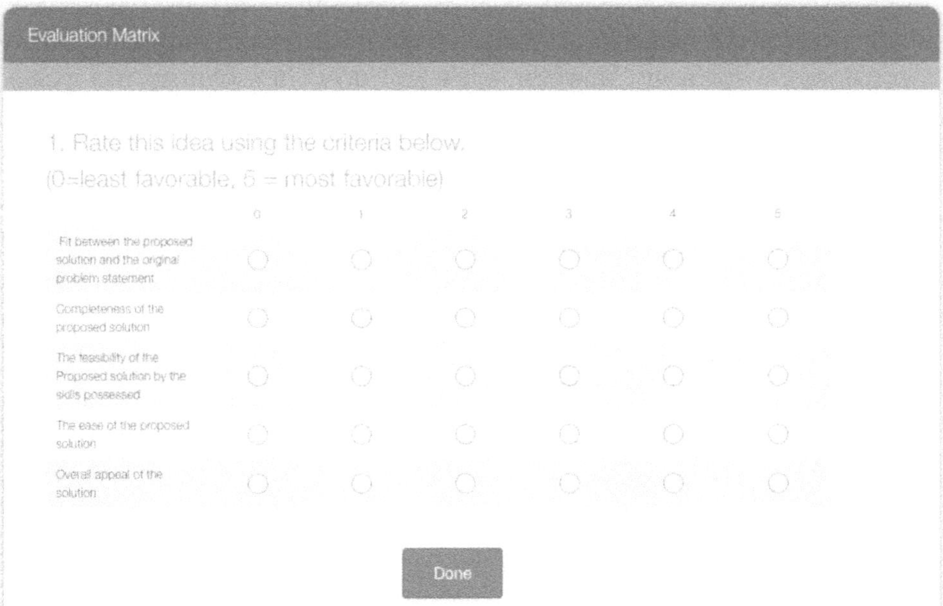

Regardless, a central place where students can share and access requirements gathered by other group members is highly beneficial. Web-based word processors (e.g., Google Docs or Microsoft Office 360) are some of the simplest and most accessible tools for document collaboration. Project teams can create and simultaneously build requirements documentation with the most up-to-date version available to all team members. The main drawback to this approach is that the document is initially unstructured. Thus, team members must agree on a format and abide by that format as they make individual contributions. Since there is no embedded mechanism for capturing feedback or comments from team members, other communication tools would have to be used in parallel.

For more complex design problems with elaborate requirements, wikis can provide a better collaborative solution. Wikis are specifically designed to allow collaborative editing and structuring of information. Wikis are particularly well suited for when a design document evolves and must be edited by multiple team members. In addition, team members can continuously review and give feedback on captured requirements.

Wiki pages can be created based on templates that reflect the nature of the project. For example, in IT projects, a wiki's pages can be pre-structured to accommodate documentation of functional requirements or use cases. Wikis also provide quick, web-based access to the latest team member contributions. As the project evolves, team members will have access to the history of not only the requirements, but also the discussion surrounding each requirement.

Beyond wikis, more advanced tools designed specifically for requirements gathering exist in the marketplace. Many of these solutions are costly and designed for corporate environments. However, a few providers offer solutions that include a free basic account suitable for student projects. Table 2 provides a sampling of requirements gathering systems.

Table 2. Examples of requirements gathering tools

Tools	Features
Wikis	Free Familiar interface, shallow learning curve
ReqView	Free version available Import documents from diverse file types Work online or offline
Gatherspace	Develop models and diagrams Report generation
iRise	Validate requirements Develop diagrams Integrate with other development environments

Phase 4 Collaborative Tools: Implementation of the Solution

In the implementation phase, students build the solution based on the requirements gathered in the previous stage. The collaborative tools used during this phase may vary greatly by discipline. However, a project web site provides an excellent means of communicating project progress during implementation. Project web sites serve as a repository for archiving and linking all aspects from each phase of the entire design project. As students review deliverables from previous phases, they reinforce their understanding of the project topic and develop a structured approach to problem-solving. The project web site provides easy access to the learning path by which they arrived at the current phase.

Within the IT discipline, the project web site can serve as point of distribution for beta versions of the artifact under design and successive releases. The website is a place for the project team to communicate the status, features, and notes regarding each new development. For users (or the instructor), the web site provides a means of gauging the project's progress and feature complexity prior to completion.

Project web sites also give instructors a view into the level of contribution from each team member. With access to a living history of the project, the instructor can assess the scope and impact of each team member's contribution on an ongoing basis. In the absence of this dynamic system, it can be difficult to ascertain the amount of effort and significance of various tasks related to the final output. Also, with visibility into the project's progress, the instructor can better ascertain when to guide project teams if intervention is necessary.

For project team members, the website serves as a hub for convenient access to all phases of the project. There are a number of free web services that provide for rapid development of visually appealing and highly functional web sites. These web services provide drag-and-drop design capabilities that make creating a web site as simple as using a word processor. Thus, students no longer need to be versed in the technical details of creating web sites to harness the collaborative power of the web to support their project. STEM students can remain focused on the project domain without having to master web site development techniques, if this is outside the scope of the assignment.

For more complex projects with a greater lifespan, can place designated project resources behind password-protected web pages, effectively creating an intranet for team members. For example, access to the discussion groups/forums and survey data could be located within members-only pages. Housing project documents and communication in a project web site provides an efficient means of accessing and storing project information beyond email attachments or texting updates to team members.

Blogs are a special feature of web sites (and many learning management systems) that merit a separate discussion. Project teams can utilize blogs as part of the web site to document the stages of implementation. Blogs can also serve as electronic journals where students post pictures, links to resources, and status updates for peers and instructors.

From the instructor's standpoint, student blogs give insight into team progress and individual intellectual growth. Blogging bestows upon each student his or her own voice and opportunity to express a unique perspective on the learning challenges and accomplishments within the project space. In addition, blogs support commenting features that can be used by the instructor to provide feedback, pose additional inquiries, or give encouragement.

From the student's standpoint, blogging provides a tool for reflection and critical thinking. In writing a blog post, students apply analytical thought, reason through hypotheses, and communicate their rationale for results. When blog posts are made viewable to peers as well as the instructor, research has shown that students are more inclined to take ownership of their own ideas and the manner in which they communicate them. As students perceive that their ideas are being evaluated by their peers their level of motivation has been shown to increase (Tseng & Tsai, 2007b). As the level of student motivation increases, students began to take ownership of learning, develop an intrinsic interest, and shift from surface to deep learning. When STEM students adopt a deep-learning approach they seek a deeper understanding by developing a holistic view of the relationship between the design tasks, technology skills, and content-related concepts (Cohen, 2002b).

In addition to the deep learning that arises from blogging and peer critique, blogging also encourages communities of inquiry and knowledge sharing. As students share ideas, challenges, and successes, mutual learning occurs. Blogging can provide more personal expression and response than traditional journaling methods. As students seek sources to support their point of view, they are more inclined to find and evaluate the merits of data sources and identify scholarly work. As students focus and strengthen their discourse, their peers benefit from a plurality of well-supported, well-reasoned perspectives. Thus, students have a richer understanding of multifaceted problems and approaches as opposed to the more myopic vantage point of a single source.

Phase 5 Collaborative Tools: Feedback and Evaluation of Solution

Once the solution has been implemented, students can receive peer or expert feedback via responses posted to project blogs or surveys in which peers or experts external to the project provide both structured and unstructured comments. When utilized among peers, making feedback anonymous encourages the most constructive and honest responses.

As discussed previously, surveys can be designed to capture highly structured or unstructured feedback. The intended goal for resulting data should drive the format for the data collection. Once the project has been implemented, there are many channels for soliciting response via survey. Surveys can be emailed or embedded in a project website or the actual project deliverable if the project involved the creation of a digital artifact. Regardless, the goal is to capture the experience of the user or client and gain insight into potential opportunities for improvement. This data can then be used to refine the project via repetition of the design cycle if time or the scope of the assignment permits.

Both the student receiving feedback and the student providing feedback benefit from participating in peer assessment as part of the final phase of design-based learning projects. When considering the merits of feedback received as well formulating well-reasoned comments, students strengthen their understanding of how to arrive at a sound solution and what comprises design excellence. The prospect of receiving feedback from peers can be just as valuable in motivating self-critique and refinement of thought.

In addition to surveys, social media can be used in the evaluation phase of DBL. Social media applications vary greatly in their interface and functionality. However, there are several common features typically found in social media that are supportive of evaluating DBL projects. First, social media provides an instant form of collective communication. Apps like Twitter allow users to create uniquely identifiable channels of communication relating to a single topic. As an example, a team might use Twitter to generate discussions around a particular feature of the project. The team can then analyze the social media stream with the goal of evaluating the user response or reception to that feature.

Using social media, project teams can have many-to-many conversations and generate "buzz" or interest in their projects. Receiving live feedback and gauging the level of interest in a proposed solution provides an authentic learning experience in discovering the practical aspects of a topic and proposed solution. A team can design the most scientifically elegant solution, but if the feedback indicates that a community of potential users/clients don't perceive this solution as needed or useful then the design may require adjustment. Even in the absence of altering the solution, students still benefit from analyzing social media content as a response to team communication about the solution. In other words, students learn not only how to refine a technical solution, but also how their communication of technical or scientific information might be improved.

There are a number of collaborative tools available to support pedagogical approaches that incorporate design-based learning. Thoughtful application of these tools can foster greater collaboration, knowledge acquisition, and critical thinking. Table 3 provides an overview of the more basic tools discussed in this chapter. The table provides a starting point for applying free and commonly available tools with a relatively shallow learning curve. As instructors and teams become more acclimated to design-based learning, experimentation with some of the newer and more advanced tools discussed in this chapter may provide additional benefits.

Table 3. Overview of basic collaborative tools

Digital Tool	Pros/Cons	Design Phase
Email	Pros: easily accessible and available to all Cons: email threads can be difficult to follow; individuals may be unintentionally dropped from group conversations	Idea generation – early brainstorming and idea introduction
Text Messaging	Pros: conversational quick communication Cons: Not well suited for group conversations or documenting	Idea generation – clarifying or refining ideas among dyads
Forums/Discussion Groups	Pros: easy to organize and structure communication around topics Cons: not as quickly assessable as email and text	Idea generation – refining and documenting ideas
Web-Based Word Processors/Office Productivity	Pros: easily accessible, simple to use Cons: unstructured, once document is formatted all users must maintain format; commenting is not built in(?)	Requirements gathering and documentation
Wikis	Pros: structure is predefined; commenting is built in Cons: more setup time required	Requirements gathering and documentation
Web-Based Surveys	Pros: easily distributed Cons: value of data and effort of analysis depends on how well questions are planned	Solution determination, feedback and evaluation of the solution
Project Web Site	Pros: can serve as a hub for all project documents and communication; can set up separate team member-only spaces Cons: can become outdated without ongoing effort to maintain	Implementation – documenting and communicating project milestones. Release new versions of the project artifact.
Blogging	Pros: easily accessible self-reflective tool for providing project updates and journaling Cons: students may require additional instruction on electronic communication etiquette and providing constructive feedback	Implementation
Social Media	Pros: can be used to generate buzz or interest in project; also provides project-related conversation streams for content analysis Cons: can be difficult to garner attention and attack focused conversations in noisy medium	Feedback and evaluation of the solution

FUTURE RESEARCH DIRECTIONS

The pace of change in educational technologies challenges researches to create persistent methodologies for assessing the impact of emerging technologies. To build a stronger theoretical framework and more robust empirical studies of technology application, it would be worthwhile to explore the development of salient features related to these collaborative systems. The structure for investigating new technologies could for example include affordances offered by the technology without overly focusing on the features, interface, or technical configuration.

CONCLUSION

When students are engaged in projects that involve real-world scenarios and authentic learning, students become more motivated to engage in deep learning. Design-based learning pedagogy provides an effective way to engage STEM students in authentic learning activities that prepare them for real-world design problems. In addition, because DBL allows students to apply creativity to generate their own design plans and specifications, students have a vested interest in project outcomes and can explore complex solution spaces. At the same time, the design process provides guidance to students on how to successfully apply their technical skills to solve problems and produce functional artifacts. These artifacts can then be evaluated and refined by students to further understand the design process cycle.

A design-based learning pedagogical approach has the potential to increase students' understanding of the design process through engaging in active and authentic learning projects. In addition, student motivation, individually and collectively in teams, is a strong determinant for deep learning and perhaps the desire to engage more fully in developing knowledge of the design process. Increased knowledge and deep learning yield more favorable design outcomes.

Design-based learning places students at the center of defining design problems and taking ownership of the process. Projects reflect issues that are uniquely relevant and topical for each team. In addition, the phases of the design process guide students through an authentic means of arriving a solution. Students progress from conceptualization to development to evaluation just as they would with design solutions to problems in the "real world."

As students move through each stage of the design process and confront unstructured inquiry and uncertain outcomes, collaborative tools can be used to add structure to communication. These tools can facilitate idea sharing, serve as knowledge repositories, and communicate solutions. Given the vast and rapidly evolving digital communication and collaboration tools available, STEM instructors and students should give careful attention to deploying the tools best suited for the tasks specific to each design phase. If applied effectively, the tools outlined in this chapter can support all phases of design-based learning in a manner that is effective yet unobtrusive to the learning process. As STEM educators find novel ways of introducing inquiry-based approaches to student-led learning, technology will remain a valuable resource. As technologies evolve, instructors must continually reassess the value of existing and emerging collaborative tools to support learning within the community of STEM students.

REFERENCES

Buchanan, R. (1992). Wicked problems in design thinking. *Design Issues*, *8*(2), 5–21. doi:10.2307/1511637

Cohen, E. (2002a). *Challenges of information technology education in the 21st century*. IGI Global. Retrieved from http://books.google.com/books?hl=en&lr=&id=Tbt3OE0UT0IC&oi=fnd&pg=PR7&dq=+%22challenges+of+information+technology+education%22&ots=g__Lx9NkCA&sig=SpuoZkU6t-sWbi5q67iYEAIJ1fQ

Cohen, E. (2002b). *Challenges of information technology education in the 21st century*. IGI Global. Retrieved from http://books.google.com/books?hl=en&lr=&id=Tbt3OE0UT0IC&oi=fnd&pg=PR7&dq=+%22challenges+of+information+technology+education%22&ots=g__Lx9NkCA&sig=SpuoZkU6t-sWbi5q67iYEAIJ1fQ

Collins, C. J., & Han, J. (2004). Exploring applicant pool quantity and quality: The effects of early recruitment practice strategies, corporate advertising, and firm reputation. *Personnel Psychology*, *57*(3), 685–717. doi:10.1111/j.1744-6570.2004.00004.x

Cope, C., Staehr, L., & Horan, P. (2002). Towards establishing the best ways to teach and learn about IT. *Challenges of Information Technology Education in the 21st Century*, 57.

Dennis, A. R., Wixom, B. H., & Vandenberg, R. J. (2001). Understanding Fit and Appropriation Effects in Group Support Systems via Meta-Analysis. *Management Information Systems Quarterly*, *25*(2), 167–192. doi:10.2307/3250928

Doppelt, Y., Mehalik, M. M., Schunn, C. D., Silk, E., & Krysinski, D. (2008). *Engagement and achievements: A case study of design-based learning in a science context*. Retrieved from http://scholar.lib.vt.edu/ejournals/JTE/v19n2/doppelt.html

Duch, B. J., Groh, S. E., & Allen, D. E. (2001). *The power of problem-based learning: a practical" how to" for teaching undergraduate courses in any discipline*. Stylus Publishing, LLC. Retrieved from http://books.google.com/books?hl=en&lr=&id=-78ZnGLRacAC&oi=fnd&pg=PR8&dq=%22Problem-based+Learning%22&ots=xnWlgDw3ka&sig=vBRvU2gMlhr8M4enSgcEfbRi51Y

Duffy, T. M., & Jonassen, D. H. (1992). *Constructivism and the technology of instruction: A conversation*. Psychology Press. Retrieved from http://books.google.com/books?hl=en&lr=&id=yS310A35FC0C&oi=fnd&pg=PR1&dq=constructivism&ots=1v4VqyqzSh&sig=dwffGNrNiTZc-3VF5omxI7ANk-Q

Dunne, D., & Martin, R. (2006). Design thinking and how it will change management education: An interview and discussion. *Academy of Management Learning & Education*, *5*(4), 512–523. doi:10.5465/AMLE.2006.23473212

Fosnot, C. T. (1996). *Constructivism. Theory, Perspectives, and Practice*. ERIC. Retrieved from http://www.eric.ed.gov/ERICWebPortal/recordDetail?accno=ED396998

Gómez Puente, S. M., Van Eijck, M., & Jochems, W. (2011). Towards characterising design-based learning in engineering education: A review of the literature. *European Journal of Engineering Education*, *36*(2), 137–149. doi:10.1080/03043797.2011.565116

Johansson-Sköldberg, U., Woodilla, J., & Çetinkaya, M. (2013). Design thinking: Past, present and possible futures. *Creativity and Innovation Management*, *22*(2), 121–146. doi:10.1111/caim.12023

Kirschner, P. A., Sweller, J., & Clark, R. E. (2006). Why minimal guidance during instruction does not work: An analysis of the failure of constructivist, discovery, problem-based, experiential, and inquiry-based teaching. *Educational Psychologist*, *41*(2), 75–86. doi:10.1207/s15326985ep4102_1

Lee, S., Koh, S., Yen, D., & Tang, H.-L. (2002). Perception gaps between IS academics and IS practitioners: An exploratory study. *Information & Management*, *40*(1), 51–61. doi:10.1016/S0378-7206(01)00132-X

Leinonen, T., & Durall-Gazulla, E. (2014). Design thinking and collaborative learning. *Comunicar, 21*(42). Retrieved from http://eprints.rclis.org/20862/

Moazzen, I., Hansen, T., Miller, M., Wild, P., Hadwin, A., & Jackson, L. (2013). Literature Review on Engineering Design Assessment Tools. *Proceedings of the Canadian Engineering Education Association*. Retrieved from http://library.queensu.ca/ojs/index.php/PCEEA/article/view/4864

Puente, S. M. G., van Eijck, M., & Jochems, W. (n.d.). A sampled literature review of design-based learning approaches: a search for key characteristics. *International Journal of Technology and Design Education*, 1–16.

Razzouk, R., & Shute, V. (2012). What is design thinking and why is it important? *Review of Educational Research*, *82*(3), 330–348. doi:10.3102/0034654312457429

Savery, J. R., & Duffy, T. M. (1995). Problem based learning: An instructional model and its constructivist framework. *Educational Technology*, *35*(5), 31–38.

Sitthiworachart, J., & Joy, M. (2003). Web-based peer assessment in learning computer programming. In *Advanced Learning Technologies, 2003. Proceedings. The 3rd IEEE International Conference on* (pp. 180–184). Retrieved from http://ieeexplore.ieee.org/xpls/abs_all.jsp?arnumber=1215052

Topi, H., Valacich, J. S., Wright, R. T., Kaiser, K., Nunamaker, J. F. Jr, Sipior, J. C., & De Vreeda, G.-J. (2010). IS 2010: Curriculum Guidelines for Undergraduate Degree Programs in Information Systems. *Communications of the Association for Information Systems*, *26*, 359–428.

Trauth, E. M., Farwell, D. W., & Lee, D. (1993). The IS expectation gap: Industry expectations versus academic preparation. *Management Information Systems Quarterly*, *17*(3), 293–307. doi:10.2307/249773

Tsai, C.-C., Lin, S. S., & Yuan, S.-M. (2002). Developing science activities through a networked peer assessment system. *Computers & Education*, *38*(1), 241–252. doi:10.1016/S0360-1315(01)00069-0

Tseng, S.-C., & Tsai, C.-C. (2007a). On-line peer assessment and the role of the peer feedback: A study of high school computer course. *Computers & Education*, *49*(4), 1161–1174. doi:10.1016/j.compedu.2006.01.007

Tseng, S.-C., & Tsai, C.-C. (2007b). On-line peer assessment and the role of the peer feedback: A study of high school computer course. *Computers & Education*, *49*(4), 1161–1174. doi:10.1016/j.compedu.2006.01.007

van Zundert, M., Sluijsmans, D., & van Merriënboer, J. (2010). Effective peer assessment processes: Research findings and future directions. *Learning and Instruction*, *20*(4), 270–279. doi:10.1016/j.learninstruc.2009.08.004

Wells, S. H., Warelow, P. J., & Jackson, K. L. (2009). Problem based learning (PBL): A conundrum. *Contemporary Nurse*, *33*(2), 191–201. doi:10.5172/conu.2009.33.2.191 PMID:19929163

Wijnen, W. (2000). Towards design-based learning. Eindhoven: Eindhoven University of Technology. Retrieved from http://w3.tue.nl/fileadmin/osc/OSC_oo/doc/OGO_brochure_1_EN.pdf

Wu, J.-H., Chen, Y.-C., & Chang, J. (2007). Critical IS professional activities and skills/knowledge: A perspective of IS managers. *Computers in Human Behavior*, *23*(6), 2945–2965. doi:10.1016/j.chb.2006.08.008

KEY TERMS AND DEFINITIONS

Authentic Learning: A pedagogical approach that encourages students to create tangible outcomes that can be examined and critiqued using the real world as a context.

Deep Learning: An approach to learning in which learner seeks to understand the theory and broader interconnectivity between concepts.

Intrinsic Interest: A desire to engage or learn based on an internal reward system.

Many-to-Many Conversations: A feature of internet-driven communication that allows participants to converse (both transmitting and receive responses) with multiple individuals simultaneously.

Peer Assessment: A process by which students evaluate and provide feedback on each other's work using criteria provided by an instructor.

Surface Learning: An approach to learning that involves simple memorization of facts with no concern for drawing larger connections.

Unstructured Problems: A problem space in which the number of potential solutions is unknown and the connection between the solution and the outcome to be produced is nonlinear and complex.

Chapter 4
Practicing Scientific Argumentation Through Social Media

Jana Craig-Hare
University of Kansas, USA

Marilyn Ault
University of Kansas, USA

Amber Rowland
University of Kansas, USA

James D. Ellis
University of Kansas, USA

ABSTRACT

The use of social media in and outside the classroom is increasing in the number of popular applications as well as pervasiveness in our culture. Teachers utilize social media to engage students, connect with experts, and expand their own professional learning. This chapter provides educators with information about the use of social media to support STEM practices. Social media can be used to engage students in active learning and problem-solving through student-posted claims and effective online questioning. Using social media supports the scientific practice of engaging in argument from evidence, as well as emulates how scientists collaborate on their own research and share research findings. Best practices and lessons learned are shared in this chapter, including a case study from a secondary science classroom and suggestions for the use of social media for educator professional learning.

INTRODUCTION

The purpose of this chapter is to provide educators with information about social media use to support STEM Learning, particularly scientific argumentation. According to Merriam-Webster's Learner's Dictionary (2016), social media are online Web

DOI: 10.4018/978-1-5225-2525-7.ch004

sites and applications "through which users create online communities to share information, ideas, personal messages, and other content (as videos)." Current sites include, for example, Facebook, Twitter, Snapchat, Instagram, and LinkedIn. The sharing of information on these sites may be as personal as the turn-taking that occurs during a dialogue or as removed as the posting of unrelated comments. As K-12 educators infuse technology into instruction, many teachers, building administrators, and district directors continue to ask for instructional resources on how to support student learning of 21st-century skills using existing and emerging digital tools. While educators may have considerable knowledge about technology integration, instructional strategies, and curriculum; they are seeking research-based and classroom-tested digital resources that they can quickly move into practice. This chapter explores the use of social media within STEM pedagogy, particularly as it can be used to support the practice of scientific argumentation. The chapter also identifies strategies to engage educators and students in teaching and learning through social media.

BACKGROUND

Social media use is on the rise. According to a recent survey by the Pew Research Center's Internet & American Life Project, 92% of teens ages 13–17 use the Internet daily, 76% use social media sites and 71% say they use more than one social media application. Seventy-one percent are Facebook users, 52% Instagram, 41% Snapchat and 33% say they used Twitter; up from 16% in 2011 (Lenhart, 2015). Chao, Parker, and Fontana (2011) write about the rise in the use of social media, stating that its impact is "so widespread and inculcated into our culture that it is futile to try to stop their [social media] influence at the classroom door" (p. 324). The 2015 Horizon Report (Johnson, Adams Becker, Estrada, & Freeman, 2015) ranked the increasing use of collaborative learning approaches as the top trend to impact K-12 education in the next three to five years (2015-2019). It states that collaborative learning approaches are a way to increase global collaboration by utilizing online technologies so students are able "to learn with others beyond their immediate environment" (p. 12). Ertmer and her colleagues (Ertmer, Ottenbreit-Leftwich, Sadik, Sendurur, & Sendurur, 2012) argue that technology "needs to be placed in the hands of students, who are encouraged and enabled to utilize it in the same ways, and for the same purposes, that professionals do -- that is, to communicate, collaborate, and solve problems" (p. 424). This perspective is echoed in the 2015 Horizon Report (Johnson et al., 2015) crediting social media apps such as Snapchat, Instagram, and Vine, for helping shift students from consumers to creators (p. 14). Social media, therefore, is a widely-used technology that has an instructional role in education, particularly science education.

The use of social media to engage students in the exchange of information and ideas is similar to how scientists conduct their work. Science is a collaborative process, during which practices, such as argumentation, can be used to present and refine claims as a result of challenges to evidence and reasoning. Argumentation, as defined by the Next Generation Science Standards (NGSS), is "a mode of logical discourse used to clarify the strength of relationships between ideas and evidence that may result in revision of an explanation" (NGSS Lead States, 2013). As a result of the NGSS, argumentation is one of the science practices that is explicitly taught during classroom instruction. This concentration on argumentation has been difficult, possibly because of a lack of instructional resources focusing on argumentation (Ault et al., 2016).

Yet understanding the practice of argumentation is key to understanding how the core ideas and crosscutting concepts in science develop. The refinement of science occurs, in part, through the social exchange process of argumentation, often occurring through social media. For example, Van Eperen and Marincola (2011) highlight the importance of scientists communicating via social media to collaborate and share findings on a range of scientific topics. Correspondingly, a 2007 survey of professionals in the field of bioinformatics (Bioinformatics, 2007) found that 77% of life scientists participated in some type of social media to advance their science, 50% said that social media was beneficial to sharing ideas with colleagues and 85% indicated that social media influenced their decision-making. In 2016, Bioinformatics asked again about the use of social media, specifically of Life Scientists and found that only 19% claim <u>not</u> to use social media. Pair these statistics of science professionals with the fact that Facebook (2016) has more than 1.79 billion monthly active users worldwide and Twitter (2016) has over 313 million, it is clear that student digital literacy in the area of science must include skills in the use of social media. Research is needed to support teachers as they pioneer the use of social media in teaching and learning (Rowland et al., 2015; Greenhow, 2011; Ertmer et al., 2012).

SOCIAL MEDIA AND STEM

Based on the NGSS, science instruction needs to utilize a three-dimensional approach consisting of core content, crosscutting concepts, and practices. The instructional use of social media can be one medium utilized to engage learners in the science practices. The practices, particularly those that engage learners in conversations about science content and crosscutting concepts when presenting an argument or defending an augment with evidence, are well suited to promote science, technology, engineering and math (STEM) literacy and the social construction of knowledge through the use of social media (National Academies of Science, Engineering, and Medicine, 2016).

Social Media

Theorizing that learning is a social activity (Vygotsky, 1978) and needs to be situated in an authentic context (Lave & Wenger, 1991), social media has the potential to engage students and teachers in the efficient and effective practice of argumentation (Mishra & Kohler, 2006). Vygotsky (1978) claimed that all learning is social. By extension, therefore, internet tools such as social media promote collaboration and information sharing and can be used in classroom settings to promote student engagement and interaction with academic content and practices (Junco, Helbergert, & Loken, 2011; Kabilan, Ahmad, & Abidin, 2010).

Social media can be defined as interactive platforms or applications with which "individuals and communities create and share user-generated content" (Kaplan, 2010). Microblogging is a form of social media or social networking where users can post text entries and follow, respond, and interact both synchronously and asynchronously with other user posts. Blogs and wikis are social media applications designed for publication of content (text or media based) and in the case of wikis, collaborative interactivity online allows multiple users to edit and post content in the same space. Other social networking sites, such as Facebook, are social spaces where users can microblog. They can also share documents, images, video, and longer posts in the form of blogs or wikis. Edmodo is the educational equivalent of Facebook, which "marries the benefits of social networking with a safe, controlled environment that educators can allow students to freely operate in" (Scott, 2012, p. 56).

The use of social media for instruction enables teachers to make learning authentic by modeling science practices in a real-world space where their peers, parents, and experts can serve as their audience. Teachers have the flexibility to integrate current events through web-links and embedded media such as images and videos. Teachers and students can also engage scientists in academic discourse, augmenting their experience and making science more accessible. Gerard, Varma, Corliss, and Linn (2011) suggest that learning is most successful when teachers utilize curriculum units with embedded technology that leverage "inquiry-oriented activities focused on distinct science concepts" (p. 439). Social media provides a medium for students to practice scientific argumentation through inquiry-oriented discourse.

The Importance of STEM Literacy Practices

At the core of student learning in the modern STEM classroom, the National Research Council (NRC) states that curricula should be rigorous and focused on critical topics while making connections through topics and performances (2013). The Committee on the Evaluation Framework for Successful K-12 STEM Education, in response to a 2011 report, (NRC, 2011) suggests that these connections might best be made through "classroom coverage of content and practices" and designated robust coverage of both content and practice as one of the 14 indicators for success. Specifically, the Common Core State Standards (CCSS) (National Governors Association Center for Best Practices Council of Chief State School Officers, 2010) and the NGSS (NGSS Lead States, 2013) emphasize the need for students to engage in the practice of science and mathematics. The practice of argumentation includes knowing how to take a more critical stance when confronted with an argument; evaluating the quality of what is read, seen or heard; and defending claims with appropriate evidence and reasoning. It has been noted that scientific practices, such as argumentation, help increase students' achievement and content knowledge by requiring them to think deeply about the content, construct their own understanding of the content, and apply it as they construct their arguments or critique those of others (Pellegrino & Hilton, 2013). Students also need to learn these skills to be part of an educated citizenry and active participants in civil discussions of issues central to our democracy. Practicing argumentation also promotes STEM literacy for students.

STEM literacy is not necessarily a content area but should be viewed as a "deictic means (composed of skills, abilities, factual knowledge, procedures, concepts and metacognitive capacities) to gain further learning," as described by Zollman (2012). The critical role of discourse in students' understanding of and learning about science has also become increasingly evident. The NRC (2012) notes that "science is fundamentally a social enterprise" (p. 27) where scientists engage in ongoing discourse with their colleagues, informally and formally, to share insights, brainstorm, and problem-solve. While the NGSS definition of argumentation includes Toulmin, Rieke, and Janik (1984) defined argumentation as "the whole activity of making claims, challenging them, backing them up by producing reasons, criticizing those reasons, rebutting those criticisms, and so on" (p. 14). Science standards now require students to engage in scientific argumentation as a critical science practice; one that supports STEM literacy.

Although the concept of literacy in science, technology, engineering, and mathematics has been defined separately by the respective professional organizations, the definitions overlap and speak to the importance of knowledge and understanding across these areas. According to Toulmin and Meghan (2007), STEM literacy is interdisciplinary, in that it bridges all four content areas and can shift student thinking toward investigating and questioning interrelated global issues. Scientific and technical knowledge is important for competitiveness in the global economy. As such, STEM literacy is a dynamic process and is critically important for students' future success. It is furthered by participants engaging in the social construction of knowledge through the exchange of information, as occurs in the process of scientific argumentation.

Social Attributes of STEM

The Framework for K-12 Science (NRC, 2012) and the NGSS present significant challenges to science teachers across the nation – intertwining the teaching and learning of science and engineering practices, crosscutting concepts, and core ideas with content-specific performance expectations. Experts in STEM education are calling for an emphasis on investigation, evaluation, argumentation, and engagement in the practices of STEM, rather than narrowly focusing on science content knowledge (Bybee, 2011; NRC, 2007; NRC, 2012). The expectations are clear: students need to learn science by doing science, and doing science is a social endeavor. Oftentimes, STEM classrooms tend to be rule-oriented with a focus on applying procedures to arrive at correct answers (Weiss & Pasley, 2004; Weiss, Pasley, Smith, Banilower, & Heck, 2003; Yackel & Hanna, 2003). Mathematics teachers, for example, report that they have students explain or justify their method for solving a problem, compare and contrast different methods for solving a problem, present solution strategies to the rest of the class, and work in small groups for less than half of the allotted class time (Banilower Smith, Weiss, Malzahn, Campbell, & Weis, 2013). Additionally, students often view a mathematical proof as a series of steps to be followed rather than a process that requires argumentation (Kamii & Dominck, 1998). Bieda (2010) found that nearly half of the opportunities that middle school students had to prove a conjecture were not supported by student justification. In these instances, Bieda found three instructional moves characterized the teacher's response to a student conjecture: 1) the teacher did not ask for justification, 2) the teacher confirmed the conjecture as valid without justification, 3) the teacher asked other students whether or not they agreed with the conjecture. Yackel and Hanna (2003) noted that students often get to high school without sufficient opportunities to engage in mathematical discourse about their own reasoning process and understanding of mathematics. A number of researchers have proposed the need for professional development to support more active, dialogic approaches to primary learning and teaching (Alexander, 2010; Mercer, Warwick, Kershner, & Staarman, 2010).

Educational researchers have long advocated for educational settings that engage students in active learning and problem solving. This includes students formulating their own claims and asking effective questions. Although questioning is key to learning, it is well documented that students infrequently ask questions in the classroom and that their questions tend to be "lower-level" (Webb, Nemer, & Ing, 2006). A strategy to engage students in more productive questioning involves engaging them in collaborative knowledge-building, with the support of new technology environments (Zhang & Sun, 2011). This involves activities such as asking a question, making a proposal, critiquing, providing assessment, agreeing, disagreeing, etc. Craig-Hare et al. (2013) found that students, when using technology, indicated a higher level of learning engagement when classroom tasks were at the higher bloom's level of creating, evaluating, and analyzing versus the lower level of applying, understanding and remembering. These "more engaged" students reported an increased feeling of being challenged, interested, in control and felt a sense of purpose in the activity. Elementary and secondary students' understanding and skills improve when peers work together on challenging, cooperative tasks (Gauvain, 2001; Littleton & Light, 1999).

Research has also shown that students who communicate and reason about mathematics problems have a greater potential for understanding the concepts that underlie the problems (Ball & Bass, 2003; Littleton & Light, 1999; Schoenfeld, 1985; Stein, Grover, & Henningsen, 1996). The ability to think about, and reflect on, the problem-solving process is an important component of learning mathematics. Communicating these ideas to classmates to cooperatively solve a mathematics problem is another step in understanding mathematics (Thomas, 2013). Other instructional methods have been identified in the research literature as a means to enhance mathematical reasoning. One focuses on metacognitive training, in which students monitor their own thinking while using appropriate information and strategies during problem-solving (Carr & Biddlecomb, 1998; Desoete, Roeyers, & Buysse, 2001; Mevarech, 1999; Mevarech & Kramarski, 1997). A second method focuses on cooperative learning (e.g., Slavin, 1996) and the importance of sharing mathematical reasoning with others as a community of learners (Ball & Bass, 2000). Instructional strategies that combined cooperative learning and metacognitive training in 8th-grade classrooms resulted in higher student problem-solving achievement than in classrooms using cooperative learning or meta-cognitive training alone (Kramarski & Mevarech, 2003). The CCSS, in Mathematics Practice #3, emphasizes that students should be able to construct arguments, critique their own or others' arguments, identify flaws in reasoning, revise their arguments based on evidence, and review their arguments and conclusions with peers (Mayes & Koballa, 2012). These practices support mathematics pedagogy based on dialogic learning.

There is significant support for the pedagogical view that learning is socially constructed through conversation (Levine & Marcus, 2010; Prestridge, 2009; Vygotsky, 1978; Warwick, Mercer, Kershner, & Staarman, 2010). Kelly (2007) contends that teaching and learning occur through discourse and interactions, that student access to science occurs through engagement in social and symbolic worlds comprising knowledge and practice of specialized communities, and that disciplinary knowledge is constructed and assessed through language. Therefore, language that is spoken as well as read is key to learning. The use of social media to engage students in STEM disciplines is, as we have stated, similar to how scientists collaborate on their research. Integrating argumentation, a key practice of science, into science instruction is a significant challenge (Alozie, Moje, & Krajcik, 2010). To address this challenge, the instructional use of social media should be linked with fostering STEM learning by designing the scaffolding, formative assessment, and self-assessment necessary to increase scientific practices and enhance STEM learning.

A number of researchers have developed computer-based resources to address the need to engage students in the social aspects of science learning. Programs such as the Web-based Inquiry Science Environment (WISE) (Linn, Clark & Slotta, 2003) and the Cognitive Apprenticeship Web-based Argumentation System (Tsai, Jack, Huang, & Yang, 2012) have successfully provided K-12 students with controlled and scaffolded environments specifically designed for the task of argumentation, supporting structured and prescriptive environments to address the evidence and reasoning associated with a claim. They do not, however, engage students in discourse with peers about science. Engaging with peers in scientific argumentation is a more authentic experience than pursuing a scaffolded program. Social media can be used to provide a flexible environment for STEM learning through discourse; one that students and teachers frequently use in their everyday lives.

Practicing Argumentation in the Classroom

British philosopher and logician Stephen Toulmin (1922-2009) first proposed his model of argumentation in his 1958 book, *The Uses of Argument*, with an updated version published in 2003. Toulmin's model of argumentation proposed a way to better understand how arguments could be used in real-world situations. The classical model of arguments was based around syllogisms in a three-statement argument of a claim, evidence and warrant/reasoning. Toulmin (1958, 2003) was unsatisfied with the absolute nature of this type of argument and felt that real-world arguments were often more convoluted and required more than just a claim, evidence and warrant/reasoning. To support this belief, Toulmin's model added three supporting components: backing, qualifiers, and rebuttal. The inclusion of reasoning and argumentation across the CCSS emphasizes the need for students to know how to take a critical stance when confronted with an argument, think logically about the relationships among concepts, evaluate a logical chain of reasoning, and defend their claims with appropriate justification and reasoning. Toulmin's model supports this type of reasoning and argumentation for real-world argumentation.

Lack of quality instruction about scientific argumentation and discourse in science classrooms has been a concern for several decades (Duschl & Osborne, 2002; Jimenez-Alexandre & Erduran, 2007; Osborne, Simon, Christodoulou, Howell-Richardson, & Richardson, 2013). With the integration of the NGSS, however, science teachers are beginning to include the practice of scientific argumentation among their classroom activities. The research on effective strategies for teaching and learning scientific argumentation is emerging, as called for in these new rigorous standards for students. Active student engagement and discourse as part of the learning process, use of direct explicit instruction, and incorporation of learning scaffolds such as graphic organizers, especially when combined with verbal descriptions and deep level questioning are just a few of the recommended instructional procedures. (Kamil et al., 2008; Pashler et al., 2007).

Ultimately, the goal of teaching argumentation is for students to this apply scientific practices to everyday challenges and develop justifiable ways to use evidence to support their claim (Lawson, 2003). A natural environment for practicing argumentation in 21st-century classrooms is in social media. Teachers often report using technology because it engages and motivates their students. Taking advantage of the emergence of social media as an engaging environment for youth supports the acceptance of this media as an educational tool. It could be argued that the social media students use every day and find highly engaging would enhance the practice of scientific argumentation and support students' ability to evaluate data, analyze claims, and propose counterarguments and rebuttals. The proliferation of and capacity surrounding the emergence of collaborative social media, including social networks and interactive spaces, have the potential to better prepare students for real-world argumentation practice by using tools with which they are already familiar while assisting teachers with the implementation of inquiry-based practices.

RESEARCH-BASED EXAMPLES OF SOCIAL MEDIA USE TO ENGAGE STUDENTS

Communication Practices: Talking Like a Scientist

Students need models for discourse, both online and face-to-face. Argumentation is a key practice for a scientist and it is well represented in the CCSS and NGSS as an integral skill for future citizens. Guided practice is one of the best ways to support communication skill building. As with any form of communication, it helps to have a common vocabulary and protocol for discourse. Over the past 15 years, researchers at the University of Kansas have developed and explored the practice of argumentation (Bulgren, Ellis, & Marquis, 2014) and the vocabulary developed by Toulmin (1958, 2003) to help students practice argumentation. A National Science Foundation exploratory research project demonstrated teachers' use of the Argumentation Evaluation Routine and associated graphic organizer to enhance students' performance in middle and secondary science classrooms (Bulgren, Ellis, & Marquis, 2014). The pilot test in the third year of the design study, during which the procedures were developed in collaboration with teachers, used an open-ended test to assess students' abilities to evaluate a scientific argument made in an article. The students were asked to identify the claim, identify and evaluate the evidence given for the claim, examine the reasoning in support of the claim, consider counterarguments, and, finally, construct and explain a conclusion about the claim. A quasi-experimental pretest-posttest design with eight experimental and eight comparison teachers was used with a total of 282 students. The quality of students' responses was assessed using a scoring rubric for each step of the argumentation process. Findings showed that students who were instructed using the Argumentation Evaluation Routine had significantly higher overall scores than students who received traditional lecture-discussion instruction, with a large effect size. Teacher satisfaction and student satisfaction and confidence levels were also reported.

This research was expanded to include the development of a multi-player game, *Reason Racer*, that helps students not only practice the vocabulary of argumentation, but also engages them in online discourse about one another's claims, evidence, and reasoning. Researchers found that the game is a compelling and engaging way to support multiple opportunities for practicing argumentation. Students who played the game at least 10 times also reported an increase in confidence and motivation to engage in science as compared to students who did not play the game (Ault, Craig-Hare, Frey, Ellis, & Bulgren, 2015). The final stage of the game is an online discussion where players must accept, reject or withhold judgment on the claim being made by the article in the scenario under study. This discourse engine resembles a Twitter stream and students reported that they enjoyed the activity. Evaluation of the discourse episodes that occurred at the end of the *Reason Racer* game showed that students could use argumentation vocabulary to critically evaluate the judgments of their peers. Since scientists use social media professionally, the research and development team involved in the game development decided to delve further into the use of mainstream social media applications to better understand how students can practice argumentation in this popular space.

Barriers for Learners

In order to better understand how argumentation can be practiced through social media, the National Science Foundation funded an exploratory project with the following research questions:

1. To determine what technological knowledge and technological professional learning supports are needed to facilitate the use of social media.
2. To determine what pedagogical knowledge and pedagogical professional learning supports do teachers require to facilitate student development of argumentation practices.
3. To determine what content knowledge and content professional learning supports are needed.
4. To investigate how students use social media to engage in scientific argumentation.
5. To investigate the extent to which students who engage in scientific argumentation within social media improve their learning about the practice of argumentation.
6. To investigate the extent to which students who engage in scientific argumentation within social media learn the core ideas in science content.

In order to explore these questions, teacher-researchers collaborated with university researchers to develop a unit about inheritance and variation of traits that focused on supporting the practice of argumentation through the use of social media. The following qualitative results represent a small portion of the study findings and focus on three emergent themes. These descriptive themes included student and teacher preferences for digital citizenship related to behavior on social media, the type of application based on the area of scientific argumentation being practiced, and the type of relevant topics and communication necessary for students to consider their activities to be worthy of their time spent. Pre and post focus groups were conducted with a population sample of 22 students across three mid-western high schools. In addition, three interviews of the teacher-researchers who participated in the three-year study contributed to these findings.

Digital Citizenship

Boyd (2014) spent eight years trying to better understand the relationship between social media and teens. She describes "networked publics" as the spaces that social media creates where teens belong and can hang out, much like malls and skate parks in the 80's and 90's, social media is the cool place to be. By bringing school and teachers to social media, there is a blending of personal and school spaces. Focus groups gave insight into students who wanted to keep their fun social media account separate from their school account. Students did not want their school feed to clutter up their personal feed and many simply preferred not to have an audience of adults in their social space. This student opinion is consistent with current policies regarding student emails and digital lives and separating personal and school accounts. Establishing ground rules with regard to the creation and/or use of social media accounts can help facilitate school-based usage and ensure that school district policies and behavioral expectations are followed.

Navigating the use of social media in science classrooms needs to be a collaborative venture between the teacher and students. Teaching students how to responsibly engage in online environments is imperative to ensuring that some of the pitfalls of social media are avoided. The persistent and visible nature of social media presents students with important decisions regarding how they present themselves online, the content that they do and do not share, as well as the networks they choose to create. Similar to brick-and-mortar settings in which youth congregate, such as drive-in restaurants, malls, and skateboard parks; social media is a public place where students gather to socialize. They, therefore, need to also learn the appropriate behavior for this environment. Purposeful conversations between students and teachers help facilitate a common understanding of protocol, decorum, and practice in a classroom or school.

The International Society for Technology in Education (ISTE) Standards for Students focus on digital citizenship; highlighting the need for students to take responsibility for their own social identity, online activities, intellectual property and digital privacy and security (ISTE, 2016). By actively engaging students in the practice of navigating the digital world, educators are preparing students for their future. No matter what social application enters the market in the coming years, connectivity and a healthy digital presence will only become more important. By practicing the use of social media in STEM courses, students will have necessary skills for academic and career advancement.

Application Preferences

During focus group discussions, students and teachers expressed mixed feelings regarding microblogging sites such as Twitter, Instagram and Snapchat and the practice of interactive scientific argumentation. Teachers found that without specific prompting and facilitation, students were not interested in back-and-forth discourse with their peers in these applications. Students indicated that the way they use these sites did not lead to natural discourse involving turn-taking and that they used texting for that type of reciprocal conversation.

Microblogging sites, such as Twitter, embolden users to post information about themselves and hope for a "retweet" or "like", depending on the application. These notifications show them that someone read their posts and responded to it. Students were willing to engage in reciprocal conversations or discourse in a microblogging site when the topic was highly engaging or personal. For example, when bioethics were used to stimulate conversation, students were more likely to engage not only in making claims and posting evidence but to voice concerns about the claims of their peers and interact with new questions, rebuttals, and counter arguments. Students said that they are more engaged by bioethical debate and are more likely to correspond with peers online, than if the discourse is contrived from less engaging, more straightforward content sharing.

Generally, however, students and teachers indicated that microblogging applications do not necessarily lend themselves to back-and-forth discourse between users. Expecting in-depth, critical dialogue between students in these applications is a misuse of their functionality based on the fact that the medium is designed for quick status updates and participants are limited to 140 characters. This is not necessarily enough space to write a convincing claim, backed by supporting evidence. It is, however, enough room to address the various components of argumentation, such as making a claim, citing a piece of evidence and responding with a rebuttal. In addition, teachers recommend that microblogging is used in biology classes as a way to disseminate information in a one-to-many strategy. For example, students could post their claim, share evidence and judgments online, and then engage in face-to-face argumentation discourse. They could use the microblogging feed to present evidence and support materials while they were engaged in conversation with their peers. Students have said that blogging sites, such as WordPress, allow more space for authoring complete arguments on scientific topics and that their threaded nature (versus Twitter's scrolling feed) offer a more formal and easier forum for back-and-forth communication. Students also indicated that blogs are "more for details and Twitter is more for [posting] thoughts."

Authentic Learning

Finally, during the focus groups, students clearly indicated that they need relevant and engaging content in order to drive them to social media sites to post for academic purposes. More importantly, relevant and engaging content was needed to get them to return and respond to their peers' postings. One student said it best, "Why would I tweet with a guy who is sitting right beside me?" When further questioned, this student explained that he would use social media sites if there was a need to go beyond the peers in his classroom and speak with students from other schools or countries, as well as experts in various fields of science. Having said that, another student in the focus group pointed out that he didn't mind tweeting live in class, even if they were arguing a current topic because it gave him time to stop, reflect and even edit his posts. Whereas if he were in class having a live, verbal conversation, he would be less likely to participate because of the social pressure that accompanies live discourse. In addition, students at three of the four research sites said that they were not used to such open-ended learning and would prefer their teacher just tell them what they need to learn and then they can fill out a worksheet or take a test to prove they learned it. Dialogic interactions in general and scientific argumentation specifically are hard work and require a level of engagement beyond knowledge attainment.

As previously mentioned, bioethics proved to be an engaging topic for students. When discourse instances were counted, per topic, bioethics produced the most "back-and-forth posts" between peers. During an activity about de-extinction, over 90% of students engaged in discourse on social media, made a claim, included supporting evidence and responded to the claim of at least one of their peers. When asked about the experience as compared to their willingness to engage in discourse on social media previously, the students responded that this topic was more fun and that they would "love to debate more" online and face-to-face in biology.

RESOURCES AND SOLUTIONS TO SUPPORT INSTRUCTION

Case Study Example: Secondary Science Classroom

Our research suggests that there is a potential for social media to support aspects of scientific argumentation including the posting of claims, evidence, and reasoning. In order to foster peer-to-peer discourse, particularly where students are offering counter arguments, rebuttals, and new questions, the content needs to be authentic and engaging. Bioethical topics have shown the highest potential for engagement and interactivity. The following case study is an example of how two secondary educators harnessed the power of bioethics to support student practice of argumentation, via social media.

High school Biology teachers Lisa and Julie were looking for a way to bridge student understanding of ecology and genetics with evolution. They chose to develop a 50-minute lesson on de-extinction using the 5E model. They started with the essential question: "Should we bring organisms back from the dead?" The following describes how they implemented the 5E Model for this lesson:

- **Engage:** They began by showing students the YouTube trailer for the *Jurassic World* movie. The trailer served as a hook because it showed the hypothetical, Hollywood version of the chaos that resulted from the genetic manipulation and rebirth of dinosaurs. After watching the video as a whole class, the teachers introduced the notion of bringing extinct animals "Back from the Dead" and asked the question: "We can, but should we?" They then introduced the 2015 National Geographic edition dedicated to de-extinction as a current event that needed further exploration (Zimmer, 2013). This article discussed how woolly mammoth DNA could be spliced with elephant DNA and the ethical dilemmas that could be sparked by that prospect. The teachers took this opportunity to remind the students of their prior exploration of genetics, inheritance, and variation of traits, as well as ecology. They asked the students to use that knowledge as they considered the essential question and ultimately prepared for the evolution unit they were beginning.

- **Engage/Explore:** In order to further support the recall of prior knowledge, teachers had the students play the woolly mammoth scenario within the free, online, multi-player game *Reason Racer,* found at ReasonRacer.com, to teach and review the various aspects of scientific argumentation. It served as a review of argumentation vocabulary and an additional source of information on the topic of de-extinction.
- **Explore:** They then engaged the class in a discussion about the questions and concerns students had about de-extinction. Students were encouraged to brainstorm their questions on the board in class or on an online poster board such as Padlet. After the list of questions was generated, the students considered the ethical, ecological, and evolutionary implications of each concern. For instance, one student wrote on the board that bringing the woolly mammoth back from the dead could disrupt the food chain, which the class indicated had implications for ethics, ecology, and evolution. Once the students determine a solid list of concerns and questions had been generated, then they were encouraged to work in pairs to research the web using their phones, computers or iPads to learn more about this current event. This research was unstructured and the students were asked to practice good digital citizenship by locating content from credible sources with good authority.
- **Explain:** As students completed their research, the teachers posted a challenge for them. They needed to make a claim regarding the essential question: "Should we bring organisms back from the dead?" Some students chose to take on the persona of congressmen, determining whether or not the United States should legalize the practice of de-extinction. Students were encouraged to make their claims on Twitter using a hashtag #advbiobftd, which the teachers created for the purpose of this lesson. Students were asked to support their claim with quality evidence. Students were encouraged to include links to the articles they used for evidence and they also needed to

explain their reasoning and share any concerns they had about their own claim. The teachers found that students needed models for how to properly create claims, use evidence and reasoning, and respond to author concerns. The teachers engaged students in practicing this skill throughout the year, as an opportunity for refinement. For instance, teachers reminded students that a good claim requires a qualifier (may, could, should) and that they needed to ensure that their evidence was backed by high-quality reasoning.

- **Elaborate:** Once the students posted their own claims, evidence, reasoning, and concerns, they were encouraged to use all of their information and understanding to refute and question their peers with counter arguments, rebuttals, and new questions. They were also encouraged to engage in back-and-forth discourse with their peers when their original post was being discussed. The teachers found it helpful to join the fray and redirect or prompt students, as needed. Students were encouraged to support their peers by ensuring that all responses were constructive and grounded in evidence with good reasoning. "You're wrong!" was offered as a non-example.
- **Evaluate:** The teachers then convened the students as a whole group to discuss the ideas and reflect on their experience. The teachers concluded the lesson by asking the students to reflect on the overall Twitter feed and cast their vote as though they were the United States congress: "Should we allow de-extinction in the U.S.?" Students were told that there was no right or wrong answer, but that as members of a democracy, their vote was important.

SOCIAL MEDIA TO ENGAGE EDUCATORS

Collaborative Professional Learning

Teenaged youth are not the only individuals who use social media. In fact, 90% of 18-29 year-olds use social media across multiple applications including Facebook, Instagram, and Twitter, followed by 77% of 30-49 year-olds and 51% of 50-64 year-olds (Perrin, 2015). Seventy percent of Facebook users report that they visit the site daily (Duggan, 2015). The median age range of public school educators in the United States is 40-49 years with 92% being under the age of 64 (U.S. Department of Education, 2016). In addition to general use, social media has become a popular option for professional learning (Robson, 2016). Teaching can be a lonely profession, particularly if an educator happens to be the only specialist in their area within the school or district. For example, in a small rural school, there may only be one fourth grade teacher or an inner-city school may only have one Advanced Placement Biology teacher. Building-based collaboration for these educators is primarily accomplished

Figure 1. Twitter feed of students discussing the legalization of de-extinction

through cross grade-level or content-area meetings. Professional learning in their specific content area is limited to what they can study, find online or acquire from sporadic in-service professional learning days or conferences.

Because of its immediacy and ability to be targeted to a specific group or content, social media has the power to break down these walls of isolation and empower educators to network with colleagues who teach exactly what they teach. By creating social spaces online that specialize in topics such as, fourth-grade instruction or

AP Biology, educators can summon answers to specific questions on a topic they plan to teach the next week or a lab that is presenting challenges (Holmes, Preston, Shaw, & Buchanan, 2013). They can access thousands of lessons plans and willing experts who want to learn from and with one another. Facebook groups, Twitter and Instagram hashtags, and the multitude of other social media applications current and yet to be released have the power to bring like-minded educators together to share ideas, brainstorm, and problem-solve.

Strategies for Using Social Media for Professional Learning

One strategy to support an educator who is new to social media in general or to a specific application is to encourage lurking. Lurking involves creating an account and browsing through the comments, hashtags, and various conversation feeds. Educators should be encouraged to take advantage of the available content that requires no reciprocation on the part of the reader. Find users who resonate, whether they share good resources, tell relatable stories or ask helpful questions, they likely

Figure 2. Facebook group where teachers are beginning to navigate social media use in the classroom

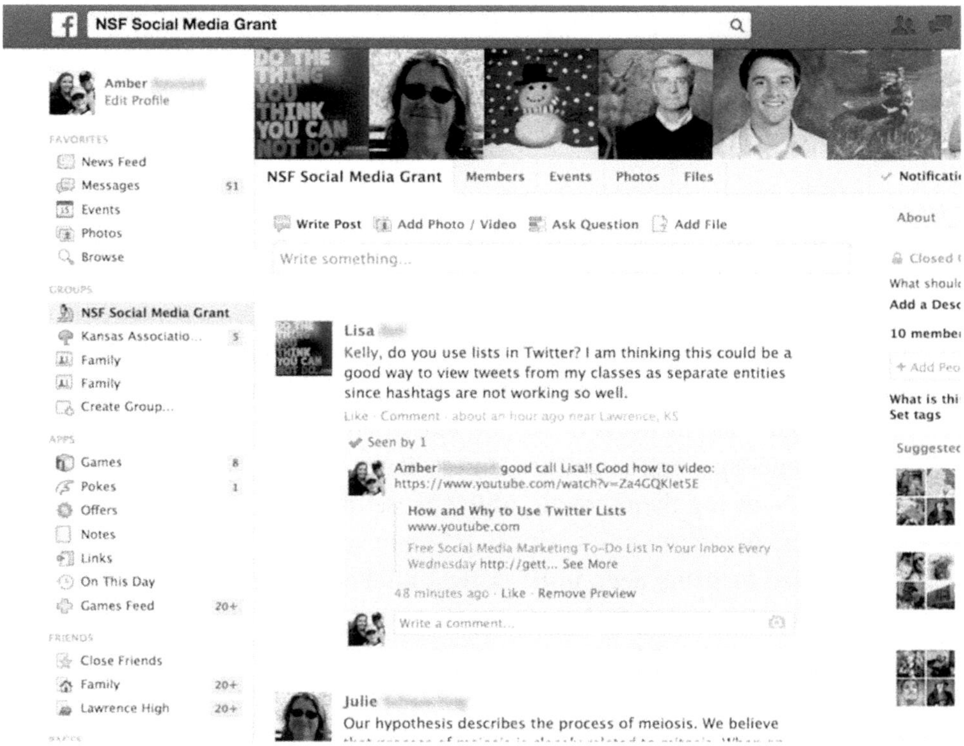

have a network of people in their "friend" list who are worthy of friending. User networks can be built by finding power users and 'friending' their friends, joining the groups they belong to and following their frequently used hashtags. Then, as educators become comfortable in a particular area of a network, they can begin to post their own ideas, receive positive feedback and become a user that others reference for growing their own networks.

Additional Tips for Educator Success:

- Maintain an ongoing understanding of the policies in place for social media use in your district. These can shift at various points throughout the year.
- Notify the building principal of your activities and confirm their support. It's better if they are aware of your usage so that they can address any concerns from the district administration, technology staff or parents. Offer them a heads-up as a professional courtesy and share classroom plans for acceptable use.
- Send a message home to parents describing student use of social media for learning purposes in your classroom.
- Establish ground rules with students for your application, hashtags and username handles. Have the students help establish the community and governing rules to engender ownership.
- Share the good stuff! When you have the opportunity to share an exciting moment from the learning taking place in your classroom, consider posting, not only the anecdote but also any links or resources that teacher friends on social media could use in their own instruction.

SOLUTIONS AND RECOMMENDATIONS

As the K-12 field of education becomes more technology-rich, teachers and researchers are continually trying to figure out how to take advantage of the affordances provided by technology and Internet access. This chapter identified specific strategies and methods to support classroom-based higher-order thinking on a social media application and students will continue to be entrenched in the digital world. Showing them how to utilize these digital resources for enhanced productivity and for learning is a critical 21st-century skill.

The idea that students need to be critical thinkers, informed consumers, and responsible digital citizens is key for college and career readiness. The ISTE Standards for Students offer guidance for supporting students to engage and thrive in a connected, digital world. Digital citizenship is among these standards, supporting students to "recognize the rights, responsibilities, and opportunities of living, learning and working in an interconnected digital world" (ISTE, 2016) with an emphasis on students acting and modeling in ways that are safe, legal and ethical. Digital citizenship is a key for digital learners to be vigilant in protecting themselves and portraying their best digital self. These student standards should be integrated into the classroom, as well as discussed and understood by educators in order to amplify learning with technology and challenge students to support their learning with technology-based resources.

As described earlier in this chapter, social media provides support for teacher professional learning. Social media applications can be used to expand peer-networks and give teachers access to the collective capacity of their profession (Dabbagh & Kitsantas, 2012; Forte, Humphreys, & Park, 2012; King, 2011). It also allows for teachers to connect outside of their profession; learning from and engaging with experts who have the potential to impact classroom learning. As teachers become more comfortable using social media to connect, collaborate and create with others, they can offer better guidance to their students on how to support their own learning with digital resources.

An increasing number of teachers are using online and blended approaches to learning, which utilize social media and online tools and resources. As such, learners are being given more autonomy over how they learn and what resources support their personal learning preferences. This further supports the need for teachers to rethink their pedagogy and curriculum in ways that support students customizing their own learning paths. Educators should continue to explore social media as a pervasive, ubiquitous application for collaboration and communication that will persist in classrooms and will continue to be a part of students' lives.

FUTURE RESEARCH DIRECTIONS

The practice of argumentation during which a person makes a solid claim, supported by quality evidence, is a model of a good citizenry. Democracy is built upon critical thinkers who can come together to solve difficult problems for their community. The ability to communicate intelligently is a skill that students must have in the digital age. Future research should explore how an online environment can support students practice of argumentation with a reduction in the social pressures that occur in face-to-face settings. Students with a disability, such as those with autism, struggle with the cognitive load required to engage in verbal argumentation in a face-to-face classroom setting. Social media may provide affordances that allow these students the time and distance to think, review, and edit their thoughts before entering into a social exchange, such as with argumentation. Research could also address whether engaging in argumentation using social media supports, rather than replaces, verbal, face-to-face discourse. Future research also needs to address the use of emerging social environments, such as virtual reality, where students can safely practice verbal communication with an avatar or artificial intelligence engine and not have to worry about what their peer or teacher might be thinking about them or their behaviors.

Finally, future research should include identifying new and emerging ways that scientists use social media. While social media has allowed scientists to communicate their research efficiently, it is critical to know how these affordances have impacted the distribution of scientific information within their profession and the general population. STEM classrooms should support students using social media in similar ways.

CONCLUSION

Social media is pervasive, with more and more people engaging in online communities through a variety of applications on a daily basis. It is also dynamic, with various applications moving in and out of style based on popularity and new applications emerging. Regardless of the application, however, students will need to be proficient, smart users of social media in order to be informed, participatory citizens. This chapter suggests that educators take advantage of the emergence of social media as an engaging environment for youth and incorporate its use as an educational tool. It also emphasizes the need to provide teachers with support to successfully use social media as a tool to support scientific argumentation and STEM practices.

The proliferation and capabilities surrounding the emergence of collaborative social media, including social networks and interactive spaces, have the potential to better prepare students for the real world by using tools with which they are already familiar while assisting teachers with the implementation of inquiry-based practice and argumentation through scaffolding and formative assessment. In order to take advantage of the opportunity provided by the affordances and prevalence of social media, teachers need pedagogical, content and technological knowledge and support to design and facilitate student learning. In addition, students need authentic and relevant reasons to invest their time and effort in the practice of scientific argumentation through social media.

Students must, in the short term, understand the application of argumentation in science. In the long term, however, citizens must independently be able to apply reasoning to claims or statements made to convince them about issues in their everyday lives, especially in evaluating the multitude of scientific claims and arguments that bombard them daily. Engaging in scientific argumentation and authentic learning tasks potentially will motivate students to pursue additional education and careers in STEM.

ACKNOWLEDGMENT

This material is based upon work supported by the National Science Foundation: Award No. DRL-1019842, The Evidence Games: Collaborative Games Engaging Middle School Students in the Evaluation of Scientific Evidence and Award #1316799, Enhancing Argumentation with Social Media: Supporting Teacher Professional Learning & Student Scientific Argumentation. Any opinions, findings, and conclusions or recommendations expressed in this material are those of the authors and do not necessarily reflect the views of the National Science Foundation.

REFERENCES

Alexander, R. (2010). *Children, their world, their education: Final report and recommendations of the Cambridge Primary Review*. Abingdon, UK: Routledge.

Alozie, N. M., Moje, E. B., & Krajcik, J. S. (2010). An analysis of the supports and constraints for scientific discussion in high school project-based science. *Science Education*, *94*, 395–427.

Ault, M., Craig-Hare, J., Ellis, J. D., Bulgren, J., Kretschmer, I., & Frey, B. B. (2016). Using Reason Racer to support argumentation in middle school science instruction. In M. J. Urban & D. A. Falvo (Eds.), *Improving K-12 STEM Education Outcomes through Technological Integration* (pp. 399–431). doi:10.4018/978-1-4666-9616-7.ch018

Ault, M., Craig-Hare, J., Frey, B., Ellis, J. D., & Bulgren, J. (2015). The effectiveness of Reason Racer, a game designed to engage middle school students in scientific argumentation. *Journal of Research on Technology in Education*, *47*(1), 21–40. do i:10.1080/15391523.2015.967542

Ball, D. L., & Bass, H. (2000). Interweaving content and pedagogy in teaching and learning to teach: Knowing and using mathematics. In J. Boaler (Ed.), *Multiple perspectives on the teaching and learning of mathematics* (pp. 83–104). Westport, CT: Ablex.

Ball, D. L., & Bass, H. (2003). Making mathematics reasonable in school. In J. Kilpatrick, W. G. Martin, & D. Schifter (Eds.), *A Research companion to principles and standards for school mathematics* (pp. 27–44). Reston, VA: National Council of Teachers of Mathematics.

Banilower, E. R., Smith, P. S., Weiss, I. R., Malzahn, K. A., Campbell, K. M., & Weis, A. M. (2013). *Report of the 2012 national survey of science and mathematics education*. Chapel Hill, NC: Horizon Research, Inc.

Bieda, K. N. (2010). Enacting proof-related tasks in middle school mathematics: Challenges and opportunities. *Journal for Research in Mathematics Education*, *41*(4), 351–382.

BioInformatics. (2007). *Scientists using social media to inform lab purchasing decisions*. Retrieved from: https://www.gene2drug.com/product/digital-marketing-to-life-scientists-social-media-and-email-best-practices

BioInformatics. (2016). *Social media and life scientists in 2016*. Retrieved from: https://www.gene2drug.com/social-media-life-scientists-2016

Boyd, D. (2014). *It's complicated: The social lives of networked teens*. New Haven, CT: Yale University Press.

Bulgren, J. A., Ellis, J. D., & Marquis, J. G. (2014). The use and effectiveness of an argumentation and evaluation intervention in science classes. *Journal of Science Education and Technology*, *23*(1), 82–97. doi:10.1007/s10956-013-9452-x

Bybee, R. (2011). Scientific and engineering practices in K-12 classrooms, Understanding A framework for K-12 science education. *Science Scope*, *35*(4), 6–11, 13.

Carr, M., & Biddlecomb, B. (1998). Metacognition in mathematics from a constructivist perspective. In D. Hacker, J. Dunlosky, & A. Graesser (Eds.), *Metacognition in educational theory and practice* (pp. 69–91). Mahwah, NJ: Erlbaum.

Chao, J., Parker, K., & Fontana, A. (2011). Developing an interactive social media based learning environment. *Issues in Informing Science and Information Technology*, *8*, 323–334.

Craig-Hare, J., Fowler, T., Landever, A., Hobin, L., Parscale, G., & Irminger, T. (2013). *Learning engagement through higher-order thinking skills and technology.* Research paper presented at the annual meeting of the International Society for Technology in Education (ISTE), San Antonio, TX.

Dabbagh, N., & Kitsantas, A. (2012). Personal learning environments, social media, and self-regulated learning: A natural formula for connecting formal and informal learning. *The Internet and Higher Education*, *15*(1), 3–8. doi:10.1016/j.iheduc.2011.06.002

Desoete, A., Roeyers, H., & Buysse, A. (2001). Metacognition and mathematical problem solving in grade 3. *Journal of Learning Disabilities, 34*(5), 435–447. doi:10.1177/002221940103400505 PMID:15503592

Duggan, M., Ellison, N. B., Lampe, C., Lenhart, A., & Madden, M. (2015). *Pew Research Center: Internet, Science and Technology*. Retrieved from http://www.pewinternet.org/2015/01/09/demographics-of-key-social-networking-platforms-2

Duschl, R. A., & Osborne, J. (2002). Supporting and promoting argumentation discourse in science education. *Studies in Science Education, 38*(1), 39–72. doi:10.1080/03057260208560187

Ertmer, P., Ottenbreit-Leftwich, A., Sadik, O., Sendurur, E., & Sendurur, P. (2012). Teacher beliefs and technology integration practices: A critical relationship. *Computers & Education, 59*(2), 423–435. doi:10.1016/j.compedu.2012.02.001

Facebook. (2016). *Stats*. Retrieved October 14, 2016, from http://newsroom.fb.com/company-info/

Forte, A., Humphreys, M., & Park, T. H. (2012, June). Grassroots Professional Development: How Teachers Use Twitter. *ICWSM*.

Gauvain, M. (2001). *The social context of cognitive development*. Guilford Press.

Gerard, L., Varma, K., Corliss, S., & Linn, M. (2011). Professional development for technology- enhanced inquiry science. *Review of Educational Research, 81*(3), 408–448. doi:10.3102/0034654311415121

Greenhow, C. (2011). Youth, learning, and social media. *Journal of Educational Computing Research, 45*(2), 139–146. doi:10.2190/EC.45.2.a

Holmes, K., Preston, G., Shaw, K., & Buchanan, R. (2013). Follow Me: Networked Professional Learning for Teachers. *Australian Journal of Teacher Education, 38*(12), 54–65. doi:10.14221/ajte.2013v38n12.4

International Society for Technology in Education (ISTE). (2016). *ISTE standards for students*. Retrieved August 2, 2016, from http://iste.org/standards

Jiménez-Aleixandre, M. P., & Erduran, S. (2007). Argumentation in science education: An overview. In *Argumentation in science education* (pp. 3–27). Springer. doi:10.1007/978-1-4020-6670-2_1

Johnson, L., Adams Becker, S., Estrada, V., & Freeman, A. (2015). *NMC Horizon Report: 2015 K-12 Edition*. Austin, TX: The New Media Consortium.

Junco, R., Helbergert, G., & Loken, E. (2011). The effect of Twitter on college student engagement and grades. *Journal of Computer Assisted Learning*, *27*(2), 119–132. doi:10.1111/j.1365-2729.2010.00387.x

Kabilan, M. K., Ahmad, N., & Abidin, M. J. Z. (2010). Facebook: An online environment for learning of English in institutions of higher education? *The Internet and Higher Education*, *13*(4), 179–187. doi:10.1016/j.iheduc.2010.07.003

Kamii, C., & Dominck, A. (1998). The harmful effects of algorithms in grades 1–4. In L. J. Morrow & M. J. Kenney (Eds.), The teaching and learning of algorithms in school mathematics: 1998 Yearbook (pp. 130–140). Reston, VA: The National Council of Teachers of Mathematics.

Kamil, M. L., Borman, G. D., Dole, J., Kral, C. C., Salinger, T., & Torgesen, J. (2008). Improving adolescent literacy: Effective classroom and intervention practices: A Practice Guide (NCEE #2008-4027). Washington, DC: National Center for Education Evaluation and Regional Assistance, Institute of Education Sciences, U.S. Department of Education. Retrieved from http://ies.ed.gov/ncee/wwc

Kaplan, A. M., & Haenlein, M. (2010). Users of the world, unite! The challenges and opportunities of Social Media. *Business Horizons*, *53*(1), 59–68. doi:10.1016/j.bushor.2009.09.003

Kelly, S. (2007). Classroom discourse and the distribution of student engagement. *Social Psychology of Education*, *10*(3), 331–352. doi:10.1007/s11218-007-9024-0

King, K. P. (2011). Professional learning in unlikely spaces: Social media and virtual communities as professional development. *iJET*, *6*(4), 40-46.

Kramarski, B., & Mevarech, Z. R. (2003). Enhancing mathematical reasoning in the classroom: The effects of cooperative learning and metacognitive training. *American Educational Research Journal*, *40*(1), 281–310. doi:10.3102/00028312040001281

Lave, J., & Wenger, E. (1991). *Situated learning: legitimate peripheral participation*. Cambridge, UK: Cambridge University Press. doi:10.1017/CBO9780511815355

Lawson, A. (2003). The nature and development of hypothetico-predictive argumentation with implications for science teaching. *International Journal of Science Education*, *25*(11), 1387–1408. doi:10.1080/0950069032000052117

Lenhart, A. (2015). *Pew Research Center: Teens, Social Media and Technology Overview*. Retrieved from http://www.pewinternet.org/2015/04/09/mobile-access-shifts-social-media-use-and-other-online-activities

Levine, T. H., & Marcus, A. S. (2010). How the structure and focus of teachers collaborative activities facilitate and constrain teacher learning. *Teaching and Teacher Education: An International Journal of Research and Studies, 26*(3), 389–398. doi:10.1016/j.tate.2009.03.001

Linn, M. C., Clark, D., & Slotta, J. D. (2003). WISE design for knowledge integration. *Science Education, 87*(4), 517–538. doi:10.1002/sce.10086

Littleton, K., & Light, P. (Eds.). (1999). *Learning with computers: Analyzing productive interaction*. Psychology Press. doi:10.4324/9780203282311

Mayes, R., & Koballa, T. R. Jr. (2012). Exploring the science framework. *Science and Children, 50*(4), 8–15.

Mercer, N., Warwick, P., Kershner, R., & Staarman, J. K. (2010). Can the interactive whiteboard help to provide dialogic space for childrens collaborative activity? *Language and Education, 24*(5), 367–384. doi:10.1080/09500781003642460

Merriam-Webster's Learner's Dictionary. (2016). Retrieved from http://www.merriam-webster.com/dictionary/social%20media

Mevarech, Z. R. (1999). Effects of metacognitive training embedded in cooperative settings on mathematical problem solving. *The Journal of Educational Research, 92*(4), 195–205. doi:10.1080/00220679909597597

Mevarech, Z. R., & Kramarski, B. (1997). IMPROVE: A multidimensional method for teaching mathematics in heterogeneous classrooms. *American Educational Research Journal, 34*(2), 365–394. doi:10.3102/00028312034002365

Mishra, P., & Koehler, M. (2006). Technological pedagogical content knowledge: A framework for teacher knowledge. *Teachers College Record, 108*(6), 1017–1054. doi:10.1111/j.1467-9620.2006.00684.x

National Academies of Sciences, Engineering, and Medicine. (2016). *Science literacy: Concepts, contexts, and consequences*. Washington, DC: The National Academies Press. doi:10.17226/23595

National Governors Association Center for Best Practices Council of Chief State School Officers. (2010). *Common Core State Standards (CCSS)*. National Governors Association Center for Best Practices, Council of Chief State School Officers.

National Research Council. (2007). *Taking science to school: Learning and teaching science in grades K-8*. Washington, DC: The National Academies Press.

National Research Council. (2011). *Successful STEM Education: A workshop summary. A. Beatty, Rapporteur. Committee on Highly Successful Schools or Programs for K-12 STEM Education, Board on Science Education and Board on Testing and Assessment. Division of Behavioral and Social Sciences and Education*. Washington, DC: The National Academies Press.

National Research Council. (2012). *A Framework for K-12 science education: Practices, crosscutting concepts, and core ideas. Committee on a Conceptual Framework for New K-12 Science Education Standards. Board on Science Education, Division of Behavioral and Social Sciences and Education*. Washington, DC: The National Academies Press.

National Research Council. (2013). *Monitoring progress toward successful K-12 STEM education: A nation advancing? Committee on the Evaluation Framework for Successful K-12 STEM Education. Board on Science Education and Board on Testing and Assessment, Division of Behavioral and Social Sciences and Education*. Washington, DC: The National Academies Press.

NGSS Lead States. (2013). *Next Generation Science Standards: For States, By States*. Washington, DC: The National Academies Press.

Osborne, J., Simon, S., Christodoulou, A., Howell-Richardson, C., & Richardson, K. (2013). Learning to argue: A study of four schools and their attempt to develop the use of argumentation as a common instructional practice and its impact on students. *Journal of Research in Science Teaching*, *50*(3), 315–347. doi:10.1002/tea.21073

Pashler, H., Bain, P. M., Bottge, B. A., Graesser, A., Koedinger, K., McDaniel, M., & Metcalfe, J. (2007). *Organizing instruction and study to improve student learning. IES Practice Guide (NCER 2007-2004)*. Jessup, MD: National Center for Education Research.

Pellegrino, J. W., & Hilton, M. L. (2013). *Education for life and work: Developing transferable knowledge and skills in the 21st century. National Research Council of the National Academies*. Washington, D.C.: National Academies Press.

Perrin, A. (2015). *Pew Research Center: Internet, Science and Technology*. Retrieved from http://www.pewinternet.org/2015/10/08/social-networking-usage-2005-2015

Prestridge, S. (2009). Teachers talk in professional development activity that supports change in their pedagogical beliefs and practices. *Teacher Development*, *13*(1), 43–55. doi:10.1080/13664530902858493

Robson, J. (2016). Engagement in structured social space: An investigation of teachers online peer-to-peer interaction. *Learning, Media and Technology*, *41*(1), 119–139. doi:10.1080/17439884.2015.1102743

Rowland, A., Ellis, J., Craig-Hare, J., Ault, M., Bradley, B., & Bulgren, J. (2015). *Social media for learning: Platform preferences, digital citizenship and authenticity*. Research paper presented at the annual meeting of the International Society for Technology in Education (ISTE), Philadelphia, PA.

Schoenfeld, A. H. (1985). *Mathematical problem solving*. New York: Academic press.

Scott, S. (2012). Go ahead. Be social: Using social media to enhance the twenty-first century classroom. *Distance Learning*, *9*(2), 54.

Slavin, R. E. (1996). Research on cooperative learning and achievement: What we know, what we need to know. *Contemporary Educational Psychology*, *21*(1), 43–69. doi:10.1006/ceps.1996.0004

Stein, M. K., Grover, B. W., & Henningsen, M. (1996). Building student capacity for mathematical thinking and reasoning: An analysis of mathematical tasks used in reform classrooms. *American Educational Research Journal*, *33*(2), 455–488. doi:10.3102/00028312033002455

Thomas, K. (2013). Changing mathematics teaching practices and improving student outcomes through collaborative evaluation. *Teacher Education and Practice*, *26*(4), 778–795.

Toulmin, C. N., & Meghan, G. (2007). *Building a science, technology, engineering and math agenda*. Washington, DC: National Governor's Association.

Toulmin, S. (1958, 2003). The uses of argument. Cambridge, UK: Cambridge University Press.

Toulmin, S., Rieke, R., & Janik, A. (1984). *An introduction to reasoning*. Upper Saddle Ridge, NJ: Prentice Hall.

Tsai, C., Jack, B., Huang, T., & Yang, J. (2012). Using the cognitive apprenticeship web-based argumentation system to improve argumentation instruction. *Journal of Science Education and Technology*, *21*(4), 476–486. doi:10.1007/s10956-011-9339-7

Twitter. (2016). *Twitter usage: Company facts*. Retrieved October 14, 2016, from https://about.twitter.com/company

U.S. Department of Education. (2016). *National Center for Education Statistics, Schools and Staffing Survey (SASS), "Public school teacher data file," 1987-88 through 2011-12*. Retrieved from: http://nces.ed.gov/programs/digest/d14/tables/dt14_209.10.asp?referrer=report

Van Eperen, L., & Marincola, F. M. (2011). How scientists use social media to communicate their research. *Journal of Translational Medicine*, *9*(199), 1–3. PMID:22085450

Vygotsky, L. S. (1978). *Mind in society: The development of higher psychological processes*. Cambridge, MA: Harvard University Press.

Warwick, P., Mercer, N., Kershner, R., & Staarman, J. K. (2010). In the mind and in the technology: The vicarious presence of the teacher in pupils learning of science in collaborative group activity at the interactive whiteboard. *Computers & Education*, *55*(1), 350–362. doi:10.1016/j.compedu.2010.02.001

Webb, N. M., Nemer, K. M., & Ing, M. (2006). Small-group reflections: Parallels between teacher discourse and student behavior in peer-directed groups. *Journal of the Learning Sciences*, *15*(1), 63–119. doi:10.1207/s15327809jls1501_8

Weiss, I. R., & Pasley, J. D. (2004). What is high-quality instruction? *Educational Leadership*, *61*(5), 24.

Weiss, I. R., Pasley, J. D., Smith, P. S., Banilower, E. R., & Heck, D. J. (2003). *Looking inside the classroom: A study of K-12 mathematics and science education in the United States*. Chapel Hill, NC: Horizon Research.

Yackel, E., & Hanna, G. (2003). Reasoning and proof. In J. Kilpatrick, W. G. Martin, & D. Schifter (Eds.), *A research companion to principles and standards for school mathematics* (pp. 227–236). Reston, VA: NCTM.

Zhang, J., & Sun, Y. (2011). Quantified measures of online discourse as knowledge building indicators. In *International conference on computer-supported collaborative learning (CSCL 2011). Proceedings, Long papers*, (p. 72). Hong Kong, China: Academic Press.

Zimmer, C. (2013). Bringing them back to life. *National Geographic*. Retrieved from http://ngm.nationalgeographic.com/2013/04/125-species-revival/zimmer-text

Zollman, A. (2012). Learning for STEM literacy: STEM literacy for learning. *School Science and Mathematics*, *112*(1), 12–19. doi:10.1111/j.1949-8594.2012.00101.x

KEY TERMS AND DEFINITIONS

Digital Citizenship: The behavior, presence, and reputation a person exemplifies online. Schools promote digital citizenship in order to empower students to be cognizant and vigilant in how they conduct themselves in a connected world.

Scientific Argumentation: The practice of using claims, evidence, reasoning, and justification to engage in the creation of meaning, knowledge, and understanding in science content.

Social Media: Online applications (such as Web sites for social networking and microblogging) through which users create online communities to share information, ideas, personal messages, and other content (such as videos). Current examples include Facebook, Twitter, Snapchat, Instagram, and LinkedIn.

Chapter 5
Wikigrams-Based Social Inquiry

Evgeny D. Patarakin
WikiVote!, Russia

ABSTRACT

This chapter presents techniques that combine the productive activity aimed at creation of collaborative stories with the research activity based on the analysis of the relationship between the participants of the productive activity. The productive activity is delivered through a wiki site Letopisi.org, where all actions of the participants are recorded in an electronic log. Log records are used as a data source to build sociograms, which constitute the basis for the research activity. The chapter describes different cycles of collaboration based on the usage of social objects. The first cycle is creation of digital story. The second cycle is social inquiry.

INTRODUCTION

Social media attracts attention of researchers and widely used in various spheres. However, in education, dynamics of social networks, extensive data on interrelations between network participants, network analysis techniques are hardly used. It comes from the widespread belief that the network data is hard to access or that the network analysis is too complicated. In this chapter, we will demonstrate that the network provides a great platform for inquiry-based learning. We present a number of simple methods of learning analytics as well as on the methods of creating static graphs (Graphvz) and dynamic models (NetLogo), which many teachers are familiar with. The source material for the research is log records. The set of log-records is a by-product of a specific learning activity, which is a digital storytelling. Cycles of

DOI: 10.4018/978-1-5225-2525-7.ch005

interleaving storytelling and research activities of the learners are interconnected by using the proposed approach. The by-product results of the learner's storytelling activities are used as a feed substrate for the network research based on the analysis of sociograms. It provides an advanced pathway to developing pedagogical means that utilize the science of complex networks as a vehicle through which students can acquire analytical skills for the network-oriented data analysis.

BACKGROUND

Science, Technology, Engineering, and Mathematics (STEM) are the driving force for worldwide economic and social advancements. The skills needed by the 21st century STEM workforce include the ability to interact with large amounts of data and the ability to understand the changing role of models. Educational systems worldwide are not keeping up with the explosion in the big data and data-driven sciences. However, exposure to these data-driven science skills is unavailable to most primary and secondary school students. Network science is a promising way to address data-intensive real-world problems (Cramer et al., 2015). Network science is an emerging, highly interdisciplinary research area that aims to develop theoretical and practical approaches and techniques to increase our understanding of natural and man-made networks(Börner, Sanyal, & Vespignani, 2007). We are surrounded by systems that are collectively called complex systems, capturing the fact that it is difficult to derive their collective behavior from knowledge of the system's components. Complex systems understanding, mathematical description, prediction, and eventually control is one of the major scientific challenges of the 21st century. Nature, science and technology are permeated with networks, and we will never understand complex systems unless we develop a deep understanding of the networks behind them (Barabási, 2016). A working definition of network science is the study of network representations of physical, biological, and social phenomena leading to predictive models of these phenomena (Committee on Network Science for Future Army Applications & National Research Council, 2005).

The network perspective allows us to address deep questions about different systems.

By postulating a friendship network in (say) a school classroom of 25 students, we have taken a theoretical step that is non-trivial. We have supposed that separate individuals are not an adequate representation, moreover that even separate dyads are insufficient; rather, that there is a unity within the classroom that makes it proper to talk of 'a' network, not 25 children or 300 dyads. To conceptualize the classroom in network terms is an implicit (and strong) claim that connectedness across individual elements is fundamentally important so that the classroom can be thought of as one 'system'. (Brandes, Robins, McCRANIE, & Wasserman, 2013, p.5)

To be successful in the 21st century, students must have a fundamental knowledge of complex networks which allows them to explore the interconnectedness of our world. Networks are pervasive across all aspects of life: biological, physical, economic, and social and our society continues to become ever more connected through the use of social media tools. This inherently interdisciplinary makes network science the perfect tool to engage students in STEM fields. Components of networks have been used as a tool for teaching math and computer science for many years. Scientists have expanded these efforts by developing and sharing materials to demonstrate how network science can successfully be used to engage students in STEM fields (Harrington, Beguerisse-Díaz, Rombach, Keating, & Porter, 2013; Sánchez & Brändle, 2014; Sheetz, Dunham, & Cooper, 2015).

As a rule, immersing school students in the field of network science begins with the study of maps that are based on data from different areas of research. For example, science maps for kids are based on the avalanche of data generated by scientific research today. These maps invite students to see, explore, and understand science (Börner et al., 2009).

We assume that introducing students and teachers to the network science may begin with mapping their own activities in the networked communities. The advantage of this approach is a network lens is used to understand situations in which students and teachers are drawn into. Consequently, the network science shows its strength in the areas of immediate experience and students and teachers become researchers of their own activity.

Inquiry-based learning attracts more attention and is expanding in Russia. Inquiry-based learning is one of educational concepts based on the constructionism philosophy. It is closely connected with learning by doing, project based learning or learning by design. The common ground for all these concepts has been the fact that they all form productive agency which there is scarce in Russia.

There is a need in contemporary society for citizens capable of acting as independent and self-sufficient agents under decentralized management. Contemporary education is intended to nurture citizens able to devise, create, use and share new products. Information and communication technologies support, facilitate and augment agent's activities. New economic models emphasize the importance of new knowledge, innovation, and the development of human capacity as the sources of sustainable economic growth. It is through education and human capacity development that individuals not only add value to the economy but contribute to the cultural legacy, participate in social discourse, improve the health of the family and community, conserve the natural environment, and increase their own agency and ability to continue to develop and contribute, creating a virtuous cycle of personal development and contribution (UNESCO, 2008).

The desired future of modern learning can be described as a situation where teachers and students are active agents that produce knowledge. Both society and education equally need participants with their own agency, because agency drives participation in the decision-making, knowledge production and collaborative learning. The concept of 'agency' reflects the individual's ability to perform an 'agent', i.e. active protagonist, the driving force behind the action (Inden, 2000). For Russia, the problem of fostering agency is particularly acute, because of the extent to which individual agency has been historically foreign to our culture (Wierzbicka, 1992). A necessity of formation participant's agency requires mentors of learning to put special attention to the problems of self-control and self-determination of behavior of participants. People have agency to the extent that they can alter their environment so that the environment affects them in adaptive ways. Schwartz (Schwartz, 1999) termed this formulation 'productive agency', because it emphasizes production through the environment. Schwartz defined productive agency as a recursive system where people take advantage of their available means to produce outwardly, and they see their ideas embodied and modified by the material or social world. Production distinguishes between situations that involve external production and those that do not and in this regard productive agency is akin to constructionism. Constructionism argues that learning occurs best when constructing a public artifact.

Constructionism shares constructivism's connotation of learning as `building knowledge structures' irrespective of the circumstances of the learning. It then adds that this happens especially felicitously in a context where the learner is consciously engaged in constructing a public entity, whether it's a sandcastle or a theory of the universe... (Papert & Harel, 1991, p.193).

Public entity or a public artifact can be not only examined or discussed but it can also be used by other people. That is to say a public artifact is inherently shareable.

Comparing the known learning communities close to the constructionism theory allows us to see that almost all of them use the idea of a cycle or a spiral of actions performed by agents over objects of actions:

- In the Globaloria Social Learning Network (Reynolds & Caperton, 2009) students perform the following actions with computer games: Play -> Plan -> Prototype -> Program -> Publish.

- In the Scratch Social Learning Network (Brennan, Hernández, & Resnick, 2009) students perform the following actions with Scratch projects: Imagine -> Create -> Play -> Share.
- In the NetLogo Modeling Commons (Lerner, 2014) NetLogo modelers perform the following actions with NetLogo Models:Create -> Run -> Share -> Comment -> Modify -> Create variations.
- In the StarLogo TNG Social Learning Network there are two linked circles (Klopfer, Scheintaub, Huang, Wendel, & Roque, 2009) and StarLogo TNG modelers perform the following actions with StarLogo Models. Research circle: Observe/Collect Data -> Generate Questions -> Test/Tinker/Play -> Observe/Collect Data. Design circle: Design -> Build -> Test/Tinker/Play -> Design.
- In the Looking Glass Social Learning Network (Kelleher & Pausch, 2007) 3D modelers perform the following actions with Alice Models: Create -> Animate -> Remix –> Share.
- In the CloudWorks Teaching Network (Galley, Conole, Dalziel, & Ghiglione, 2010) teachers perform the following actions with clouds (objects on the base of CompendiumLD map) Find -> Share -> Discuss.
- In the WebGrid Social Inquiry Network (Gaines & Shaw, 2012) researchers perform the following actions with grids (objects on the base of Repertoir Grid Test): Display -> Cluster -> Map -> Crossplot -> Matches -> Compare.

The circular consequence of agent's actions affecting an artifact is close to the theory of object sociality and its application to network communities organization defined by Engeström (Engeström, 2005). We draw in particular on the work of Engeström in terms of the concept of 'social objects' and his arguments for the importance of social objects as the key mediating artifacts. Engeström drawing on the work of Knorr-Cetina (Knorr-Cetina, 1999) puts forward an argument for the need to adopt an approach to social networking based on object oriented sociality. He provides examples of successful social networking sites built around social objects. three out of five for the inquiry-based organizational principles of design for the social networks singled out by Engeström deserve special attention as they are compulsory for learning communities:

- Clearly define the social object your service is built around.
- Define the verbs that users perform on the objects, so that is it clear what the site is for.
- Make the objects shareable.

The third principle (Make the objects shareable) means that an object created by one of the participants of the social network may be used or modified in any way by other participants of the social network.

In all the above-mentioned communities, there is a certain social object – a game, a story, a model or other 'virtual chips' over which agents perform their actions. Objects or actions can be different, but all agents are permitted to carry actions over one and the same object. If agents perform actions over one and the same object, they become indirectly connected by the social object. The links are similar to those between movie actors, or scientists creating an article, or Wikipedia editors working on a page together. The structure of these social connections is the subject of social network analysis and learning analytics. We agree with the definition of learning analytics as a collection of methods that allow learners to understand what is going on in a learning scenario (Schneider et al., 2012).

Learning analytics and diagrams methodology help to analyze and discuss situations that develop during a network collaboration in different domains. Analysts in this field have established Social Network Analysis (SNA) as an empirical method to study the ties between actors in the network (Wasserman & Faust, 1994). SNA uses various concepts to evaluate different network properties like centrality, connectivity, cliques, etc. A sociogram is a powerful analysis tool, helping researchers identify points of interest and other structural properties that otherwise would not be obvious in numeric data. The use of graphic imagery to represent relational information was key to the birth of sociometry. Depictions of social networks were invented by social scientist Jacob L. Moreno in 1934 (Moreno, 1934). As Freeman notes Moreno was the first to propose using sociograms to reveal important structural information about social linkage patterns and specified general rules for constructing sociograms (Freeman, 2000). We make maps not just of the physical world but also of our social worlds (Mehra et al., 2014).

The study of dynamic networks greatly benefits from visualizations that can illustrate ideas and concepts not immediately visible in a static sociogram. Moody and others' research illustrates the need to visualize how networks develop and change over time (Moody, Mcfarl, & Bender-demoll, 2005). Among the tools developed in the complexity field, agent-based modeling and network analysis are very important in sustaining the process of bringing complexity to bear on the policy world. The combination of the two methods can increase enormously the potential of complexity-based policies (Fontana & Terna, 2015). Since agent-based modeling is inherently dynamic, the problems with static networks are overcome naturally. Agent-based modeling permits the desired richness of behaviors and attributes that might bridge the gap between agent-nodes and the real world.

Termites With Logs

It was mentioned in the previous section that actors perform actions over same objects becoming interlinked via this social object in many various situations. Simple actions that students perform over objects of actions in learning communities are very much alike the procedures performed by turtles in a famous Termites model (Resnick, 1997): search-for-chip -> find-new-pile -> put-down-chip. As Resnick (1997, p.76) wrote, each individual termite should obey the following rules:

If you are not carrying anything and you bump into wood chip, pick it up.

If you are carrying a wood chip and you bump into another wood chip, put down the wood chip you're carrying.

Each turtle performs a sequence of steps in procedures search-for-chip find-new-pile put-down-chip which leads to the result when chips scattered randomly over the screen are gradually gathered into one round pile. Termites model seems most optimal because it contains chips as objects for collaboration. Experimenting on this model we can get a deeper understanding of collaboration phenomenon. Let's imagine that termites record their labor actions over chips in a log. I.e. if an actor performs a meaningful action on a chip, he leaves a record on this in the log. To understand how we can benefit from collateral records we have modified the original text of Termites model by adding new variables and rules. We have inserted a variable called list WIKILOG where turtles make records of their actions. There also have been made some additions to the procedures *search-for-chip* and *put-down-chip.*

Procedures

```
to search-for-chip
ifelse pcolor = yellow [
ifelse 0 = [pagenum] of patch-here [
let newpage 1 + length pages set pages lput newpage pages
set t-pagenum newpage
set wikilog lput (se [who] of self newpage "create") wikilog ]
[set t-pagenum [pagenum] of patch-here ]
set pcolor black set color orange fd 20 ]
[ wiggle search-for-chip ]
end

to put-down-chip
```

```
ifelse pcolor = black
[
ask patch-here [set pagenum [t-pagenum] of myself] set wikilog
lput (se [who] of self [t-pagenum] of self "edit") wikilog
set pcolor yellow
set color white
set t-pagenum 0
get-away ]
[ rt random 360
fd 1
put-down-chip ]
end
```

A procedure to define links between actors having carried the same chip has been added to the model. It uses log records.

Procedures

```
to-report edits
report filter ["edit" = item 2 ?] wikilog
end

to logs_to_sociogram
ask patches [set pcolor 0]
foreach edits [
let friend1 item 0 ?
let p1 item 1 ?
let friend2 first first filter [(p1 = item 1 ?) and ("create" =
item 2 ?)] wikilog
if friend1 != friend2 [
ask turtle friend1 [ create-friendbond-with turtle friend2 ]]]
repeat 8 [layout-spring turtles links 1 5 7 ]
end
```

The interface for the modified model is shown in Figure 1. Logs_to_sociogram button switches all the chips off the screen and only shows linkage between agents carrying one and the same chip.

Figure 1. Interface of termites with logs model

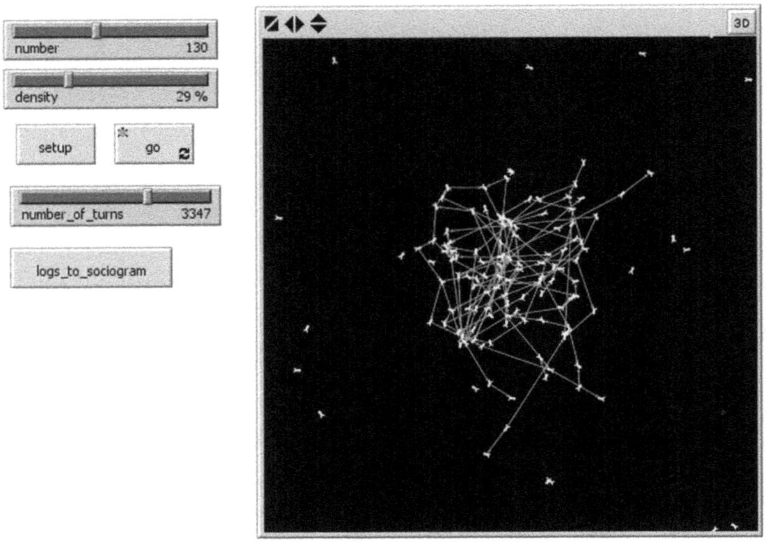

Termites with Log Model was implemented in NetLogo language 5.2 and its source code is available on the Internet http://modelingcommons.org/browse/one_model/4749

This simplest model for sociograms creation based on the actions with social objects can be applied to various collaboration situations. In all cases when actors perform actions over shareable objects, the log for these actions can help build sociograms as well as conduct social research. This statement was used when we designed a learning social network Letopisi.org, which combines cycles of creative and research learning.

Letopisi.org: Creative and Exploratory Cycles

Initially the network Letopisi.org was regarded as a field for creative productive learning where learners create common stories about their teachers, schools and cities. As a medium for collaboration at the start of the project in 2006 we chose wiki. The majority of wiki software (also known as a wiki engine) is free and open source software and is developed collaboratively. MediaWiki is the most popular of these and is also used to run Wikipedia, the world's largest online encyclopedia, as well as many other (non-profit) educational projects.

We draw attention to the fact that wiki is a very simple and very quick way to create and deploy a new page in the network. In the same time, attention falls from the area that it strengthens with the interaction of wiki pages and the collaborative efforts. What has been created by one person, can be further used by other people. We can collect articles from the finished blocks, the same way as the program collects from Lego blocks. Lego blocks or a Wiki article are useful and powerful not because they are simple but because they determined the exact template, and they are always compatible with each other. Thanks to the template and transclusion mechanism, with MediaWiki it is possible to use articles as "building bricks" and from these we can assemble larger "building blocks" (Patarakin & Visser, 2012).

A wiki is the simplest online database that could possibly work. Users of wiki do not need to know the programming commands of the hypertext language. The text of any article or page of recorded items is interpreted by the program as hypertext. The special wiki checking agent looks through the texts of all the pages in search of patterns before they get to the browser. If a match is found, the agent checks whether a page of that name already exists in the database; in that case, it makes a reference on that page - if such a page does not exist then reference is made to the creation of a new page.

This example shows us how wiki can be used to elaborate a project in a variety of ways. We may select those words/concepts which we want to use as a basis and about which we wish to learn more, MediaWiki syntax imply that we include the key words in two square brackets. Following up on the example given we can use [[Sedum telephium]] or to [[Pustinskiy nature reserve]]. If a botanist has already written and placed an article about [[Sedum telephium]] in wiki, the reference will work immediately and will lead us to the text. If s/he intends to write an article in the future, there will be a deferred reference which can be activated in the future. The beauty of wiki is that we do not have to bother a botanist with questions about how to find the file containing his or her article on the Sedum telephium or to ask a geographer about the name of a swamp. We just abide by the way wiki works —the article is named and has a reference. And if a geographer then writes another article about the swamp in question, the link becomes activated automatically – nothing else has to be done.

In Russian education MediaWiki is represented first of all by Letopisi.org project and its regional clones in several teaching colleges and universities. Letopisi - http://Letopisi.org is the national educational project with international participation and has continued for more than ten years. Currently Letopisi.org is a huge multimedia archive and an experimental site, where everyone can find useful information and make an experiment with modern methods of collective storage, retrieval, editing and classification of texts, photos, audio, video. The leading idea of the Letopisi concept is that the collaborative network activity and the network cooperation of the learning agents are aimed at the creation of various types of learning products which in general could be marked by a widely recognized term "digital story". The role of narration and the meaning of the practice of narration for various fields of activity were stressed by Bruner (Bruner, 2003) When the computer technologies started to be used multiple forms of digital storytelling emerged. More recent research (Gee, 2004; Grobstein, 2005) allowed for widening our understanding of the variety of the modern digital story modifications. The basic scheme of the concept is the following: the digital story and the constituent elements of it could be used by other participants of the collaborative activity in creation of new stories.

In the Letopisi.org Social Learning Network students perform the following actions with wiki pages:

Read -> Create -> Edit -> Connect -> Share -> Read

Read stories created by others -> create new page -> edit page -> put pages together into a story -> share the new story

This is a base cycle of actions in Letopisi where key objects are pages over which actions are performed. The learning design within the frame of the given concept structures not only the conditions for a separate learning agent activity within a limited time and space interval but also determines conditions for mid-term interrelation of the learning agents and the exchange of the activity products as well as the conditions for the development and long-term evolution of the whole collaborative activity system based on the selection of the most important stories.

In the focus of the learning design is not only the learning activity of a separate learning agent related to the creation of an individual product, but also a system of relations between all elements. Thus, the subject of the learning design represents not only social projecting of the system "agent-tools-product-outcome" in which digital stories are created, but also projecting of a system in which the interrelation and evolution of these digital stories occur as well as the interrelation and co-evolution of the digital stories authors who are the learning agents. Using the metaphor of chess school, we can state that the subject of the learning design is not just the playing board upon which a single game is being performed but also the whole network of relations which is formed "around the board" between the participants of the collaborative activity.

The MediaWiki environment in which the modern collaborative activity is carried out allows us to follow the links which occur between the agents and objects of activity. As a rule, modern socio-technical systems in which the collaborative activity is carried out, store the full history of all activities. In general terms, this history can be presented as a record of a chess or go game, consisting of many moves. Each move in the socio-educational project as a game contains three required elements:

Agent ID| Object ID| Type of an action (Writing, Editing, Voting)|

Every action of an agent towards an object leads to the formation of a link between them. If the agents perform action over one and the same object they become agents of the collaborative activity, indirectly linked with one another by the mutual object of activity. The collaborative activity network could be presented as the bipartite graph combining agents with objects of collaborative activity.

There is a variety of software tools that has been developed to support analysis of network structures. In our work, we have chosen a tool that allows us to easily identify the stable patterns in the field of collaborative work, as well as to track and analyze the dynamics of sociograms, to select the participants among stable patterns for their targeted management.

Our first learning analytics application is based on Graphviz - open source graph visualization software. Graph visualization is a way of representing structural information as diagrams of abstract graphs and networks. The Graphviz layout programs take descriptions of graphs in a simple text language, and build diagrams in useful formats. The use of GraphViz diagrams allows us to conduct express analysis of the position at the field of the collaborative network activity.

We have created a simple application for social network analysis that help to detect and visualize cliques of densely connected articles and editors in wiki - Collaboration Diagram or WikiGram - https://www.mediawiki.org/wiki/Extension:Collaboration_Diagram

The use of wikigrams allows us to conduct express analysis of the position at the field of the collaborative network activity, to determine the key units of the position – authors and articles, with which the majority of other units are linked, to evaluate the stability of a community according to the number of authors involved into collaborative editing of the collaborative pages.

On the basis of this application we have developed the technique of the discussion of collaboration. With this application we have identified the typical configurations in the field of a collaborative work (single, paired figures, small groups, cliques, clusters, giant components, binder key players).

Our second learning analytics application was implemented in NetLogo language 5.2 and its source code is available on the Internet as Dynamic Wikigram Model http://modelingcommons.org/browse/one_model/4769

The use of NetLogo allowed us to present the collaboration as a dynamic diagram in which each actor can interact with each other tens of thousands of actors. The created model used NetLogo features such as breeds and agentset. Interface of Dynamic Wikigram Model presented in Figure 2.

Details of methodology, technology and a range of applications are best illustrated for the purposes of this chapter by a brief example as presented in the next section.

Figure 2. Interface of dynamic Wikigram model

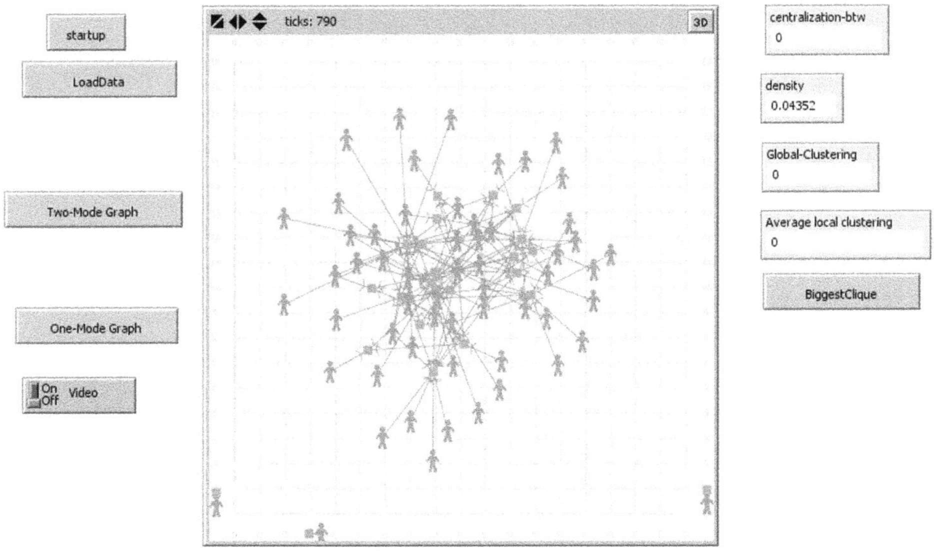

Wikigrams in Action

To illustrate wikigrams reflecting collaborative editing let us have a look at an example of a page and a category. Each page among working tabs has a wikigram tab. By clicking this tab a user opens a special page where Graphviz creates a bipartite graph – wikigram (Figure 3). The participants are shown as little figures and pages image represents an internet wiki page. The diagram is interactive, and you see the name of an author or a page when mouse on image on the diagram. If you click on the images, a relevant page for the author or the article opens.

Graphviz interprets the program, the code of which is placed under the diagram. The first lines determine the layout of a wikigram. Those lines with an arrow symbol –> are interpreted as links between units.

```
<graphviz>
digraph W {
layout = "neato";
node[fontsize="10", fontcolor="blue", nodesep=2, shape="none",
style=""] ;
edge [arrowhead=normal, arrowsize=0.4,len=2.5];
node [URL="/index.php?title=\N"] ;
"User:T Pirog" [tooltip="T Pirog"] ;
"User:T Pirog" [label ="", image="/CollaborationDiagram/user_
```

Figure 3. WikiGram of page

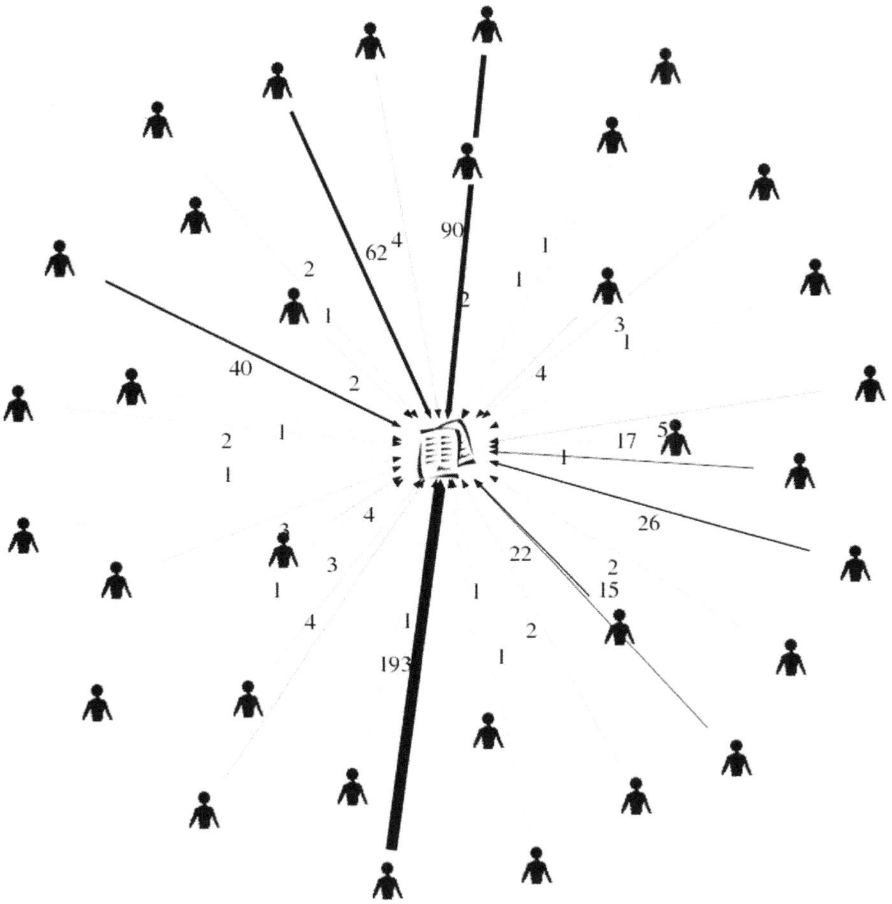


```
icon.png" ];
"UbiPlace" [image="/CollaborationDiagram/article_icon.png"
tooltip="UbiPlace" label=""];
"User:T Pirog" -> "UbiPlace" [ penwidth=1.18 label=4] ;
"User:Aygun Azizova A." -> "UbiPlace" [ penwidth=1.47 label=5]
;
"User:Ljudmilla Rozhdestvenskaja" -> "UbiPlace" [ penwidth=0.30
label=1] ;
</graphviz>
```

A wikigram is automatically built not only for a separate page, but for groups of pages under one category.

Categories, a software feature of MediaWiki, provide automatic indexes that are useful as tables of contents. You can categorize pages and files by adding one or more Category tags to the content text. These tags create links at the bottom of the page that take you to the list of all pages in that category.

As a rule, co-editors unite under one category those pages which are their collaborative work within one project. Thus, a category wikigram reflects interaction within a creative project (Figure 4).

Any participant can take graphs initial text and use it for personal experiments on a separate designated page. Available actions are:

Figure 4. Category wikigram

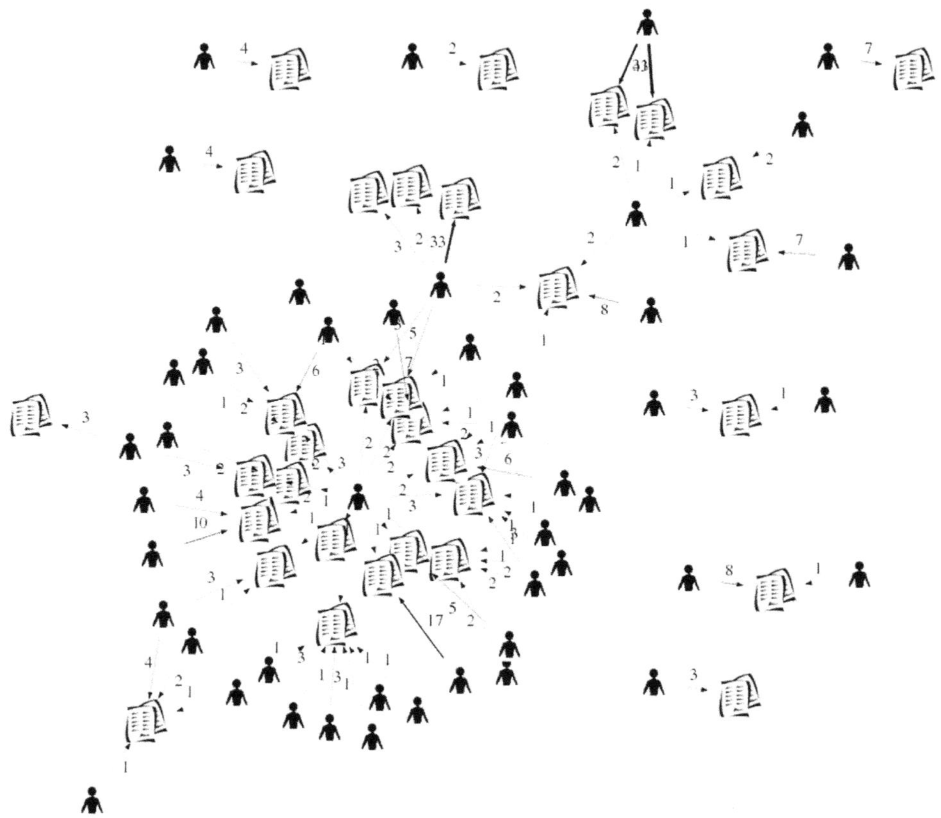

- Trying different layouts (neato, circo, sfdp, fdp) and choosing an optimal one for showing links between units in a way when links are not crossed. For example, layout = “circo”
- Changing diagram size for smaller or bigger to make objects clearly visible on a screen. For example, size=”10,10”
- Setting sizes for edges and arrows. For example, edge [arrowhead = normal, arrowsize = 0.4, len = 1.5]
- Changing other parameters – permit or forbid unit overlap. For example, overlap = false.

A more complicated and interesting task is connected with deletion of some units or breaking a graph into clusters. A researcher can visually identify a unit with the highest central concentration, and then delete all the lines bearing it in a word processor. The following figures show the example of such task. Complete project wikigram is shown in Figure 5. A key actor is in the center of the wikigram.

Figure 5. Complete project wikigram

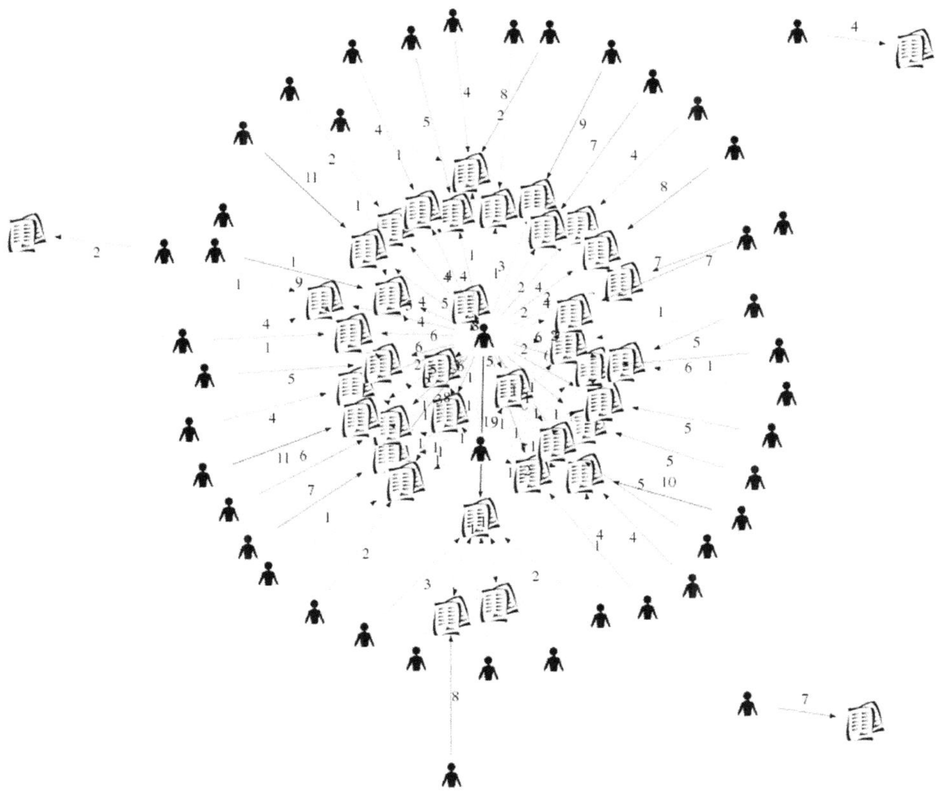

Figure 6. Wikigram after removal of a key actor

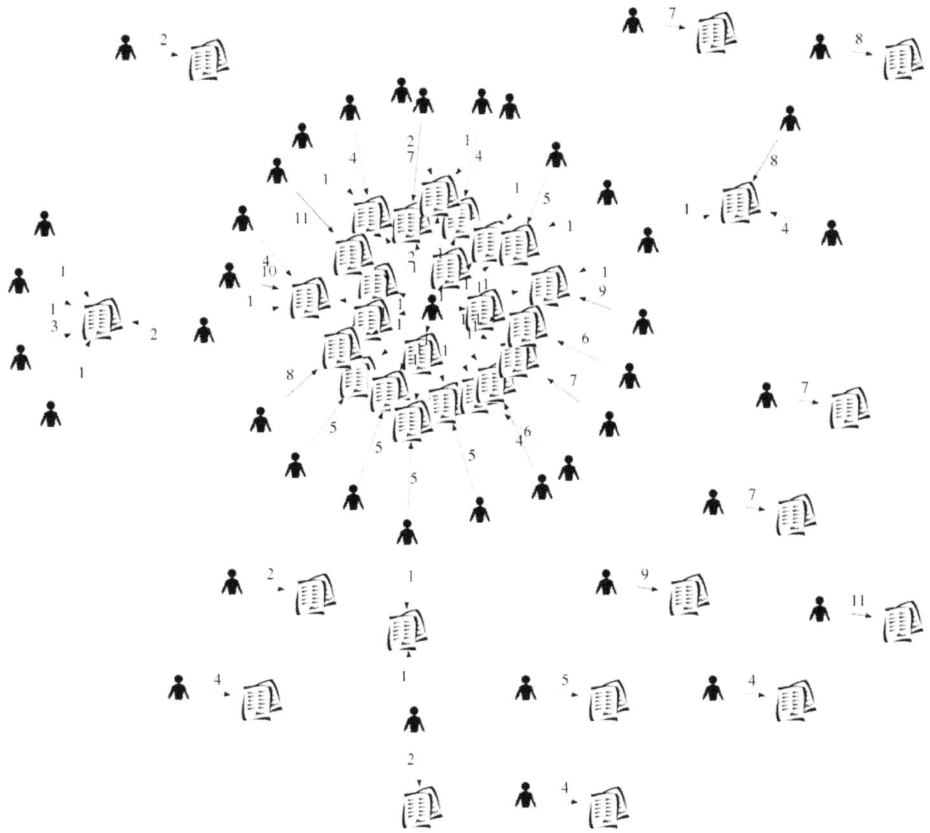

After removing from the graph all the rows, which mentions a key actor, the wikigram is transformed to the following form (Figure 6).

The next possible task is connected with the comparison and evaluation of wikigrams of different categories. To make it easier for teachers and students to compare and evaluate wikigrams we have created evaluation criteria shown in Table 1.

Dynamic Wikigram Model can be used as a simple and illustrative tool of analysis. During his/her research, a student uses the technique of dynamic agent-based sociograms to:

- Trace how and based on what objects' links between participants of collaborative production activity are formed.
- Identify key players and stable biggest cliques, which serve as cores that support the operation of network communities.

Table 1. Wikigram evaluation criteria

Criterion	High level	Medium level	Low level
Productivitya	A wikigram contains a variety of objects created by participants.	Participants have created digital objects. The number of objects is approximately equal to the number of participants.	In the project very few digital objects have been created. The number of objects is much fewer than the number of participants.
Action frequency	A graph shows that objects have been modified frequently (more than 5 times) by authors and editors	Some objects have been enhanced by the authors and other participants 3-4 times.	Objects created by participants have never been altered since being created and published in the system.
Connectedness	All actors and objects are connected in one graph.	Participants and objects are connected into small groups of components (3-4).	The graph of the project is broken into a number of non-connected components.
Cohesion	A wikigram depicts one cohesive clique, in which all the participants are connected to each other via pages.	Participants form several small groups.	There are no groups on the wikigram which proves lack of interaction.
Stability	A wikigram shows several key players, whose linking provides stability for collaboration.	A wikigram contains 2-3 main players, whose removal will lead to breaking the net into separate components.	A wikigram has one key player in charge of all informational processes. Removal of this unit terminates the net.

Key players are those elements in the network that are considered important, in regard to certain criteria. One of the most popular criteria is the betweenness centrality of vertex. The betweenness of an edge is the number of these paths running through it. It is clear that, when a graph is made of tightly bound clusters, loosely interconnected, all shortest paths between nodes in different clusters have to go through the few interclusters connections, which therefore have a large betweenness value. A node with high betweenness centrality is responsible for connecting many pairs of nodes via the best path, and deleting that node should cause many pairs of nodes to be more distinctly. The idea behind betweenness centrality is that being in between actors makes actor powerful because he may be able to control the flow of information between them (Borgatti, 2006). Nodes with high betweenness centrality are often called key-players. To calculate the betweenness centrality of a Netlogo turtle, you take every other possible pairs of turtles and, for each pair, you calculate the proportion of shortest paths between members of the pair that passes through the current turtle. The betweeness centrality of a turtle is the sum of these. Top ten key players are determined in NetLogo as

sublist reverse sort-one [norm-betweenness] users 0 9

Key players are important in themselves and as nodes that connect communities. Qualitatively, a community is defined as a subset of nodes within the graph such that connections between the nodes are denser than connections with the rest of the network (Radicchi, Castellano, Cecconi, Loreto, & Parisi, 2004). The detection of the community structure in a network is generally intended as a procedure for mapping the network into a tree. For the divisive class of algorithms one starts with the whole graph and iteratively cuts the edges. The crucial point in a divisive algorithm is the selection of the edges to be cut. Girvan and Newman have introduced an "edge betweenness algorithm" where the selection of the edges to be cut is based on the

Figure 7. Wikigram with key player in the center

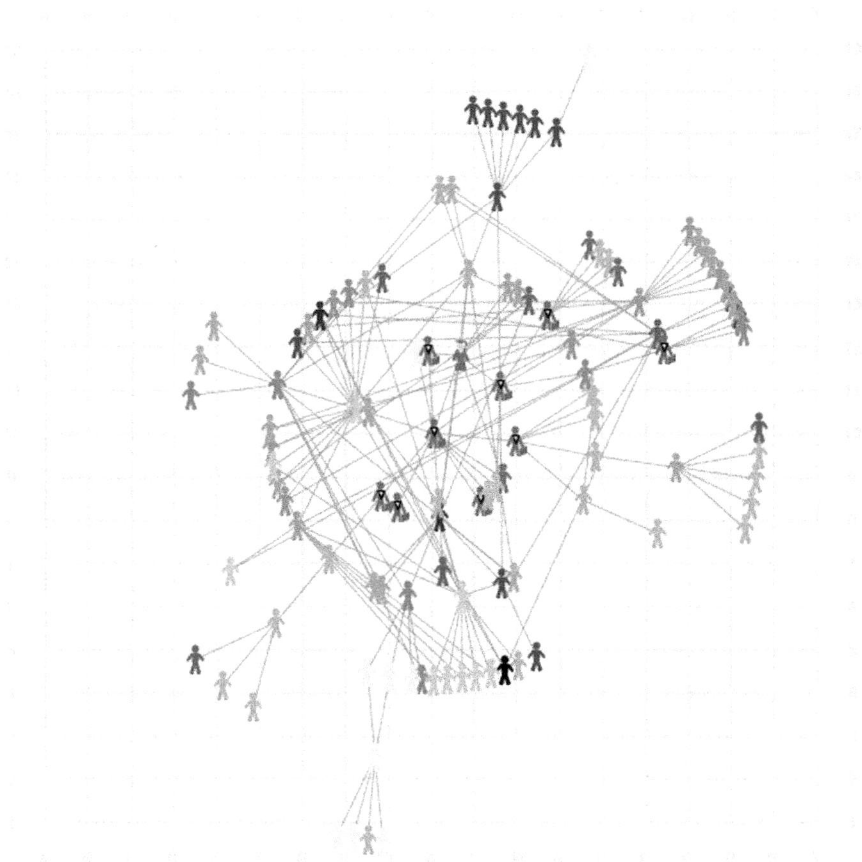

value of their edge betweenness centrality (Girvan & Newman, 2002). The single step of the edge betweenness algorithm consists in the computation of the edge betweenness for all edges in the graph and in the removal of those key players with the highest score. The iteration of this procedure leads to the splitting of the network into disconnected subgraphs, until the whole graph is divided in a set of isolated nodes. At each step of the edge betweenness algorithm NetLogo model created the sociogram, which was used to discuss the role of a key players (Figure 7).

A clique is a subset of a network in which every node has a direct link to every other node. A maximal clique is a clique that is not itself contained in a bigger clique. Cliques containing more than N members are determined in NetLogo as

Figure 8. Wikigram of maximal cliques

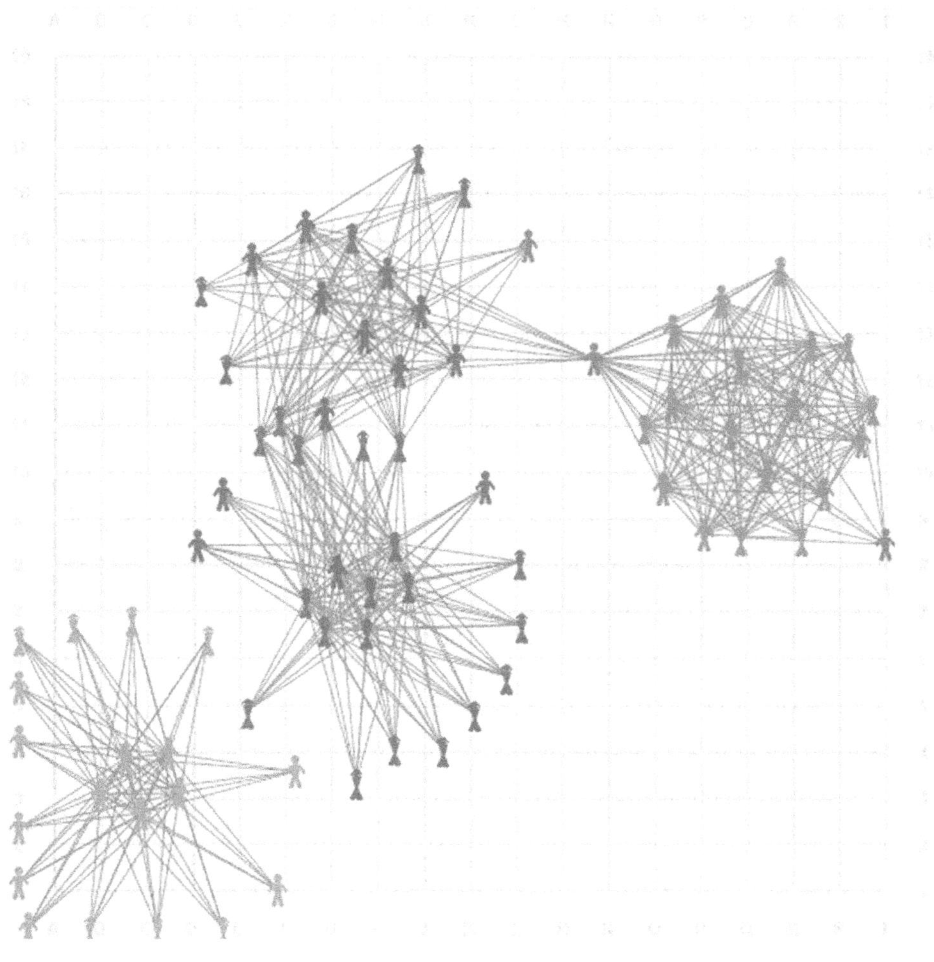

nw:maximal-cliques [if (count ?) > N

FUTURE RESEARCH DIRECTIONS

The projected system of the collaborative network activity gives extra possibilities not only for the productive activity but also for the analysis and reflection on the processes inside the system. The tools should open the possibility to assess from the network point of view both the position of each participant and the degree of the whole system development as the learning network. The analysis of the activity of each participant inside the acting community combined by mutual story or mutual game, allows us linking the act of activity and the development of one participant with the development of the whole community.

The point of the co-creative projects is not only the created product of the project, but also the social structure itself. The value is in the creation process of this social structure during collaborative work and the formation history of this structure. Usually the social structure and the history of its formation are hidden for the participants. In best cases the subject for participants' discussion is the number of created objects and the number of comments and ratings. Meanwhile a social structure is a significant characteristic in many respects determining the success of collaboration. We believe that visualization of the social structure can support the process of group reflection.

Reflexivity is defined as "the extent to which group members overtly reflect upon the group's objectives, strategies, and processes, and adapt them to current or anticipated endogenous or environmental circumstances" (West, 1996, p. 559). West distinguished between task reflexivity and social reflexivity. In teams, high in social reflexivity, team members often reflect and deal effectively with collaboration problems, and thus display a good quality of relationships and friendly attitudes. Social reflexivity is associated with the social functioning part of a team, deals with interpersonal relation, strengthens collaboration among team members, and therefore leads to better performance (Carter & West, 1998). While various studies have found that social reflexivity has significant positive relationship with team outcomes (Gurtner, Tschan, Semmer, & Nägele, 2007; Schippers, Den Hartog, & Koopman, 2007), little is known regarding the mechanisms underlying group social reflexivity in network collaborative projects. We start from the hypothesis that social reflection can be triggered by sociograms.

It's possible to use the approach and the applications described in this chapter not only inside Wiki or in learners' research activity, but for a wider field of productive activity to form productive agency. Productive agency formation isn't limited by learning or research communities where students work over project creation of simply educational value. This field of productive agency formation includes socio-pedagogical projects where citizens interact with each other, and artifacts are the documents influencing educational policies.

We have successfully used NetLogo wikigram application to support several socio-educational projects, where the number of participants was several thousand people.

CONCLUSION

Inquiry-based learning is most often connected with supervision and modeling of interaction within real flocks, swarms, anthills or termite nests. Digital technology development is now opening the door to research what Foucault defined as "a cet immense fourmillement de traces verbales" (vast swarm of verbal traces) (Foucault, 1969, p. 35). This being said, verbal traces, which turn into the subject of analysis, appear over the course of production of digital stories. The network of relationships between wiki pages collaborative editors is mapped in the same way as the network of relationships between dolphins, ants and termites. This does not mean that we suggest that our students study the process of creating digital stories in the same way as they would study building an ant-hill or a termitary. But we draw their attention that their own collaboration in creating digital stories is based on the same systemic rules as in forming flocks, building ant-hills or termite nests.

Do we really want to think of students as ants, each following the same simple rules in an almost mindless fashion? Certainly not. Students in a classroom are, of course, much different than ants in a colony—or roots on a walking tree. But in all of these cases, the behaviors of the overall system arise, often in unexpected ways, from interactions among the parts of the system. (Resnick, 2003, p.58)

We attract students' attention to the network science in the field of their own activity. In this field, they are the actors between whom digital stories edited pages mediated connections emerge. As a result, we get to unite the cycles of productive and research activities within one social network. The simple application for social network analysis described in this work is an experimental step to form a culture of open network research. The method of defining and visualizing the group participants, connected by editing the same articles, allows us to observe the groups connected by editing wiki-pages without leaving the wiki-media. The application and analysis methods under consideration serve as guidelines and can be used as invitation to conduct research. We have offered a methodology of wikigram demonstration on a gameboard with an alphabetical and numerical notation, which allows users to have discussions and analyze quality data on groups of participants of wiki-projects. Real cases analysis shows that the technology and methodology of wikigrams allows users to analyze and discuss situations which occur over the course of collaborative network activity.

REFERENCES

Barabási, A.-L. (2016). Network Science. Cambridge University Press. Retrieved from http://barabasi.com/networksciencebook/

Borgatti, S. P. (2006). Identifying Sets of Key Players in a Social Network. *Computational & Mathematical Organization Theory*, *12*(1), 21–34. doi:10.1007/s10588-006-7084-x

Börner, K., Palmer, F., Davis, J. M., Hardy, E., Uzzo, S. M., & Hook, B. J. (2009). Teaching children the structure of science. In *IS&T/SPIE Electronic Imaging* (pp. 724307–724307). International Society for Optics and Photonics. Retrieved from http://proceedings.spiedigitallibrary.org/proceeding.aspx?articleid=1334930

Börner, K., Sanyal, S., & Vespignani, A. (2007). Network science. *Annual Review of Information Science & Technology*, *41*(1), 537–607. doi:10.1002/aris.2007.1440410119

Brandes, U., Robins, G., McCranie, A., & Wasserman, S. (2013). What is network science? *Network Science, 1*(1), 1–15. https://doi.org/10.1017/nws.2013.2

Brennan, K., Hernández, A. M., & Resnick, M. (2009). Scratch: creating and sharing interactive media. In *Proceedings of the 9th international conference on Computer supported collaborative learning* (vol. 2, pp. 217–217). International Society of the Learning Sciences. Retrieved from http://dl.acm.org/citation.cfm?id=1599503.1599576

Bruner, J. S. (2003). *Making stories: law, literature, life*. Cambridge, MA: Harvard University Press.

Carter, S. M., & West, M. A. (1998). Reflexivity, Effectiveness, and Mental Health in BBC-TV Production Teams. *Small Group Research*, *29*(5), 583–601. doi:10.1177/1046496498295003

Committee on Network Science for Future Army Applications, & National Research Council. (2005). *Network Science*. The National Academies Press. Retrieved from https://www.nap.edu/catalog/11516/network-science

Cramer, C., Sheetz, L., Sayama, H., Trunfio, P., Stanley, H. E., & Uzzo, S. (2015). In G. Mangioni, F. Simini, S. M. Uzzo, & D. Wang (Eds.), NetSci High: Bringing Network Science Research to High Schools (pp. 209–218). Complex Networks, VA: Springer International Publishing. Retrieved from http://link.springer.com/chapter/10.1007/978-3-319-16112-9_21

Engeström, J. (2005). *Why some social network services work and others don't — Or: the case for object-centered sociality*. Retrieved from http://www.zengestrom.com/blog/2005/04/why-some-social-network-services-work-and-others-dont-or-the-case-for-object-centered-sociality.html

Fontana, M., & Terna, P. (2015). *From Agent-based models to network analysis (and return): the policy-making perspective* (Department of Economics and Statistics Cognetti de Martiis. Working Paper No. 201507). University of Turin. Retrieved from https://ideas.repec.org/p/uto/dipeco/201507.html

Foucault, M. (1969). *L'Archéologie du savoir*. Gallimard.

Freeman, L. C. (2000). Visualizing Social Networks. *Journal of Social Structure*, (1). Retrieved from http://www.cmu.edu/joss/content/articles/volume1/Freeman.html

Gaines, B. R., & Shaw, M. L. G. (2012). Sociocognitive Inquiry. In I.-H. Ting, T.-P. Hong, & L. S.-L. Wang (Eds.), *Social Network Mining, Analysis, and Research Trends: Techniques and Applications* (pp. 35–55). IGI Global. doi:10.4018/978-1-61350-513-7.ch003

Galley, R., Conole, G., Dalziel, J., & Ghiglione, E. (2010). *Cloudworks as a "pedagogical wrapper" for LAMS sequences: supporting the sharing of ideas across professional boundaries and facilitating collaborative design, evaluation and critical reflection*. Presented at the 2010 European LAMS & Learning Design Conference, Oxford, UK. Retrieved from http://oro.open.ac.uk/22328/

Gee, J. P. (2004). *What video games have to teach us about learning and literacy*. Palgrave Macmillan.

Girvan, M., & Newman, M. E. J. (2002). Community structure in social and biological networks. *Proceedings of the National Academy of Sciences of the United States of America*, *99*(12), 7821–7826. doi:10.1073/pnas.122653799 PMID:12060727

Grobstein, P. (2005). Revisiting Science in Culture: Science as Story Telling and Story Revising. *Journal of Research Practice*, *1*(1).

Gurtner, A., Tschan, F., Semmer, N. K., & Nägele, C. (2007). Getting groups to develop good strategies: Effects of reflexivity interventions on team process, team performance, and shared mental models. *Organizational Behavior and Human Decision Processes, 102*(2), 127–142. doi:10.1016/j.obhdp.2006.05.002

Harrington, H. A., Beguerisse-Díaz, M., Rombach, M. P., Keating, L. M., & Porter, M. A. (2013). Commentary: Teach network science to teenagers. *Network Science, 1*(2), 226–247. https://doi.org/10.1017/nws.2013.11

Inden, R. B. (2000). *Imagining India. C.* Hurst Co. Publishers.

Kelleher, C., & Pausch, R. (2007). Using Storytelling to Motivate Programming. *Communications of the ACM, 50*(7), 58–64. doi:10.1145/1272516.1272540

Klopfer, E., Scheintaub, H., Huang, W., Wendel, D., & Roque, R. (2009). The Simulation Cycle: Combining Games, Simulations, Engineering and Science Using StarLogo TNG. *E-Learning and Digital Media, 6*(1), 71–96. https://doi.org/10.2304/elea.2009.6.1.71

Knorr-Cetina, K. (1999). *Epistemic cultures: how the sciences make knowledge.* Harvard University Press.

Lerner, R. M. (2014). *Agent-Based Modeling as a Social Activity.* Northwestern University. Retrieved from http://gradworks.umi.com/36/69/3669272.html

Mehra, A., Borgatti, A., Soltis, S., Floyd, T., Halgin, D. S., Brandon, O., & Lopez-Kidwell, V. (2014). Imaginary Worlds: Using Visual Network Scales to Capture Perceptions of Social Networks. In *Contemporary Perspectives on Organizational Social Networks* (Vol. 40, pp. 315–336). Emerald Group Publishing Limited. Retrieved from http://www.emeraldinsight.com/doi/abs/10.1108/S0733-558X%282014%290000040016

Moody, J., Mcfarl, D., & Bender-demoll, S. (2005). Dynamic Network Visualization. *American Journal of Sociology, 110*(4), 1206–1241. doi:10.1086/421509

Moreno, J. L. (1934). *Who Shall Survive? A New Approach to the Problem of Human Interrelations.* Nervous and Mental Disease Publishing Company. doi:10.1037/10648-000

Papert, S., & Harel, I. (1991). Situating Constructionism. In *Constructionism* (pp. 193–206). Ablex Publishing Corporation. Retrieved from http://www.papert.org/articles/SituatingConstructionism.html

Patarakin, E., & Visser, L. (2012). New Tools for Learning - The Use of Wiki's. In L. Visser, Y. L. Visser, R. Amirault, & M. Simonsom (Eds.), Trends and Issues in Distance Education 2nd Edition International Perspectives (pp. 287–299). Information Age Publishing.

Radicchi, F., Castellano, C., Cecconi, F., Loreto, V., & Parisi, D. (2004). Defining and identifying communities in networks. *Proceedings of the National Academy of Sciences of the United States of America, 101*(9), 2658–2663. doi:10.1073/pnas.0400054101 PMID:14981240

Resnick, M. (1997). *Turtles, termites, and traffic jams: explorations in massively parallel microworlds*. MIT Press.

Resnick, M. (2003). Thinking Like a Tree (and Other Forms of Ecological Thinking. *International Journal of Computers for Mathematical Learning, 8*(1), 43–62. doi:10.1023/A:1025632719774

Reynolds, R., & Caperton, I. H. (2009). Comparison of Middle School, High School and Community College Students' Wiki Activity in Globaloria-West Virginia: (Pilot Year-two). In *Proceedings of the 5th International Symposium on Wikis and Open Collaboration* (p. 29:1–29:2). New York, NY: ACM. https://doi.org/10.1145/1641309.1641350

Sánchez, A., & Brändle, C. (2014). *More network science for teenagers*. Retrieved from http://arxiv.org/abs/1403.3618

Schippers, M. C., Den Hartog, D. N., & Koopman, P. L. (2007). Reflexivity in Teams: A Measure and Correlates. *Applied Psychology, 56*(2), 189–211. doi:10.1111/j.1464-0597.2006.00250.x

Schneider, D., Class, B., Benetos, K., Lange, M., Schneider, D., Class, B., … Lange, M. (2012). *Requirements for learning scenario and learning process analytics*. Presented at the World Conference on Educational Multimedia, Hypermedia and Telecommunications. Retrieved from http://edutechwiki.unige.ch/en/Learning_process_analytics

Schwartz, D. L. (1999). The productive agency that drives collaborative learning. In I. P. Dillenbourg (Ed.), *Collaborative learning: Cognitive and computational approaches* (pp. 197–218). Elsevier Science/Permagon.

Sheetz, L., Dunham, V., & Cooper, J. (2015). Professional development for network science as a multi-disciplinary curriculum tool. In *2015 IEEE Integrated STEM Education Conference (ISEC)* (pp. 178–182). https://doi.org/10.1109/ISECon.2015.7119919

UNESCO. (2008). ICT Competency Standarts for Teachers: Policy Framework. Paris: UNESCO.

Wasserman, S., & Faust, K. (1994). *Social Network Analysis: Methods and Applications* (1st ed.). Cambridge University Press. doi:10.1017/CBO9780511815478

West, M. A. (1996). Reflexivity and work group effectiveness: A conceptual integration. In M. A. West (Ed.), *Handbook of work group psychology* (pp. 555–579). Chichester, UK: Wiley.

Wierzbicka, A. (1992). *Semantics, Culture, and Cognition: Universal Human Concepts in Culture-Specific Configurations*. Oxford University Press.

KEY TERMS AND DEFINITIONS

Clique: A subset of a network in which every node has a direct link to every other node.

GraphViz: An open source graph visualization software.

Key Players: Those elements in the network that are considered important, in regard to certain criteria.

Learning Analytic: A collection of methods that allow learners to understand what is going on in a learning scenario.

Log-File: A file of the transactions between a system and the users of that system.

MediaWiki: The most popular of wiki engines, which is also used to run Wikipedia.

NetLogo: A multi-agent programmable modeling environment.

Network Science: An emerging, highly interdisciplinary research area that deals with theoretical and practical approaches and techniques of analysis, synthesis and behavior of natural and man-made networks.

Network Analysis: An empirical method to study the ties between actors in the network.

Sociogram: A powerful analysis tool that provides researchers with the ability to identify specific points of interest and other structural properties.

Wiki: The simplest online database that could possibly work.

Section 2

Inquiry-Based Teaching of STEM in Digital Era

Chapter 6
Inquiry-Based Science Education and the Digital Research Triad

Dina Tsybulsky
Tel Aviv University, Israel & Mofet Institute, Israel

Ilya Levin
Tel Aviv University, Israel

ABSTRACT

The chapter deals with a new research field that has arisen at the intersection of scientific experiment and emerging digital technologies. The classical triad of experimental research 'subject-instrument-object' and its implementation in science education are in the focus of the chapter. The triad is studied in its evolution to a so-called digital triad corresponding to the experimental science of digital society. In the digital triad, each of the three components are transformed. The knowing subject - researcher is transformed to the digital scholar; the experimental instrument is transformed on the base of emerging cloud and mobile technologies; the research object comprising hybrid natural-artificial components emerges. The digital transformations of the experimental research triad and educational practices based on the digital triad are manifested in a number of pioneer inquiry-based projects analyzed in the chapter.

DOI: 10.4018/978-1-5225-2525-7.ch006

INTRODUCTION

Global society has transitioned into the Digital Era; this shift represents a revolution in human history – the so-called digital revolution (Dewandre, 2011; Ess, 2015; Floridi, 2014; Turner, 2006; Yuan, 2013). This revolution relates to fundamental principles of humanity. It has changed peoples' understanding of their place in the world, as people no longer merely consider humanity a part of the nature, but also a part of the artificial world that humans have created. The digital revolution also changed peoples' perception of society. Digital society is a hyper-connected one, in which people may have hundreds of acquaintances living in distant parts of the planet, with whom they exchange information. Presently, moreover, people have unlimited and ubiquitous access to desirable information, which, in turn, becomes personalized and context-cognizant. Indeed, our world has become one of information abundance. This state, in contrast to the previous state of information scarcity, comprises one important characteristic of digital society (Ganascia, 2015).

Indeed, the digital era has led humanity to take on a new perspective on its surrounding environment. People are gaining an awareness of the fact that we live not only in a 'real' environment, but also in a virtual space. Such a 'twofold reality' has led to emerging technologies of augmented and mixed reality, which enable new forms of knowing the surrounding world. This phenomenon manifests another characteristic of digital society – a blurred distinction between reality and virtuality (Ess, 2015).

Within digital society, people have begun perceiving themselves as "informational organisms (inforgs), mutually connected and embedded in an informational environment (the infosphere)" (Floridi, 2014, p. 94). The concept of inforg includes not only humans but also specific informational artifacts that are able to communicate with people and even demonstrate elements of social behavior. The emerging cyber-physical systems (CPS) comprise an example of such informational artifacts, since they are hybrid systems that can be considered neither purely artificial nor purely natural. Indeed, as advanced computer-embedded systems, CPSs demonstrate one of the characteristics of digital society whereby the boundaries between people, artifacts, and nature are blurring (Ess, 2015).

Obviously, the above transformations involving such fundamental features of the human being (the emergence of the informational abundance, transforming ways of observation of the world, and a changing view on the nature of surrounding objects) could not possibly leave unchanged key components of human culture such as scientific inquiry. Needless to say, scientific experiments are changing with emerging technologies; this change contributes to an epistemological breakthrough and to the construction of knowledge (Ganascia, 2008).

In this Chapter, we address the ways in which scientists obtain new knowledge and how these new methods affect science education. We focus on the classical triad of experimental research, 'subject-instrument-object,' and study the triad in its evolution into a so-called *digital triad* corresponding to digital society. We show that the digital triad parts from the classical triad in three ways: (1) it places the subject in the role of a new type of researcher – a digital scholar and an informational organism; (2) the digital triad's emerging experimental instruments are based on mobile, wearable and mixed-reality technologies; and (3) a new type of digital objects has emerged – namely, hybrid natural-artificial objects. We apply this proposed interpretation of the digital triad to study present transformations in inquiry-based science education. We also discuss how digital transformations are manifested in certain educational ventures by analyzing a number of pioneer projects.

The Chapter is organized as follows. In the Background section, some philosophical issues connected with the Nature of Technology and the Nature of Science are discussed. In section "The research triad", the experimental research triad and its evolution are presented. Then, the detailed description of the digital triad and the interrelations 'subject-instrument-object' in digital triad are discussed. The section "The digital triad in science education" is devoted to the digital triad in inquiry-based science education. A number of educational projects that utilize the digital triad are presented. Each project demonstrates changes in a specific component of the triad.

BACKGROUND

The Nature of Technology in the Experimental Science Context

In this Chapter, we focus on experimental science that generally accumulates knowledge by controlled experimentation involving manipulation of the study objects (as per Cleland, 2002; Cartwright, 1999). Traditionally, experimental science is considered to aim towards the generation of reliable knowledge of the world. According to such an approach, technological instrumentation plays a secondary, auxiliary role in experimental science. However, the development of experimental science has been intricately interwoven with the development of technology. Indeed, experimental science extensively relies on technological equipment, and resultantly, experimental research often contributes to technological innovations (Radder, 2009). Moreover, substantial conceptual similarities exist between the realization of experimental methods and that of technological processes, most significant of which is the implied possibility and necessity of manipulating and controlling natural objects. As such, scientific experimentation undergoes fundamental changes in the digital epoch. To understand the nature of experimentation and the manner in which it is affected by the digital era, we must comprehend the nature of technology and its interrelations with experimental science.

The most popular view on the role of technology in science considers technology as instrumental, as a tool enhancing scientific experimentation by improving human sensing capacities. Martin Heidegger (1954) criticized the instrumental view of technology as inadequate, in that it overlooks technology's essence. Gilbert Simondon (1958) explained the fallacy of the pure instrumental perception of technology and pointed out that a technological object is a component of so-called technicity. The technicity includes, along with the object, also human technical activities (work) (Simondon, 1958).

Questions about the nature of technology were profoundly explored throughout the twentieth century. Remarkable philosophers such as Lewis Mumford, Oswald Spengler, José Ortega y Gasset, Martin Heidegger, Gilbert Simondon yielded breakthroughs in our understandings of technology's role and nature. However, the impact of their notions remained restricted to the narrow circle of the academic community. At the level of broad social consciousness, the instrumental perception of technology remained intact; ultimately, technology is considered a tool.

The alternative, non-instrumental view of technology was unpopular in the industrial era, when technology was perceived dominantly as an 'applied science' (Gil-Peretz et al., 2005). This perception began changing with the advent of the digital epoch. When mobile, wearable, cloud, and other technologies began emerging in all spheres of life – and with the advent of concepts such as context awareness, social awareness, and ubiquitous computing – the instrumental perception of technology has become overly simple and inadequate.

Alongside the instrumental approach, another classical approach to technology is a so-called autonomous one (Peters, 2006), according to which, technology evolves by its own internal principles independently of surrounding world. One consequence of the autonomous approach was the perception of artificial technological objects as inorganic ones which may even contradict to human culture. Simondon (1958) argued that there is no conflict between technical artifacts and human culture. Humans and technical artifacts coexist. Moreover, "...humans are the conductors in the world-orchestra of technical artifacts" (Simondon, 1958, p. 11). Additionally, Simondon defines an artifact not just as an entity but as an evolution process. He "used the word 'objet' for indicating a process rather than a device or machine" (De Vries, 1998). Such dynamic and evolving perception of the artificial object, which is traditionally considered as a static and unchangeable, is highly impactful.

Another traditional approach to technology is the reflective one. The "reflective philosophy of technology sees as its task the study, analysis and evaluation of technology and its relation to society and the human condition" (Brey, 2014, p. 129). Those holding the reflective perspective consider technology to be something created by professionals and unchangeable by users. In contrast, supporters of the constructive philosophy of technology advocate "the development of philosophical approaches to guide and transform the practices of those actors in society that are responsible for the development, regulation and use of technology" (Brey, 2014, p. 129). Technological constructivism emphasizes the global qualities of emerging technologies – their openness, configurability, and availability. Indeed, constructivism is connected to the 'science-as-technology' notion: according to that concept, the old dictum: 'science discovers, technology invents' should be replaced with, 'science discovers because it invents' (Lelas, 1993). As such, scientific theory becomes an instrument of design, while technology itself becomes the object of research.

We argue that in digital society, experimental science will evolve against the background of the *non-instrumental, non-autonomous* and *constructive* approach to the nature of technology. Moreover, our analysis of important changes in technology and in the philosophical understanding of its role in the experimental science taking place within digital society reveals that to complete the general picture, a particular emerging epistemological issue must be taken into account.

Namely, the need exists to rethink the concept of scientific collaboration and its role in science. The commonly acknowledged status of scientific collaboration as one of the key characteristics of contemporary science is based on the two following conditions:

1. *The significance of scientific collaboration as an integral part of research.* Presently, scientific collaboration is not only considered a way of distributing knowledge, but also constitutes knowledge itself (Nielsen, 2013). It is becoming a knowledge-making activity, which includes different types of communication: "communication among scientists is what makes knowledge possible" (Nielsen, 2013, p. 2081).
2. *The phenomenon of data abundance followed by the emergence of a new phenomenon of Data Intensive Science.* Recently, a number of new methods of collection and analysis of scientific data have yielded substantial progress in research practices. Philosophers of science and many scientists have noted the recent qualitative progress in the methodology of scientific research (Hey, Tanslev, & Tolle, 2009). They characterize this progress as leading to a new 'data-intensive science' research paradigm.

The emerging research field positioned at the intersection of scientific experiment and digital technologies will inevitably include the fundamental topics of both the Nature of Technology and the Nature of Science. In the present section, we claim that for studying this emerging field one needs fresh philosophic approaches. Namely, the non-instrumental, non-autonomous and the constructive approaches of Nature of Technology were ahead of their time in the past, and the communication-centered approach of Nature of Science has been proposed quite recently. We consider that such approaches should be used together and are essential for studying the digitalization of experimental science.

THE RESEARCH TRIAD: SUBJECT-INSTRUMENT-OBJECT

As noted above, any traditional experimental scientific research comprises three fundamental components: (1) a research subject (researcher); (2) an instrument of research (experimental equipment); and (3) an object of research (natural objects, processes, or phenomena). This section focuses on the classical experimental triad, *subject-instrument-object.*

In the classical research triad, the only technological component is the second one, the instrument. The other two components are natural. Philosopher Luciano Floridi in his notion of 'technology as in-between' (Floridi, 2013), proposed a scheme of three orders classifying the relations between scientific research and technology. Under this scheme, the classical triad aligns with a 'scheme of the first order' ('nature-technology-nature'). Floridi considers technology to be an entity located between two others: (1) an interacting user, comprising the main actor in any technological phenomenon; and (2) what Floridi calls 'natural affordances,' which can be natural objects, processes, or phenomena. The second order in Floridi's scheme, 'nature-technology-technology,' corresponds with experiments in which a natural object is replaced by objects comprising an artificial portion. In the third and last order in his scheme ('technology-technology-technology'), the remaining non-technological component, the experimenter him or herself, becomes an artificial, automated knowing subject – something close to an artificial intelligent entity. In this last order, the humans are no longer in the scheme. The 'in-between' model provides just three possible schemes. Floridi proposes no scheme of the fourth order because "such a chain can always be reduced to a series of triples, each of which will be either of first-, of second-, or of third-order" (Floridi, 2014, p. 32).

In this Chapter, we apply the 'in-between' concept to study the digitalization of experimental science. We consider the research triad of the second order as a triad of the digital epoch, in contrast to the triad of the first order, which characterizes the epoch preceding the digital one. In other words, we distinguish two epochs in development of the experimental science triad: pre-digital and digital.

The pre-digital epoch relates to the longest period of history of science – from the emergence of empirical science at the end of XVI century through the beginning of the XXI century, when society's digital transformations commenced.

The digital epoch is the time when digital technologies become ubiquitous in human life and when these new emerging technologies demonstrate their qualitative difference from habitual technologies. This is an era in which society in general, and experimental science in particular, begins to change significantly. In conjunction, the non-instrumental and constructive approaches to technology become topical.

Notably, the transition from the pre-digital to the digital epoch was not instantaneous. A transition period emerged from the middle of the twentieth century (when information technologies just started to be used as experimental research instruments) through the present (when we witness the ubiquitous character of the digital technologies and their integration in all spheres of scientific research). The transition period has formed the first generation of researchers who used digital technologies in their research. Moreover, it provided both practical and theoretical basics of such integration.

The Research Triad in the Pre-Digital Epoch

The pre-digital epoch was characterized by experimental facilities, based on either analogous amplification of energy or transformation of energy. Various instruments (including scientific ones) that function on these principles played (and keep playing) an important role in human life. Technological apparatus such as the microscope, telescope, compass, timer, and other means of observation and measurement have become integral components of human culture throughout the span of history.

We argue that pre-digital technology naturally corresponds to humans' physical nature. Though the functional complexity of such instruments varies broadly, overall, humans consider these instruments to be quite natural and generally understandable. Instruments of the pre-digital epoch operate according to human notions of function; namely, according to analog principles. Such devices are relatively simple since they are based on linearity and continuity and facilitate processes of learning. The contemporary physicist John Barrow, in his book, "New Theories of Everything," notes:

Simple linear phenomena can be analyzed in pieces. The whole is nothing more than a sum of its parts according to this approach. Thus, we can understand something about a system without understanding everything about it. Non-linear chaotic systems are different. They require a knowledge of the whole in order to understand the parts because the whole amounts to more than the mere sum of its parts. (Barrow, 2007, pp. 147-148).

Social consciousness yielded greater awareness about the tight relations between technology and science. Such awareness resulted from the great progress in instrumentation based on energy transformation, which, in turn, created a view of technology as an applied science (Gil-Perez et al., 2005).

The pre-digital epoch – the longest stage of scientific technology development – forms the basis for understanding the processes of scientific research. As discussed above, the pre-digital triad reflects the first-order scheme of the 'in-between' approach. This traditional experimental research triad comprises: (1) the subject – a classical researcher who uses instrumental facilities for observations and experimentations; (2) the instrument – experimental facilities based on analogous amplification and/or transformation of energy; and (3) the object – natural objects/phenomena.

When defining the pre-digital epoch triad, it is not sufficient just to define the entities comprising the triad (subject, instrument and object). Clarifying interrelations among the entities are needed. The relation "subject-instrument" is clearly instrumental, since the instrument is a technological artifact, which functioning is based on physical principles and energy transformation. The relation "instrument-object" is mediated by a subject during an experimental research process. Both the research process and the instrument are components of the "technicity" (Simondon, 1958), the practical sphere of the experimental research.

As we discuss below, the digital triad has transformed each of the components, creating a more complex formula of scientific experimentation.

The Research Triad in the Transitional Period

The digital triad did not replace the traditional research triad instantly; rather, a transitional process emerged alongside developments in digital technology. Thus, we can point to a transitional period between classical scientific research and the digital reality in which scientific inquiry is conducted at present.

The transition period in the history of the research triad commenced when instruments began functioning on the principle of transforming information, in addition to transforming energy. The first devices of this kind were digital computers invented in the middle of the twentieth century. In the 1950s, computers had no relation to scientific research. They were used mostly for various calculations. This situation changed in the early 1960s, when the first mini-computers based on transistors were invented. Those computers were: (a) relatively miniature, and thus could be physically placed in an experimental laboratory; and (b) were capable of working under a 'real time' regime, which enabled them to be connected to the experimental equipment.

The next serious step in the scientific use of informational technologies was the invention of the microprocessor, followed by the creation of a personal computer in the mid-1970s. Since then, scientists have nearly universally come to appreciate the efficiency of microprocessor-based devices. Microprocessors have become an integral part of any innovative scientific equipment. "In this day and age very few real experiments are conducted without employing the latest technology – sophisticated measurement instruments supplying large amounts of accurate data to a computer for storage, analysis and display" (MacKenzie, 1988, p.13).

Simultaneously, innovative computer literacy skills become mandatory components of scientists' background. This trend has become even more noticeable in the mid-1990s, following the emergence and broad expansion of the Internet. Scientists have finally obtained essential informational support in their research.

Notwithstanding the intensive development of informational technologies and their dissemination in scientific research, the role of technology in experimental science did not change during the transition period. Technology remained just a means intended to improve a scientist's work by taking on some routine of the scientific research and by intensifying the scientist's work. Correspondingly, interrelations between components of the triad are not being changed in the transition epoch. Using a terminology introduced by Floridi, one can call the transition period as the period of technology enhancing but not augmenting the technology (Floridi, 2014, p.89).

The situation changed again around 2005, with the advent of the digital epoch. Indeed, the change was followed by intensive development of informational technologies. Notably, we must consider transformations in scientific research of the digital epoch as a part of the global transformations of human society in this era.

THE RESEARCH TRIAD IN THE DIGITAL EPOCH

As noted in the Introduction, the main symptomatic transformations of the digital epoch are the following: (a) a blurred distinction between reality and virtuality; (b) further blurry distinctions between people, nature, and artifacts; and (c) a reversal from scarcity to abundance of information. In this section, we define the research triad of the digital epoch and demonstrate how the components of the triad reflect the era's social transformations. For simplicity, we refer to the research triad of the digital epoch as the *digital triad*.

The Experimental Researcher of the Digital Triad

The first component of the triad, the experimental subject/researcher, has altogether transformed in the digital epoch. This is not to say that the classical scientist is extinct. Merely, alongside researchers of the classical type, a new type of researchers has appeared.

First of all, such new researchers are inhabitants of the digital society; they are inforgs living in active harmony with the informational environment of digital society. We call such new researchers 'researcher-inforgs'. Their scientific activities are digitally inspired, including intensive collaboration and data exchanges with colleagues, experimental research based on analysis of 'big data,' use of digital curation, and other emerging scientific undertakings.

Moreover, the professional identities of such researchers are expressed in the virtual space. Their collaboration with their colleagues and research assistants takes place mainly online. Indeed, science nowadays is increasingly collaborative (Nielsen, 2013; Olson, Zimmerman & Bos, 2008). Whereas in the past, scientific collaboration required that scientists share the same physical place, contemporary communication technologies enable cooperation among scientists from distant institutions and different disciplines. Thus, scientific research has become both data-intensive and highly dependent on intensive communications. Many web-based resources supporting experimental research become available. The new researchers of the digital epoch are able to integrate raw data received from different resources into their own research, applying big data analysis and network analytic methods. As a result, their research success depends on how well they collaborate with colleagues and apply technologies in emerging areas including databases, visualization, and cloud computing.

The new researcher-inforg reflects one of the global transformations of digital society: a reversal from scarcity to abundance of information. Indeed, informational abundance enables the researcher's transformation from a traditional knowing subject into a new researcher who utilizes a plurality of sources and applies informational technologies for comparing, transforming, analyzing, and demonstrating results.

Cloud computing is among the key technologies supporting the above-mentioned innovations in scientific research and supporting the researcher-inforg. Mell (2011, p. 2) defines cloud computing as "a model for ubiquitous, convenient, on-demand network access to a shared pool of configurable computing resources (e.g., networks, servers, storage, applications, and services) that can be rapidly provisioned and released with minimal management effort or service provider interaction." Floridi considers cloud computing as the first "graceful step out of the idea that computers need humans" (as cited in Miller, 2010, p. 4). Cloud computing resources are remote and independent from their users. The fact that both personal and public data is stored in the cloud changes fundamental views about ownership, accessibility, and usefulness of information and knowledge. These changes are critical in transforming the researcher into the researcher-inforg.

Experimental Instruments of the Digital Triad

The second component of the digital triad – the instrument – is also changing significantly. As is the case with the researcher, the emergence of instruments of a new nature coexists with the ongoing use of traditional experimental equipment.

Several scholars have proposed classifications of scientific instruments. Radder (2009) offered a distinction between instruments that represent a natural property by measuring its value (e.g., a device that registers blood pressure), instruments that create phenomena that do not exist in nature (e.g., a laser), and instruments that closely imitate natural processes in the laboratory (e.g., an Atwood machine, which mimics processes and properties of falling objects).

Measurement in particular has changed in the digital epoch, with the emergence of a so-called sensor revolution. The emerging smart sensors are not merely measurement devices in the traditional sense, which can transfer a value of a specific parameter from an object to the researcher by performing analog to digital transformation. The new sensors, interconnected online, are adapted to receive and/or process information from a cloud memory and other resources, so as to provide the researcher with the desirable meaningful data beyond just rough values.

The kind of scientific instruments that create non-existing phenomena can be upgraded by emerging technologies that provide computer simulations of the non-existing phenomena. A number of scholars even distinguish computer simulation and modeling from the empirical research, considering such modeling and simulation a substantive experimental research paradigm in its own right (Hey et al., 2009).

The instruments imitating natural processes utilize the emerging digital facilities and form a new type of instruments – mixed-reality devices. Mixed reality lets the user observe the real world while also recognizing in it some (believable) virtual objects. The mixed reality device anchors those virtual objects to a point in real space, making it possible to treat them as real.

Among the emerging technologies supporting the above-mentioned innovations are mobile and wearable devices, as well as ubiquitous computing. Such technologies include context-aware computing, ubiquitous wireless networks, smart objects, and location-based systems. Ubiquitous computing involves the most natural possible interaction between a user and a computer, geared towards the ultimate aim of the user not even realizing that s/he is interacting with a system (Kwok, Cheng, Ho-Shing Ip, & Kong, 2011).

Such new instruments based on mobile and wearable technologies, have transformed to comprise a new entity. The new entity, being different from the habitual instrument, does not just enhance it, but (and which is especially significant) - augments it. Indeed, new instruments that include virtual components, reflect one of the transformations of the digital epoch – a blurred distinction between reality and virtuality.

The Research Object of the Digital Triad

We argue that the third component of the digital triad – the research object – is no longer purely natural. The new object of research may be comprised of an artificial portion, which is mainly a digital one. In other words, the object becomes a hybrid one with functions partially based on natural laws and partially grounded in the technological specification of the digital devise embedded within it.

One of the clear manifestations of the new object of research is emergence of a cyber-physical system (CPS). CPSs are hybrid natural/artificial systems that humans create by providing systems with an ability to behave socially. The emergence of the CPS is an important cultural phenomenon of digital society (Levin, 2016). CPSs comprise an example of objects that cannot be attributed to either nature or technology, since they contain both natural and artificial components.

Indeed, CPSs' integrated computational and physical capabilities enable interaction with humans through many new modalities. The ability to interact with and expand the capabilities of the physical world through computation, communication, and control is a key enabler for future technology developments (Baheti, 2011). CPSs represent a specific branch of complex technical systems' confluences of knowledge and computing technologies; networking and informing; and physical artefacts and engineered systems – towards operating and servicing in human and social contexts. CPSs are artifacts with characteristics that allow them to be inhabitants of the newly recognized infosphere – alongside humans (Vroom & Horvath, 2014).

Humans behave as informational beings not only in cyberspace, but also (even more so) in reality. 'Reality' changes permanently by accommodating more and more other informational entities, which, in turn, may also serve as artifacts. Awareness of such a previously unexpected phenomenon is an important feature of the digital epoch. People interact with means of digital communication socially. This interaction takes place even if people do not perceive the communication means, programs, and media agents as partners in their frame of their socialization.

CPSs reflect a key transformation of digital society: a blurred distinction between people, nature, and artifacts. As Bernadette Bensaude-Vincent and Bill Newman (2007, p.2) note, "instead of opting for an absolute distinction of quality between the artificial and the natural, one should accept only a gradual distinction of degree."

In this section, we have shown that all three components of the digital triad are different from their corresponding ones in classical experimental science. To sum, in the digital triad: (1) the subject is a researcher-inforg; (2) the instrument is based on emerging mobile, wearable, and mixed-reality technologies; and (3) the research object involves hybrid natural/artificial systems.

Interrelations Subject-Instrument-Object in Digital Triad

The relation Subject-Instrument of the Digital Triad is a relation between the researcher-inforg and the instruments based on the above-mentioned emerging technologies. This relation can be clarified by using the Simondon's concept of "trans-individualization" (1958). Particularly, mobile and wearable devices provide a bright example of the trans-individualization. Indeed, each device is absolutely equal to all others as a figure (form) but, being totally individualized, absolutely different as a ground (content).

The relation Instrument-Object of the Digital Triad is not a simple relation between the above-mentioned augmented instrument and said hybrid natural/artificial system. This relation is sophisticated since it includes a researcher as a mediator. Obviously, such relations are not yet studied enough, and undoubtedly represent a rich field for a highly topical research.

THE DIGITAL TRIAD IN SCIENCE EDUCATION

In the previous sections, the research triad was considered as a model of the experimental scientific research which is traditionally plays a central role in science education. Namely, a school science laboratory constitutes a classical and highly important educational environment. Thus, an idea to utilize the digital research triad in class of the Digital society looks highly justified. In turn, in a science class, the research triad is reflected in inquiry-based science education. In this section, we consider students as subjects of the triad that engage in inquiry-type investigational activities.

The research triad has been evolving alongside a number of reforms in science education. First, the triad appeared at the time of the 'Sputnik' reforms, when Schwab (1960; 1962) suggested that science must be taught in a manner that is consistent with the way science operates – namely, "in an inquiry format" (Barrow, 2006, p. 266). Since then, a number of policy documents emphasize that students should develop the abilities necessary to conduct inquiry (AAAS, 1993); have a fundamental understanding about specific characteristics of scientific inquiry (NRC, 2000); and engage in scientific practices and methods of investigation used by scientists (NRC, 2012).

As such, the concept of scientific experimentation for the sake of learning is considered as inseparable part of science education for over 60 years. Below, we discuss the implications of the digital triad for inquiry-based science instruction.

The Research Triad in Inquiry-Based Science Class

Historically, in the pre-digital epoch, the research triad in a science class was shaped as Floridi's first order scheme. Students *(subjects of the triad)* conducted hands-on investigation of natural phenomena (*object of the triad)* in a lab or field, by utilizing equipment (*instruments of the triad*) based on energy amplification or transformation. This type of investigation still comprises a significant and inseparable part of the science curriculum in various countries (Abd-el-Khalick et al., 2004).

The transition period (between pre-digital and digital epochs) in a science class was characterized by emerging of Microcomputer-Based Labs (MBL) – namely, labs utilizing various probes and sensors for performing various hands-on inquiries. A number of successful educational inquiry-based projects emerged at the turn of the century: The Global Lab Project (Berenfeld, 1999; 2010), the CoVis Project (Pea et al., 1997; Pea, 2002), and the GLOBE Project (Penuel & Means, 2004). Empirical research of these projects focused mainly on the *instrumental* component of the triad and on the effect of innovative inquiry activities on students' outcomes (such as their learning of scientific concepts, inquiry skills, and understanding of the nature of science). Specific aspects of MBL technology such as real-time graphing (Nicolaou, Nicolaidou, Zacharia, & Constantinou, 2007; Pierri, Karatrantou, & Panagiotakopoulos, 2008); collaborative data visualization (Pea, 2002; Penuel & Means, 2004); and collaborative data acquisition, sharing, and analysis (Berenfeld, 2010) were reported. As a next stage, MBL technologies became broadly applied in different combinations of physical and virtual labs. These initiatives included sequentially combined labs; blended physical and virtual parts of labs with selected affordances associated with certain learning objectives; labs run in parallel, such that virtual and physical labs are available at the same time; and connected physical and virtual labs whose findings are compared. Combining physical and virtual labs became a popular and highly acceptable educational practice, providing students with a profound understanding of certain scientific topics (Chao, Chiu, DeJaegher, & Pan, 2016; De Jong, Linn, & Zacharia, 2013).

Since science classes began functioning through the digital triad, most studies evaluating the phenomenon focused on the instrument component. Two systematic reviews of the use of mobile technologies in education have indicated that the majority of the studies on the matter focused on designing systems, followed by evaluating the effectiveness of such systems (Crompton, Burke, Gregory & Grabe, 2016; Wu et al., 2012).

One of the important MBL innovations in the digital epoch is mixed reality technology (Concord Consortium, 2012). Indeed, different types of the inquiry activities based on mixed-reality technologies were introduced into science education, including augmented reality lab/field inquiry utilizing real phenomena augmented by virtual images and augmented virtual labs with embedded virtual phenomena in real, physical space and time (Chao, Chiu, DeJaegher, & Pan, 2016). Such inquiry enables students to perform scientific experiments that they cannot easily perform in the real world (Klopfer & Squire, 2008), to visualize concepts such as airflow or magnetic fields, and to experience events by displaying virtual elements over real objects (Dunleavy, Dede, & Mitchell, 2009; Wu et al., 2013).

As noted above, mobile and wearable technologies also comprise some of the key developments of the digital epoch, and have already been broadly integrated into inquiry-based science learning. A number of studies report using mobile and wearable technologies as measurement devices within science education (e.g. Gonzales et al., 2015; Gómez-Tejedor, Castro-Palacio, & Monsoriu, 2014; Hochberg, Gröber, Kuhn, & Müller, 2014; Kuhn & Vogt, 2013). Measurement with such technologies can be performed by using a variety of sensors, utilizing either standard applications or ad-hoc ones. Bracelet products such as Fitbit, Garmin, and Striiv enable the measurement of motion and vital signs, giving rise to the 'quantified self' phenomenon (Van't Hooft, Swan, Cook, & Lin, 2007). These products provide wireless connectivity, on-board analytics, and interfaces for hands-free feedback that provide educators with a wide range of opportunities in conducting experiment-based training.

Yet the use of mobile and wearable technologies is not limited to their measurement features. The use of mobile technology in science education has been studied also in the context of scaffolding, location-aware functionality, visual/audio representations, digital knowledge-construction tools, and digital knowledge-sharing mechanisms (Zydney & Warner, 2016). A number of characteristic features of wearable technologies (e.g. Google Glass) were studied in this regard: interactive sight and manipulation, including virtual representations (simulations) of physical phenomena simultaneously with corresponding virtual phenomena (Lukowicz et al., 2015); recording and providing a feedback (Coffman & Klinger, 2015), communication and distribution of resources (Bower & Sturman, 2015); and hands-free access and first-person view (Wu et al., 2014).

Regarding the *subjects* of the digital triad (students), studies have focused on the expanding scope of students' inquiry activities via online communication, use of big data, and participation in forums and networks. The main focus of studies to date has been on technology use in collaborative, inquiry-based science learning, and the general conclusion seems to be that "using collaborative technologies can help support collaborative inquiry within science education and can help model how scientists use technology" (Donna, 2013, p. 2). Likewise, scholars noted that "collaboration is very natural to real scientific inquiry, so the introduction of collaboration into the learning process brings the learning environment closer to real inquiry" (Van Joolingen, De Jong & Dimitrakopoulou, 2007, pp. 114-115). Specifically, studies of instrumentation of the digital triad in the classroom have investigated issues involving collaborative inquiry argumentation (Laru, Järvelä, & Clariana, 2012; Raes, Schellens, & De Wever, 2014); and brainstorming, reflective writing, feedback, interacting dialogue (Donna & Miller, 2013; Hwang, Tsai, Chu, Kinshu, & Chen, 2012).

Notably, the *object* component of the digital triad has been studied to a much lesser extent; only one educational project seems to have been reported about in this regard – a so-called Connected Gardening project (Zuiker & Wright, 2015), which utilizes cyber-physical systems as objects of the inquiry. We expect inquiry into this field to expand in upcoming years as use of the digital triad in inquiry-based learning continues to grow.

The Digital Triad in Educational Projects: A Few Case Studies

In this section, we analyze three educational inquiry-based science projects in the context of the digital triad. We trace how specific emerging technologies implement the above-mentioned transformations in science education.

The first project involves a location-based, augmented reality environment with a five-step guiding mechanism custom-developed so as to guide students in sharing knowledge during inquiry-based learning (Chiang, Yang, & Hwang, 2014). In this project, fourth-grade students from an elementary school in North Taiwan studied the ecological environment at a nearby pond. Specifically, students physically observed the habitats and morphology of selected aquatic plants by using a location-based Augmented Reality (AR) system based on a five-step activity series: 'ask,' 'investigate,' 'create,' discuss,' and 'reflect.'

At the 'ask' stage, the students were capable of: (1) using custom-tailored devices to take photos in order to discover and to observe aquatic plants; (2) expressing their observations and inquiries through various designs; (3) adding text and supplementing pictures; (4) sharing pictures and text messages through the location-based AR interface.

At the 'investigate' and 'create' stages, the students could browse information that they and their peers shared in posts on the targeted platform. They could also activate the text or pictures shared by their peers to learn more or use verbal descriptions to understand the knowledge that the sharer of this content wished to convey. They could then send textual questions and visual or verbal descriptions to study or enhance the data provided by their peers.

At the 'discuss' and 'reflect' stages, many students formed subgroups in chat rooms to focus on a particular topic and became engaged in communication and discussion. Through the platform, the students could ask questions, participate in discussions, and think cooperatively, which led them to discuss additional questions and to make additional inquiries.

This project clearly illustrates changes that take place at the *subject* component of the digital triad, associated with a high level of inquiry-oriented student collaboration. Such effective collaboration proved possible due to the abundance and availability of scientific knowledge regarding to the learning topic. Notably, this knowledge is highly relevant to the object of inquiry, while also reflecting the personalities of the research subjects (the students). Through such inquiry-based learning processes, students form their own online personalities as researcher-inforgs – researcher connected by a network. New technologies have enabled this innovative learning activity – above all, the emerging communication technologies and cloud technologies playing a revolutionary role in today's scientific inquiry.

The second project is the 'G-Physics' program, which utilizes a Google Glass application for a standard high school acoustics experiment: determining the relationship between the tone frequency generated by hitting a glass filled with water and the amount of water in the glass (Lukowicz et al., 2015). The project requires that students conduct experiments interactively, that they manipulate the theoretical representation of the relevant phenomena, and, simultaneously, that they interact with real-world phenomena. The inquiry activities in this project mainly involve the *instrumental* component of the digital triad. The use of Google Glass as a virtual component in the study enables research regarding multiple objects by observing and manipulating them in reality and by enhancing the inquiry virtually.

Finally, the Connected Gardening Project (Zuiker & Wright, 2015) reflects the changes of interest in the *object* component of the triad. In this project, a fourth-grade class organizes and refines its garden plot using observations of the physical environment and evaluations of data from a networked digital probe. In this way, the probe and the plot together form at least one specific cyber-physical system that visualizes and provides access to some parameters of the garden's soil, water, and sunlight. In turn, the data further allows students to display the garden system according to its underlying physical markers of plants growth, as well as to experience the patterns and processes involved in dynamically balancing the garden system.

This program is based on the ability of the plants, by means of the various sensors they were attached to, to transfer data both between the objects and from the objects to the cloud memory. Hence, the main feature of the project is the fact that the objects (the plants in the garden) are no longer the purely natural entities they would have been in the pre-digital epoch. As CPSs, these objects are hybrid natural-artificial systems that are able to transfer a substantial amount of data. Moreover, the objects demonstrate some basic social behavior. The emergence and use of such hybrid systems as the objects of scientific inquiry is unprecedented in science education.

FUTURE RESEARCH DIRECTIONS

The proposed triad-based approach relates both to the foundations of experimental science, and to science education in digital epoch. In both of them, there are a number of highly desired research directions. The proposed idea of the triad evolution from a pre-digital to a digital period is just the very beginning of a deeper and wider study of the triad and its evolution. The proposed model has to be studied in the fields of: Nature of Technology, Nature of Science and Science Education.

In the fields of the Nature of Technology and the Nature of Science, the future study will relate to each component of the triad (researcher-inforg, augmented instrument and hybrid system). Approaches of such philosophers as Simondon, Rabardel, Floridi, Brey, De Vries, Feenberg can be useful in this research. In particular, Simondon's "trans-individualization", his theory of the technological artifact and his perception of cybernetics may be highly applicable.

In the field of Science Education, the future study may include the research of a new, digital pedagogics based on the digital research triad. We hypothesize that the digital triad is a new technological and educational entity, which must be considered as not a straightforward enhancement of the traditional triad. Consequently, a number of emerging pedagogical phenomena can be detected and studied.

CONCLUSION

This Chapter discusses some of the research that has arisen at the intersection of areas of scientific experiment and emerging digital technologies. This intersection represents a fascinating new field that deals with the nature of experimental science, the nature of digital technology, and the interrelations between experimental science and digital technologies.

Axiomatically, the *subject-instrument-object* triad comprises the core foundation of experimental science. The intrinsic presence of technological components as integral parts of the triad has transformed the triad in the new digital environment, in turn stimulating the study of these changes. We have shown that the experimental digital triad is different from the classical triad of the pre-digital epoch. Each component of the digital triad is enriched by particular features that are specific to the digital epoch: intensive communication as a feature of the research subject, virtualization as a feature of the research instrument, and not merely physical but rather hybrid features of the emerging research objects. Changes in each component of the triad reflect fundamental transformations of the digital epoch.

Our interpretation of the digital triad paves the way for understanding digitalization of experimental science in general. This understanding is based on the non-instrumental and constructive approach to the nature of the technology, as well as on awareness of the significant role of communication in scientific research.

The digital triad is integrated into science education in the form of inquiry-based learning activities. New hyper-connected students (researchers), digital instruments, and hybrid research objects are gradually introduced into educational institutions. A number of successful, innovative educational projects confirm this transformation.

The main contribution of our Chapter lies in its proposed approach towards the analysis of transformations in experimental science. Applying this approach in analyzing the digitalization of contemporary scientific research will yield understandings regarding both experimental science and science education. Moreover, the proposed approach has both theoretical and practical significance. This new method provides a theoretical framework for the study of digital transformations in experimental science and inquiry-based science education, while also serving as a practical tool for the assessment of innovative educational practices.

REFERENCES

Abd-El-Khalick, F., Boujaoude, S., Duschl, R., Lederman, N. G., Mamlok-Naaman, R., Hofstein, A., & Tuan, H. L. (2004). Inquiry in science education: International perspectives. *Science Education*, *88*(3), 397–419. doi:10.1002/sce.10118

American Association for the Advancement of Science (AAAS). (1993). *Benchmarks for Science Literacy*. Oxford University Press.

Baheti, R., & Gill, H. (2011). Cyber-physical systems. *The Impact of Control Technology*, *12*, 161–166.

Barrow, J. D. (2007). *New theories of everything: the quest for ultimate explanation*. Oxford, UK: Oxford University Press.

Barrow, L. H. (2006). A brief history of inquiry: From Dewey to standards. *Journal of Science Teacher Education*, *17*(3), 265–278. doi:10.1007/s10972-006-9008-5

Bensaude-Vincent, B., & Newman, W. R. (2007). *The artificial and the natural: an evolving polarity*. MIT Press.

Berenfeld, B. (1999). The Internet in Our Classrooms: Teaching Tomorrow's Skills for Tomorrow's World. In *Science and Environment Education. Views from Developing Countries* (pp. 215–235). Washington, DC: The World Bank.

Berenfeld, B. (2010). Leveraging technology for a new science learning paradigm. In University Leadership: Bringing Technology-Enabled Education to Learners of All Ages. Cambridge, MA: Learning International Networks Consortium (LINC).

Bower, M., & Sturman, D. (2015). What are the educational affordances of wearable technologies? *Computers & Education*, *88*, 343–353. doi:10.1016/j.compedu.2015.07.013

Brey, P. (2014). From reflective to constructive philosophy of technology. *Journal of Engineering Studies*, *6*(2), 129–136.

Cartwright, N. (1999). *The Dappled World: A Study of the Boundaries of Science*. Cambridge, MA: Cambridge University Press. doi:10.1017/CBO9781139167093

Chao, J., Chiu, J. L., DeJaegher, C. J., & Pan, E. A. (2016). Sensor-Augmented Virtual Labs: Using Physical Interactions with Science Simulations to Promote Understanding of Gas Behavior. *Journal of Science Education and Technology*, *25*(1), 16–33. doi:10.1007/s10956-015-9574-4

Chiang, T. H. C., Yang, S. J. H., & Hwang, G. J. (2014). Students online interactive patterns in augmented reality-based inquiry activities. *Computers & Education*, *78*, 97–108. doi:10.1016/j.compedu.2014.05.006

Cleland, C. (2002). Methodological and epistemic differences between historical science and experimental science. *Philosophy of Science*, *69*(3), 474–496. doi:10.1086/342455

Coffman, T., & Klinger, M. (2015). Google glass: using wearable technologies to enhance teaching and learning. *Proceedings of Society for Information Technology & Teacher Education International Conference*, 1777-1780.

Concord Consortium. (2012). Framing Mixed-Reality Labs. *Newsletter*, 2014.

Crompton, H., Burke, D., Gregory, K. H., & Gräbe, C. (2016). The Use of Mobile Learning in Science: A Systematic Review. *Journal of Science Education and Technology*, *25*(2), 149–160. doi:10.1007/s10956-015-9597-x

De Jong, T., Linn, M. C., & Zacharia, Z. C. (2013). Physical and virtual laboratories in science and engineering education. *Science*, *340*(6130), 305–308. doi:10.1126/science.1230579 PMID:23599479

De Jong, T., Sotiriou, S., & Gillet, D. (2014). Innovations in STEM education: The Go-Lab federation of online labs. *Smart Learning Environments*, *1*(1), 3. doi:10.1186/s40561-014-0003-6

De Vries, M. J. (2008). Gilbert Simondon and the dual nature of technical artifacts. *Techné: Research in Philosophy and Technology*, *12*(1), 23–35.

Dewandre, N. (2011). The Sustainability Concept: Can We Stand Between Catastrophism and Denial? In C. Jaeger, J. D. Tàbara, & J. Jaeger (Eds.), *European Research on Sustainable Development* (pp. 29–34). Springer Berlin Heidelberg. doi:10.1007/978-3-642-19202-9_4

Donna, J., & Miller, B. G. (2013). Using cloud-computing applications to support collaborative scientific inquiry: Examining pre-service teachers' perceived barriers towards integration. *Canadian Journal of Learning and Technology*, *39*(3), 1–17.

Dunleavy, M., Dede, C., & Mitchell, R. (2009). Affordances and limitations of immersive participatory augmented reality simulations for teaching and learning. *Journal of Science Education and Technology*, *18*(1), 7–22. doi:10.1007/s10956-008-9119-1

Ess, C. (2015). The Onlife Manifesto: Philosophical Backgrounds, Media Usages, and the Futures of Democracy and Equality. In L. Floridi (Ed.), *The onlife manifesto* (pp. 89–111). Springer-Verlag GmbH. doi:10.1007/978-3-319-04093-6_14

Floridi, L. (2013). Technologys in-betweeness. *Philosophy & Technology*, *26*(2), 111–115. doi:10.1007/s13347-013-0106-y

Floridi, L. (2014). *The fourth revolution: How the infosphere is reshaping human reality*. OUP Oxford.

Floridi, L. (2015). *The onlife manifesto*. Springer-Verlag GmbH. doi:10.1007/978-3-319-04093-6

Ganascia, J. G. (2008). "In silico" Experiments: Towards a Computerized Epistemology. *Newsletter on Philosophy and Computers*, *7*(2), 11–15.

Ganascia, J. G. (2015). Views and Examples on Hyper-Connectivity. In L. Floridi (Ed.), *The Onlife Manifesto* (pp. 65–85). Springer International Publishing.

Gil-Pérez, D., Vilches, A., Fernández, I., Cachapuz, A., Praia, J., Valdés, P., & Salinas, J. (2005). Technology as Applied Science. *Science & Education*, *14*(3-5), 309–320. doi:10.1007/s11191-004-7935-0

Gómez-Tejedor, J. A., Castro-Palacio, J. C., & Monsoriu, J. A. (2014). The acoustic Doppler effect applied to the study of linear motions. *European Journal of Physics, 35*(2), 025006. doi:10.1088/0143-0807/35/2/025006

González, M. Á., González, M. Á., Martín, M. E., Llamas, C., Martínez, Ó., Vegas, J., & Hernández, C. (2015). Teaching and Learning Physics with Smartphones. *Journal of Cases on Information Technology, 17*(1), 31–50. doi:10.4018/JCIT.2015010103

Heidegger, M. (1954). The question concerning technology. *Technology and Values: Essential Readings*, 99-113.

Hey, T., Tansley, S., & Tolle, K. M. (2009). The fourth paradigm: data-intensive scientific discovery (Vol. 1). Redmond, WA: Microsoft Research.

Hochberg, K., Gröber, S., Kuhn, J., & Müller, A. (2014). The spinning disc: Studying radial acceleration and its damping process with smartphone acceleration sensors. *Physics Education, 49*(2), 137–140. doi:10.1088/0031-9120/49/2/137

Hwang, G. J., Tsai, C. C., Chu, H. C., Kinshuk, K., & Chen, C.-Y. (2012). A context-aware ubiquitous learning approach to conducting scientific inquiry activities in a science park. *Australasian Journal of Educational Technology, 28*(5), 931–947. doi:10.14742/ajet.825

Klopfer, E., & Squire, K. (2008). Environmental Detectives—the development of an augmented reality platform for environmental simulations. *Educational Technology Research and Development, 56*(2), 203–228. doi:10.1007/s11423-007-9037-6

Kuhn, J., & Vogt, P. (2013). Smartphones as experimental tools: Different methods to determine the gravitational acceleration in classroom physics by using everyday devices. *European Journal of Physics Education, 4*(1), 16–27.

Kwok, R., Cheng, S., Ho-Shing Ip, H., & Kong, J. (2011). Design of Affectively Evocative Smart Ambient Media for Learning. *Computers & Education, 56*(1), 101–111. doi:10.1016/j.compedu.2010.08.015

Laru, J., Järvelä, S., & Clariana, R. B. (2012). Supporting collaborative inquiry during a biology field trip with mobile peer-to-peer tools for learning: A case study with K-12 learners. *Interactive Learning Environments, 20*(2), 103–117. doi:10.1080/10494821003771350

Lelas, S. (1993). Science as technology. *The British Journal for the Philosophy of Science, 44*(3), 423–442. doi:10.1093/bjps/44.3.423

Levin, I. (2016). Cyber-physical Systems as a Cultural Phenomenon. *International Journal of Design Sciences and Technology, 22*(1), 67–80.

Lukowicz, P., Poxrucker, A., Weppner, J., Bischke, B., Kuhn, J., & Hirth, M. (2015). Glass-physics: Using Google Glass to Support High School Physics Experiments. *Proceedings of the 2015 ACM International Symposium on Wearable Computers*, 151-154. doi:10.1145/2802083.2808407

MacKenzie, I. S. (1988). Issues and Methods in the Microcomputer-Based Lab. *Journal of Computers in Mathematics and Science Teaching*, *7*(3), 12–18.

Mell, P., & Grance, T. (2010). The NIST definition of cloud computing. *Communications of the ACM*, *53*(6), 50.

Miller, K. W., & Voas, J. (2010). Ethics and the Cloud. *IT Professional*, *12*(5), 4–5. doi:10.1109/MITP.2010.129

National Research Council (NRC). (2000). *Inquiry and the national science education standards*. Washington, DC: National Academy Press.

National Research Council (NRC). (2012). *A framework for K-12 science education: Practices, crosscutting concepts, and core ideas*. Washington, DC: The National Academies Press.

Nicolaou, C. T., Nicolaidou, I. A., Zacharia, Z. C., & Constantinou, C. P. (2007). Enhancing fourth graders' ability to interpret graphical representations through the use of microcomputer-based labs implemented within an inquiry-based activity sequence. *Journal of Computers in Mathematics and Science Teaching*, *26*(1), 75.

Nielsen, K. H. (2013). Scientific communication and the nature of science. *Science & Education*, *22*(9), 2067–2086. doi:10.1007/s11191-012-9475-3

Olson, G. M., Zimmerman, A., & Bos, N. (2008). *Scientific collaboration on the Internet*. The MIT Press. doi:10.7551/mitpress/9780262151207.001.0001

Pea, R. D. (2002). Learning science through collaborative visualization over the Internet. In N. Ringertz (Ed.), *Nobel Symposium: Virtual museums and public understanding of science and culture*. Stockholm, Sweden: Nobel Academy Press.

Pea, R. D., Gomez, L. M., Edelson, D. C., Fishman, B. J., Gordin, D. N., & O'Neill, D. K. (1997). Science education as a driver of cyberspace technology development. In K. C. Cohen (Ed.), *Internet links for science education* (pp. 189–220). New York, NY: Plenum. doi:10.1007/978-1-4615-5909-2_12

Penuel, W. R., & Means, B. (2004). Implementation variation and fidelity in an inquiry science program: Analysis of GLOBE data reporting patterns. *Journal of Research in Science Teaching*, *41*(3), 294–315. doi:10.1002/tea.20002

Peters, M. A. (2006). Towards philosophy of technology in education: Mapping the field. In *The international handbook of virtual learning environments* (pp. 95–116). Springer Netherlands. doi:10.1007/978-1-4020-3803-7_3

Pierri, E., Karatrantou, A., & Panagiotakopoulos, C. (2008). Exploring the phenomenon of change of phase of pure substances using the Microcomputer-Based-Laboratory (MBL) system. *Chemistry Education Research and Practice*, *9*(3), 234–239. doi:10.1039/B812412B

Radder, H. (2009). The philosophy of scientific experimentation: a review. *Automated Experimentation, 1*(2), 2-9.

Raes, A., Schellens, T., & De Wever, B. (2014). Web-based collaborative inquiry to bridge gaps in secondary science education. *Journal of the Learning Sciences*, *23*(3), 316–347. doi:10.1080/10508406.2013.836656

Schwab, J. J. (1960). Inquiry, the science teacher, and the educator. *The School Review*, *68*(2), 176–195. doi:10.1086/442536

Schwab, J. J. (1962). The teaching of science as enquiry. In J. J. Schwab & P. Brabdwein (Eds.), *The Teaching of Science*. Cambridge, MA: Harvard University Press.

Simondon, G. (1958). *Du mode d'existence des objets techniques*. Paris: Aubier.

Turner, F. (2006). From counterculture to cyberculture. Stewart Brand, the whole earth network, and the rise of digital utopianism. Chicago: University of Chicago Press.

Van Joolingen, W. R., De Jong, T., & Dimitrakopoulou, A. (2007). Issues in computer supported inquiry learning in science. *Journal of Computer Assisted Learning*, *23*(2), 111–119. doi:10.1111/j.1365-2729.2006.00216.x

Van't Hooft, M., Swan, K., Cook, D., & Lin, Y. (2007). What is ubiquitous computing? In M. Van't Hooft & K. Swan (Eds.), *Ubiquitous computing in education*. Mahwah, NJ: Lawrence Erlbaum Associates.

Vroom, R. W., & Horvath, I. (2014). *Cyber-physical augmentation: An exploration. In* Tools and methods of competitive engineering. *Digital Proceedings of the 10th International Symposium on Tools and Methods of Competitive Engineering TMCE 2014*.

Wolf, F., Hobby, R., Lowry, S., Bauman, A., Franza, B. R., Lin, B., & Kolker, E. (2011). Education and data-intensive science in the beginning of the 21st century. *OMICS: A Journal of Integrative Biology*, *15*(4), 217–219. doi:10.1089/omi.2011.0009 PMID:21476844

Wu, W. H., Jim Wu, Y. C., Chen, C. Y., Kao, H. Y., Lin, C. H., & Huang, S. H. (2012). Review of trends from mobile learning studies: A meta-analysis. *Computers & Education*, *59*(2), 817–827. doi:10.1016/j.compedu.2012.03.016

Yu, J., Dong, K., & Zheng, Y. (2016). VLAB-C: Collaborative Virtual Laboratory in Cloud Computing and Its Applications. In *Big Data Applications and Use Cases* (pp. 145–174). Springer International Publishing. doi:10.1007/978-3-319-30146-4_7

Yuan, E. J. (2013). A culturalist critique of online community in new media studies. *New Media & Society*, *15*(5), 665–679. doi:10.1177/1461444812462847

Zuiker, S. J., & Wright, K. (2015). Learning in and beyond school gardens with cyber-physical systems. *Interactive Learning Environments*, *23*(5), 556–577. doi:10.1080/10494820.2015.1063512

Zydney, J. M., & Warner, Z. (2016). Mobile apps for science learning: Review of research. *Computers & Education*, *94*, 1–17. doi:10.1016/j.compedu.2015.11.001

KEY TERMS AND DEFINITIONS

Cyber-Physical Systems: A specific branch of complex technological systems comprising knowledge and technologies of computing, networking and communication, and those of physical artifacts and engineered systems towards operating and servicing in human and social contexts.

Data Intensive Science: The fourth paradigm of science after the three known paradigms of empirical, theoretical, and computational science. It is a data-driven style of science, where ICT is used to manage, analyze, and share data.

Digital Era: A period in human history characterized by the shift from traditional industry to economy based on ICT.

Digital Society: A society corresponding to the Digital Era in human history. It is the ICT dependent society, where the creation, distribution, use, integration and manipulation of information becomes the main economic, political, and cultural activity.

Experimental Science: A branch of science based on the empirical science research paradigm, which utilizes the gathering knowledge obtained by controlled experimentation, in which natural phenomena are manipulated in order to discover the unknown phenomena or test a known one.

Inquiry-Based Science Teaching: A plurality of methods and environments, which allow students to consider problems and situations scientifically by actually being involved in research process.

Nature of Science: An academic field that deals with methodological, epistemological, and ontological issues of science, including science knowledge, science limits, tools, products of science and human elements of science.

Nature of Technology: An academic field dealing with the essence of technology, its inherent biases, the questions how individuals and society direct, react and changed by technology as well as its implications for education.

Research Triad: A theoretical model of the experimental scientific research, which comprises: (1) a research subject (researcher); (2) an instrument of research (experimental equipment); and (3) an object of research.

Scientific Inquiry: The diverse ways in which scientists study the natural world and propose explanations based on the evidence derived from their work. It defines in terms of two dimensions: (1) scientific ways of knowing and (2) a way of learning science.

Chapter 7
Beyond Angry Birds™: Using Web-Based Tools to Engage Learners and Promote Inquiry in STEM Learning

Isha DeCoito
Western University, Canada

Tasha Richardson
University of Toronto, Canada

ABSTRACT

This chapter explores digital web-based tools for engaging learners and promoting inquiry-based STEM learning. Specifically the authors analyze a selection of technological supports in STEM education, including remote laboratories and simulations, within the context of inquiry based teaching and learning in physics. Teaching physics through inquiry continue to create high levels of anxiety amongst elementary school teachers, which in turn influences their pedagogical choices and limits the possibility of spontaneous events arising from student exploration in the classroom. The authors maintain that teachers will require professional development opportunities to work within the Technological Pedagogical Content Knowledge (TPACK) framework to ensure that they are able to select from a broad spectrum of technological supports. The authors highlight the potential of web-based digital tools to promote inquiry-based STEM learning, and engage both teachers and students, thus potentially improving attitudes toward teaching and learning STEM content through digital technologies.

DOI: 10.4018/978-1-5225-2525-7.ch007

INTRODUCTION

New realities of the 21st century demand individuals with different competencies than those considered appropriate for success in the past. The shift in our society's growing reliance on technology mandates that education emphasize technological/digital literacy including: digital competence (skills, concepts, approaches, attitudes), digital usage (professional/discipline application), and digital transformation (innovation/creativity) (Belshaw, 2011). Improving digital literacy underpins not only a nation's capacity to provide individuals and groups with equity of access to social opportunity; it is a necessity for participation in the digital economy (Warsh, 2007). Research suggests that models of education are inadequate for addressing the challenges and opportunities facing 21st century learners (Barron & Darling-Hammond, 2008; Perkins, 2009). Consequently, education must change. Milton (2015) maintains that surface changes in education will not equip students for the 21st century and that change is needed at the core of educational practice. A shift must occur from the traditional view of educational practice to a transformative view, one in which "learning is a social process, with students and teachers working in partnership with each other and with experts beyond school, supported by digital technologies" (p. 9). Moreover, this shift must aim to incorporate technologies in schooling in a manner that digress from disciplinary experts' determinations of what and how students should learn – a classic perspective which has resulted in challenges for educators as they continue to search for strategies to effectively address the development of skills reminiscent of the preferred learning styles of today's students. Thus, in technology enriched environments teachers and textbooks are not the keepers of knowledge, as students are provided with new possibilities for pursuing inquiry and engaging in knowledge construction.

Overall, the research evidence over the last forty years on the impact of digital technologies (e.g., Integrated Learning Systems (Parr & Fung, 2000), interactive whiteboards (Higgins, Beauchamp, & Miller, 2007), mobile and handheld technologies (Cheung & Hew, 2006) and virtual learning environments (Passey & Higgins, 2011)) on learning consistently identifies positive benefits (Higgins, Xiao, & Katsipataki, 2012). Despite the educational value of web-based technologies, very few educators use them in any substantive way in teaching and learning. Technological tools, when used judiciously, have the potential to impact teaching and learning in STEM disciplines (DeCoito & Richardson, 2016). One specific STEM discipline, physics, continues to garner the least positive attitudes through educational experiences thus resulting in the highest levels of anxiety amongst elementary teachers. This in turn influences their pedagogical choices and limits the possibility of spontaneous events arising from student exploration in the classroom, as they feel ill equipped to troubleshoot (Czerniak & Haney, 1998). Thus, teachers will require professional development opportunities that focus on digital literacies. Specifically, how to navigate the Technological Pedagogical Content Knowledge (TPACK) framework (Koehler & Mishra, 2009), since development of teachers' TPACK is critical to effective teaching with technology. It is not enough to simply know that remote labs or simulations exist, but rather it is necessary for teachers to understand how to incorporate these technologies into their classrooms to provide the best learning opportunities for their students.

In this chapter the authors explore the potential of digital web-based tools for engaging learners and promoting inquiry-based STEM learning. Specifically, the authors aim to analyze a selection of technological supports in STEM education, including simulations and remote laboratories, within the context of supporting teaching and learning in physics. The following questions guide the analysis:

1. How might these web-based tools promote inquiry-based STEM learning, in terms of helping students develop domain knowledge and inquiry skills?
2. How should teachers and/or learners be supported in developing their use of digital technology to ensure it improves learning?
3. What kinds of modeling and scaffolding should educators provide to help learners engage in this process?
4. What promises and challenges do these technologies hold in terms of transforming education practice?

In summary, this chapter has the potential to make significant contributions to our understanding of how web-based digital tools i) promote inquiry-based STEM learning, and ii) engage both teachers and students, thus potentially improving attitudes toward teaching and learning STEM content through digital technologies in the classroom.

BACKGROUND

Promoting Inquiry Based Learning and Enhancing 21st Century Learning Skills

The preparation of young people for lifelong learning in a 21st century knowledge-based information society has become an increasingly important objective of educational systems worldwide (Dagienė, 2011). A primary challenge for education is the transformation of students' learning processes in and out of school and to engage student interest in gaining 21st century skills. Lemke (2004) reported a link between 21st century skills and academic achievement, making the case for incorporating teaching activities that help students develop 21st century skills, including: *Learning and Innovation Skills*; *Life and Career Skills*; *Information, Media, and Technology Skills*; and *Knowledge and Skills in Core Subjects* (National Research Council, 2011; Partnership for 21st Century Skills, 2009), as illustrated in Figure 1.

Despite the fact that the aforementioned skills have been around for a very long time, currently the challenge is, according to Bolstad, Gilbert, McDowall, Bull, Boyd, and Hipkins (2012), "how to achieve a shift that creates a more coherent educational ecology that can support what is known about good learning *and* that can accommodate new knowledge about learning and, importantly, new *purposes* for learning in a changing world". The acquisition of 21st century skills in teaching and learning will require a shift in what we teach, how we teach it, the tools we use and how we educate, train, nurture, and retain our teachers. It is undeniable that we cannot change how our students learn unless our teachers are equipped to teach in new ways.

Figure 1. Framework for 21st century learning
(Partnership for 21st Century Skills, 2009)

Content Knowledge and 21st Century Themes

- English, Reading or Language Arts; World Languages; Arts; Mathematics, Economics, Science; Geography; History; Government and Civics

Learning and Innovation Skills

- Creativity and Innovation; Critical Thinking and Problem Solving; Communication and Collaboration

Information, Media and Technology Skills

- Information Literacy; Media Literacy; ICT Literacy

Life and Career Skills

- Flexibility and Adaptability; Initiative and Self Direction; Social and Cross-Cultural Skills; Productivity and Accountability; Leadership and Responsibility

Inquiry in the Classroom

When taught well, inquiry is a powerful tool in the classroom and is considered a careful balance between freedom and structure as teachers provide a space for students to make decisions and to have ownership of the process. Its success hinges on planning in advance the learning goals, focusing on what is most important, and providing scaffolding for the learning experience (Forbes, 2009). According to the National Research Council (2000), regardless of what type of activity teachers use in their classrooms, for the activity to be "inquiry" it must i) focus on a question that can be investigated; ii) gather evidence during the investigation; iii) explain the evidence; iv) connect the explanation to the larger community; and v) communicate conclusions that are both effective and justifiable (Settlage, 2007, p. 465). Researchers (Banchi & Bell, 2008; Martin-Hansen, 2002) have described student scientific inquiry on a continuum, demarcated by the extent of the students' involvement in the investigation design and implementation and are typically delineated as four forms of inquiry: *confirmation inquiry* includes teacher questions, prescribed procedures, and teacher-presented results; while *structured inquiry* focuses on teacher questions, prescribed procedures, and student results. *Guided inquiry* involves teacher questions, student-designed procedures, and student results; and finally, *open inquiry* incorporates student questions, student-designed procedures, and student results.

Even though scientific inquiry is regarded as an important learning goal by both science teachers and the science education community, the reality in schools is that teachers lack a practical pedagogical framework to inform their inquiry-based instruction (Bell, Smetana, & Binns, 2005). This makes teaching inquiry-based science challenging and necessitates science teachers become immersed in ample opportunities to develop their pedagogical design capacity (Ball & Cohen, 1999; Barron & Darling-Hammond, 2008; Forbes, 2009). Czerniak (1990) and Swackhamer, Koellner, Basile and Kimbrough (2009) found that teachers who are efficacious are more likely to incorporate student-centred and inquiry-based activities into their classrooms. With inquiry based science teaching, teachers create learning environments that allow students to uncover scientific phenomena and engage in constructing their own understandings of science concepts. With careful planning and structure, students can be more independent in investigating the world around them while engaging in powerful inquiry processes. With the rise of digital technologies there is an increase in the amount and diversity of resources available, which poses additional challenges for teachers to navigate as they develop curriculum around inquiry practices. In this chapter the authors explore a number of simulations and remote laboratories, as examples of technological advancements which provide opportunities that allow students to enhance 21st century learning skills as they work collaboratively with their peers to create, share, refine, and exchange knowledge and ideas. For this reason, we should consider what kinds of support teachers need so that they feel confident to select from a broad spectrum of options available for students.

Teacher Professional Development

Simply introducing technological tools and infrastructure into schools will not trigger beneficial and meaningful educational change. Moreover, technology cannot be effective in the classroom without teachers who are knowledgeable about both the technology itself and its implementation to meet educational goals, that is, teachers who are technologically literate. There are several issues that may arise when implementing digital literacies in the classroom. Among these are choices about which one(s) to use, how effective the choices are in reinforcing learning, and the role of these digital literacies in shifting from an instruction paradigm, which is teacher focused, to a learning paradigm, which is student focused (Cox, 2008). Research studies (Sasseville, 2004; Zhang & Martinovic, 2008) revealed that teachers' levels of self-efficacy (Bandura, 1997; Skaalvik & Skaalvik, 2008) and attitudes toward digital technologies determined whether or not, and how, they used information and communications technology (ICT) in their classrooms.

Teacher professional development is absolutely essential if technology is to be used effectively. Traditional one-time teacher training workshops have not been effective in helping teachers to feel comfortable using technology or to integrate it successfully into their teaching. Professional development should entail initial preparation/training (pre-service); workshops, seminars, and short courses (in-service); and ongoing pedagogical and technical support for teachers as they address their daily challenges and responsibilities. Training and on-going inquiry-based approaches seem the most successful professional development targets for teachers, thus implying that support should go beyond using technology and focus on effective pedagogical use of technology to support teaching and learning goals (Higgins et al., 2012). Therefore, it can be said that, while technology use in the classroom is increasing, improving teachers' digital competencies through the application of technology should remain the goal. The impact of technology and the changing face of curriculum, as well as the accompanying changes in the roles of teachers can no longer be ignored; roles must be reconceived in order to engage learners in many decisions about their learning (Bennett, 2002; Bolstad et al., 2012). In the report, *Teachers' Use of Educational Technology in U.S. Public Schools: 2009,* Gray, Thomas, and Lewis (2010) highlighted the fact that 17% of all public schools teachers reported that their students often developed, ran demonstrations, models or simulations, with 25% doing so in mathematics/computer science, and science (Table 1). Livingstone's study (2012) highlighted the fact that teachers' integration of

Table 1. Percent of teachers reporting frequency of student use of educational technology

Characteristic	Solve problems, analyse data, perform calculations		Conduct experiments or perform measurements		Develop and present multimedia presentations		Create art, music, movies, or webcasts		Develop or run demonstrations, models, or simulations		Design and produce a product	
	R	S/O	R	S/O	R	S/O	R	S/O	R	S/O	R	S/O
All public school teachers	23	45	25	25	25	42	24	25	23	17	17	13
Secondary level	25	46	27	30	24	53	24	32	26	25	20	17
Math/ computer science, science	20	61	27	45	29	40	23	17	27	25	17	11

*S/O – sometimes or often; R – rarely Source: National Center for Education Statistics, U.S. Department of Education. (2010)

digital supports into their classrooms differed; with 30% in science at the elementary level, and 10% in science at the secondary level. Unfortunately, virtual field trips and simulations are some of the least used technologies for teachers (Uluyol & Sahin, 2016), with the majority of technologies being used to support teacher presentation in one form or another (Livingstone, 2012; Ruggiero & Mong, 2015).

Achieving changes associated with the integration of technology in the overall learning environment will require professional development around teachers' technological and pedagogical content knowledge (TPACK, see Mishra & Kohler, 2006), which describes teachers' knowledge and skills about technology (newer and digital) integration, as illustrated in Figure 2. Shulman's (1986, 1987) descriptions of pedagogical content knowledge (PCK) informs the TPACK framework. PCK is a type of knowledge that is unique to teachers, and is based on the manner in which teachers relate what they know about teaching (PK) to what they know about what they teach (CK). Thus, TPACK incorporates how teachers' understanding of educational technologies and PCK interact with one another to produce effective teaching with technology. At the core of TPACK is a transactional relationship between content, pedagogy, and technology knowledge (Kohler, Mishra & Yahya, 2007). Therefore, TPACK is an intersection of all forms of teachers' knowledge – technological knowledge (TK), pedagogical knowledge (PK), and content knowledge (CK) applied while teaching (Hofer & Grandgenett, 2012).

In order for teachers to shift towards using a TPACK framework, professional development will be required for a number of reasons. First, for teachers of core subjects who are responsible for teaching students towards state examinations,

Figure 2. TPACK framework
(adapted from Mishra & Kohler, 2006)

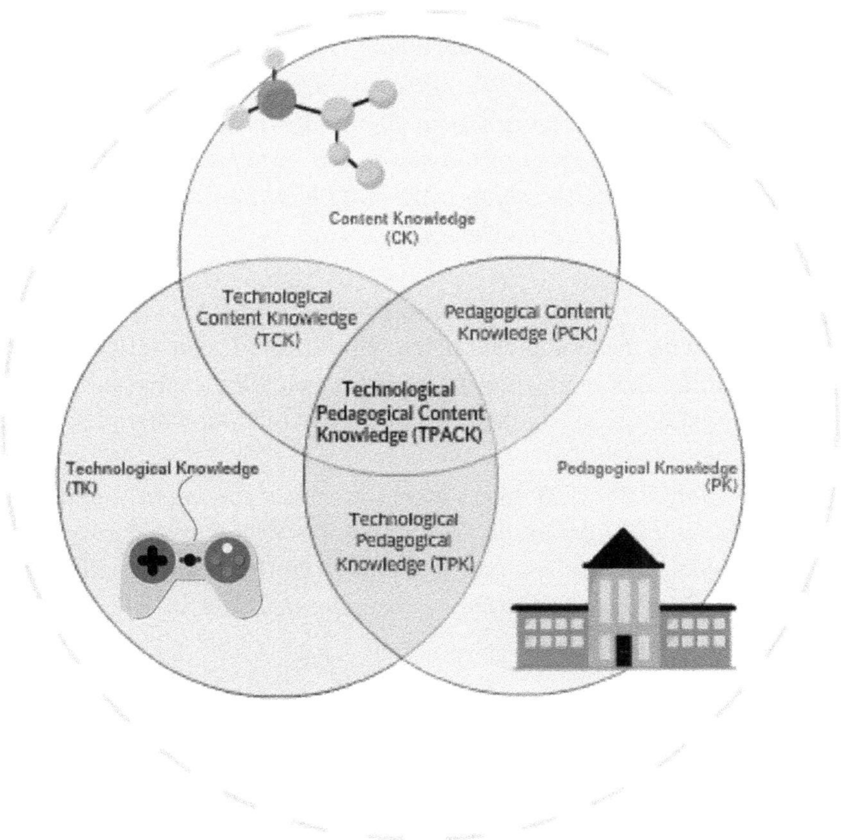

traditional methods are shown to be effective in that preparation, so there is little impetus to shift their pedagogical decisions, especially when faced with challenges to incorporating technologies into the classrooms, such as file size, connectivity to the Internet, or the time teachers are required to devote to learning how to use the technology (Livingstone, 2012; Muehrer, Jenson, Friedberg & Husain, 2012; Nielsen, Miller & Hoban, 2015; Seiter, 2008), to name a few. Thus, professional development needs to be geared towards helping teachers understand the benefits of using the available technologies in order to facilitate a shift away from more traditional, teacher-centred pedagogies. For example, a longitudinal study (Richardson & DeCoito, 2017) exploring the impact of a STEM outreach program designed for middle school students revealed factors, such as access to technology, professional support, and teacher readiness, that work to constrain the use of a TPACK framework in the classroom, as exemplified in the following teacher quote:

I would need to take workshops or something. That's the problem. I don't know. Some of the meters and stuff that [the outreach program] use, I'm not sure how to use them... Yeah, I mean computer, sure, fine, scales, calculators, spring scales. These kinds of things, but some of the newer, like electronic type tools we don't have currently. We don't have robotics. We don't have that kind of stuff currently. So that's definitely an area for improvement.

Furthermore, teachers in the STEM study (Richardson & DeCoito, 2017) reported being inundated with technological choice with little time for reflection, and the lack of opportunities that model how to integrate technologies that support inquiry and 21st century skills, as illustrated by a teacher in the following quote:

The technology part is tricky, because there is so much coming at you all the time. It's hard to keep up. I just got Apple TV. I've got this Whiteboard. I have iPads and laptops and people give you "oh, here's 37 things that you can do with them" and it's like, give me five things ... two maybe, because that's on top of everything else you're doing.

Early exposure to STEM does not guarantee that students will pursue careers in STEM related fields. The Natural Sciences and Engineering Research Council of Canada (NSERC, 2010) reported that of those students who graduated from high school in 1997, only 9.2% went on to earn a bachelor's degree in Natural Science and Engineering in 2001. What a TPACK framework allows teachers to provide in their classroom is effective preparation for both students entering STEM careers as well as those who choose other career paths, as it enriches students' digital literacies. Therefore, if there is a way for teachers to effectively combine teaching strategies that support students to master concepts that will benefit them (or even encourage them) in their understanding of STEM, then this is an area that warrants further exploration.

Digital Literacies

Technology, including anything that aids student learning, can shift the learning experience from teacher-centered to student-centered and increase inquiry activities when it is placed into the hands of students. There is increasing access and reliance on technologies such as the Internet, and as such, educators need to provide opportunities for students to develop their digital literacies. Cornell University (2009) describes digital literacy as "the ability to find, evaluate, utilize, share, and create content using information technologies and the Internet." In today's classrooms, opportunities to enhance students' digital literacies take multiple forms: engaging in online discussions within virtual classrooms, Web 2.0 technologies (emphasis on user generation), research on class topics, construction of digital timelines, websites, apps, and presentations, use of simulations and remote laboratories, and playing digital video games (DVGs). Despite these inclusions supporting learning, not all develop students' digital literacies in the same manner or to the same extent.

Consider the more recent inclusion of popular DVGs, such as Angry Birds™ (McConnell & Dickerson, 2013), Super Mario Bros.™ (Nordine, 2011), or Spore™ (Wright, 2007) in the classroom. While DVGs can be used effectively to increase student understanding of content and context (Carr & Bossomaier, 2011; Chen, Wang, & Lin, 2015; DeCoito & Richardson, 2016; Rondon, Sassi, & de Andrade, 2013), they are often used as a way to increase student motivation and improve student attitudes toward subject matter by providing an engaging and familiar virtual context for a teacher to plug into their lessons on science and its theory (Annetta, Frazier, Folta, Holmes, Lamb, & Cheng, 2013; Cai et al., 2006; Dominguez et al., 2013). This is an incredibly powerful and effective use of DVGs by teachers, however the task ultimately lies with the teacher to provide or guide activities to make scientific theories less abstract and to refine and develop student understandings (Sherin, 2006). In this chapter, the authors focus on less frequently used technologies that can potentially assist teachers in bridging the conceptual gaps that may otherwise remain: simulations and remote laboratories.

SIMULATIONS AND REMOTE LABORATORIES IN PHYSICS EDUCATION

Teaching and Learning Physics

The Natural Sciences and Engineering Research Council's (2010) *Women in Science and Engineering in Canada* report revealed, with the exception of Quebec, that students in grade 12 are least likely to study physics. This is problematic because physics is often the gate-keeper to further studies in STEM related fields in post-secondary education (Trumper, 2006). Physics may, in part, be viewed as such because it is shown to increase critical thinking skills (Quitadamo, Kurtz, Cornell, Griffith, Hancock, & Egbert, 2011), and is often considered a precursor for developing deeper understandings of content knowledge in other disciplines. For example, physical laws govern how chemicals behave, and the way chemicals behave determines biological processes and pathways (Bybee et al., 2006). For students who opt out of further studies in physics, they are not just opting out of a course of study, but rather they are also opting out of a host of potential careers associated with the natural sciences and engineering.

There has been much research devoted to exploring why physics is the least popular science. Among the more common themes are physics' dependence on mathematics (Hansson, Hansson, Juter, & Redfors, 2015; Korpershoek, Kuyper, & van der Werf, 2015), its abstract nature, the tensions between theoretical and naive understandings (diSessa, 1993; Taber, 2008), and its alignment with males and masculinity (Carlone, 2004; Sadler, Sonnert, Hazari & Tai, 2012). In this chapter the authors have selected two themes in physics to explore in depth: (1) the level of abstraction, and (2) the reliance on mathematics.

Abstract Concepts

When considering any given day in our life, we interact with the physical world in a multitude of ways, and it is through these interactions that we construct our everyday understanding of physics (Taber, 2008). These understandings allow us to make predictions about future interactions, and in general guide us through the physical world. Part of the challenge for educators is that these constructions in physics are often erroneous, and helping students to sort through their constructions requires focused attention on students' ideas (Andrews, Leonard, Colgrove, & Kalinowski, 2011; diSessa, 2006). Often, teachers make pedagogical choices that are more teacher centred (Markic, Eilks, & Valanides, 2008), and if these methods are devoid of activities, students are left with a feeling that physics is a difficult, abstract subject (Johnston & Ahtee, 2006; Kapucu, 2014; Redish, Saul, & Steinberg, 1998). Additionally, teachers tend to use problem solving to focus on mathematical manipulations involved in physics rather than contextual understanding (Duit, Niedderer, & Schecker, 2007)

The challenge for teachers is to shift to teaching methods that provide multiple opportunities for students to focus on difficult concepts in a way that maximizes opportunities to challenge students' constructed physics explanations from their daily experiences, or phenomenological primitives (p-prims) (diSessa, 1993), as well as engage and motivate them. Despite some teachers using problem solving to help make concepts less abstract (Hansson et al., 2015; Pask, 2003; Redish & Kuo, 2015), this approach assumes a certain level of mathematical competency on the student's part.

Reliance on Mathematics

Physicists use mathematics to explain and predict phenomena, and as such, it is an integral part of understanding physics (Duit et al., 2007; Hansson et al., 2015). For some students, using mathematics to work through their misconceptions is helpful (Sherin, 2006), and in many classrooms, teachers of physics emphasize the mathematical manipulations as they teach (Bøe & Henriksen, 2013). However, for those students who are academically weak in mathematics, unless they are able to experience physics in an alternate way, they may find learning physics confusing and frustrating. This is especially true when physics is taught devoid of context, and when the focus is on equations associated with physical phenomena (Wegener, Mcintyre, McGrath, Savage, & Williamson, 2012).

Thus, teachers are challenged to make pedagogical choices that provide multiple opportunities for students to focus on difficult concepts in a way that minimizes the challenges, maximizes opportunities to challenge p-prims, extend and support mathematical understanding, as well as engage and motivate students in their formal education. Educators also recognize their role in nurturing students' science interest and in helping alleviate their science anxiety. As teachers face very real restrictions with respect to equipment and time, when using traditional approaches it becomes increasingly difficult to meet the needs of diverse learners in the classroom. The authors contend that incorporating technological supports into lessons would provide resources for teachers to integrate inquiry activities, support student learning in science, and could potentially alleviate teachers' anxiety and improve their self-efficacy.

Technology Integration

Some examples of technological supports include remote laboratories and simulations, amongst others. Remote laboratories, such as FAR Labs™, differ from simulations in that they allow students to conduct a laboratory exercise with real equipment, except the equipment is accessed through the Internet and activated remotely. Simulations differ from remote laboratories in that students are able to access them in real-time (other than connectivity issues) without waiting their turn. Simulations, such as the Physics Education Technology (PhET™) Interactive produced by the University of Boulder, Colorado, allow students to work with a single concept to help make that concept less abstract and more concrete. This can either be a single simulation that essentially animates a concept for the student or can be interactive, thus allowing the student to change parameters and witness the effects of those manipulations. In both instances of remote laboratories and simulations, the benefits to the classroom teacher include promoting inquiry and more experiential learning for students to solidify their understanding of STEM concepts. Learners are able to engage with a single concept without concern for experimental time, usually devoted to gathering, setting up and dismantling equipment; access to otherwise costly, fragile, or inaccessible equipment; and the removal of safety hazards. Additionally, in the case of simulations, there is access to learning experiences that are otherwise not available in the classroom. The authors discuss in greater detail the differences between remote laboratories and simulations, with concrete examples, and strategies for incorporating them effectively in the classroom to promote inquiry-based STEM learning.

Simulations

What Is a Simulation?

The National Research Council (2011) describe a simulation as a non-contextual presentation of information which minimizes the number of distractions from the primary learning goal, thus allowing the student to focus on a single phenomenon in order to understand what factors impact its behaviour. Simulations are effective ways to demonstrate what cannot be seen (Perkins et al., 2006), and are effective tools for challenging preconceived explanations and facilitating a shift in students' epistemological beliefs (Psycharis, 2013). Even though there are many very good videos available through the Internet, they are not the same as simulations. Simulations support a student driven question and answer approach. Students can ask "how does this factor influence …?" or "if I change this, what will happen?" or "how can I get the phenomenon to behave the way I want it to?" and seek the answers through their exploration of the tool.

Examples of Simulations in Physics Education

There are an increasing number of digital simulations available through the Internet that can be used in the classroom. These are generated not only by institutions and private companies, but also by educational practitioners who are making use of powerful software available, such as Physion (http://physion.net/), Comsol (https://www.comsol.com/comsol-multiphysics) or Algodoo (http://www.algodoo.com/). Simulations range in scope (specific to general), flexibility (number of parameters the user is able to manipulate), accessibility (fees attached), and collaboration opportunities (teacher resource sharing). Table 2 illustrates a sample of physics simulations commonly used in the classroom and available through the Internet.

An example of a simulation on the Photoelectric Effect is represented in Figure 3. In this simulation students are able to manipulate the wavelength and intensity of light, target metal, and the retarding potential (shown as current). The simulation can generate graphs for the user, thus the relationship can be seen immediately.

Table 2. Simulations used in physics education

Simulation	Platform	Cost	Grades*	Curriculum	Prepared Materials	Community Sharing
Gizmos (Explore Learning)	Mac PC	Y	ES	General science.	Y	Y
Interactive Physics (Design Simulation Technologies)	Windows	Y	SP	A wide range of topics is available. This program is an offshoot from an initiative to assist engineering students.	Y	Y
PhET™ Interactive Simulations (University of Colorado)	Windows Mac Linux	N	ESP	A wide range of topics is available in both classical and modern physics.	Y	Y
Science of Everyday (3M)		N	ES	Application focus.	Y	N

*elementary = E; secondary = S; post-secondary = P

Figure 3. PhET simulation of Photoelectric Effect

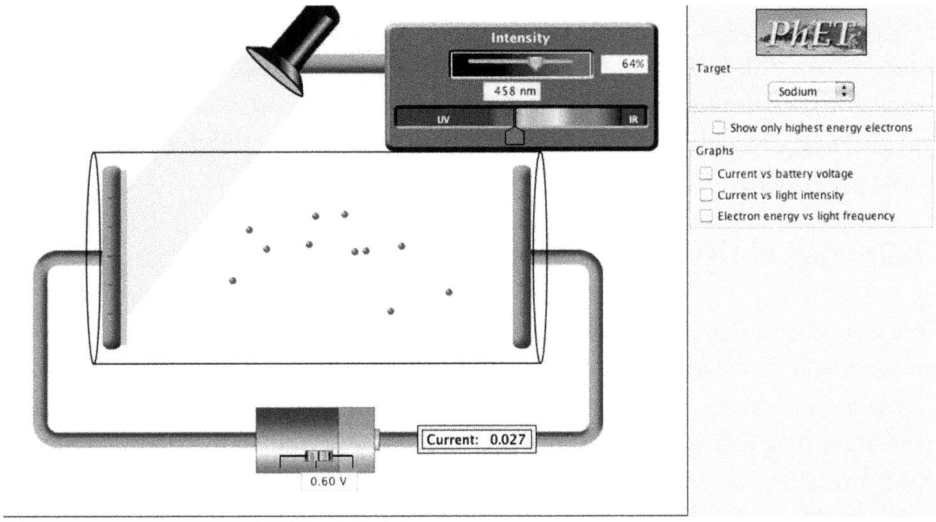

Included, and what is especially significant of simulations (Perkins et al., 2006) is that the photoelectrons are shown as they leave the target and make their way to the anode. This detail cannot be directly observed; hence the simulation emphasizes the theoretical aspects of the phenomenon in question, thus rendering the abstract more concrete.

Advantages of Using Simulations

The literature identifies several advantages of using simulations to support science learning, including improved student belief in their understanding of physics (Perkins et al., 2006; Psycharis, 2013), homework support, opportunities for teachers to track students' misconceptions (Lee, 2011), equivalent or better understanding of physics concepts compared to students who participated in hands on activities (Steinberg, 2000), and opportunities to explore laboratory conditions without the limitations and restrictions that exist in a real environment (Finkelstein et al., 2005). To this last point, Marshall and Young (2006) recommend that simulations be used primarily as a tool to explore concepts.

In the case of physics, part of both its elegance and its challenge for students is the well-established interconnection of physics and mathematics (Karam, 2015; Redish & Kuo, 2015). Some simulations provide the best of both worlds for students in that they remove the challenges associated with doing multiple calculations, and do them rapidly so that students can see how values change during a simulation (as is the case in PhET™ simulations when vector values can be activated during the simulation), where others, such as Modellus, are able to extend students' mathematical understanding by delving into more advanced concepts needed for later studies in physics (Psycharis, 2013).

Challenges of Using Simulation

Research shows that while there are many benefits attached to simulations, they must be used with care as students tend to shift their focus and attention. When the simulation is non-intuitive (such as its input parameters and constraints), students spend more time focused on interpretation of the interface rather than the phenomenon under investigation (Marshall & Young, 2006). Additionally, students shift the responsibility of finding an answer to the computer as they wait passively to be "told" by the computer why phenomena react the way they do (Steinberg, 2000). This can lead to a shift in focus on the student's part in terms of learning how to interpret computer feedback rather than designing and implementing a scientific investigation (Marshall & Young, 2006).

Remote Laboratories

While technologies can bring many facets to the student experience, hands on laboratories are still generally considered the best option. The primary goals of laboratory activities include (Singer, Hilton, & Schweingruber, 2005, p. 76-77):

- Enhancing mastery of subject matter;
- Developing scientific reasoning;
- Understanding the complexity and ambiguity of empirical work;
- Developing practical skills;
- Understanding the nature of science (NOS);
- Cultivating interest in science and interest in learning science;
- Developing teamwork abilities.

The above goals are also applicable to laboratory activities that are remotely accessed via the computer. Maziewski, Dobrogowski, and Zablotskii (2007) argue for the inclusion of remote laboratories and maintain that they should be part of every teacher education program. In most cases, teachers employ laboratory activities that lie somewhere in the middle of the inquiry spectrum. Of course, not all inquiry activities will support the development of students' digital literacies or 21st century learning skills in the same way. For this reason, teachers should consider purposefully which activities they use in their classrooms.

What is a Remote Laboratory?

Remote laboratories (also known as remote control laboratories or remote experiments) refer to laboratory equipment that can be accessed through the Internet to perform a specific experiment about a specific phenomenon. Remote laboratories allow students to alter some predetermined aspect of the experimental set-up and a camera captures and broadcasts the results through the Internet to the observing student. The student can then interact with the laboratory equipment in real time (unless the equipment is already in use). First established to help accommodate students in distance education programs, universities, such as the Spanish University of Distance Education, are developing more experiments that can be done from home (De la Torre, Sanchez, Dormido, Sanchez, Yuste, & Carreras, 2011), with more remote laboratories being developed around the world, and with increasing numbers of students accessing them worldwide (Maziewski et al., 2007).

Table 3. Remote laboratories freely accessible to the public

Remote Laboratory	Fee	Curriculum	Data gathering	Prepared documents
Freely Accessible Remote Laboratories (FARLabs) (Australian Government)	Free	Nuclear, Environment, Structure	Graphical data	Teacher documents available with registration
Remotely Controlled Laboratory (RCL) (Universität der Bundeswehr München)	Free	Modern Physics (e.g., electron diffraction, Millikan's experiment, Photoelectric effect)	Students make measurements from display; screen captures of data	Available discussion questions and background information

A Few Examples of Remote Laboratories in Physics

There are a growing number of remote laboratories accessible through the Internet. While some are only available for those registered in distance education programs, (for example, FisL@bs from the Spanish University of Distance Education, NetLab from University of South Australia, or Practical Experimentation by Accessible Learning (PEARL) from The Open University, UK), there are a few others that are available to the general public (Table 3), such as the Freely Accessible Remote Laboratories (FARLabs) and the Remotely Controlled Laboratory (RCL) (Schauer, Ozvoldova, & Lustig, 2009).

At present, the RCL showcase 12 experiments available to the general public that connect to the Ontario Physics curriculum (e.g., the Wave Nature of Light and Modern Physics). One example, the Photoelectric Effect lab in Figure 4, provides background theory, information about the laboratory set-up, discussion questions to guide teachers and students, and access to equipment that allow students to observe the effects of changing both light intensity and wavelength, and the resulting energy of the photoelectrons emitted from the target. The remote laboratory itself is very simple; the student interacts with the laboratory equipment in real time by selecting a filter (to limit intensity of light) and the wavelength of light (to irradiate the metal surface), and the resulting electric potential difference can be read from the display (shown on the left).

Some Advantages of Using Remote Laboratories

Remote laboratories provide more access for students to engage in experiments (Hatherly, Jordan, & Cayless, 2009) that may otherwise be missed due to safety concerns (such as radioactive labs that can be accessed through FARLabs), financial concerns connected to sensitive equipment (such as the Millikan experiment accessed

Figure 4. Remotely control labs - photoelectric effect
(Universität der Bundeswehr München)

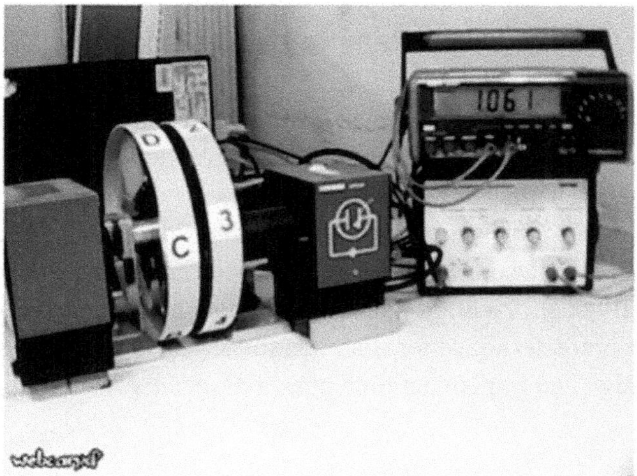

Voltmeter is switched on automatically.
Voltage is displayed in unit Volt.

Table of filters

grey filters		spectral filters	
A	T = 100 %	1	366 nm
B	T = 75 %	2	405 nm
C	T = 50 %	3	436 nm
D	T = 25 %	4	546 nm
E	T = 0 %	5	578 nm

through RCL) (Grober, Vetter, Eckert, & Jodl, 2007; Schauer et al., 2009), or other reasons. Remote laboratories provide the opportunity for multiple trials, and each trial returns the results specified by the student from a real context (De la Torre et al., 2011). Hatherly et al. (2009) also found that the experience students gain through engaging with remote laboratories can translate to greater confidence in the lab, resulting in less guidance being required during hands-on laboratories.

In addition to the above noted benefits, there are some interesting outcomes associated with using remote laboratories, including focusing on i) the concept of learning as an activity, ii) the context of the learning (Hmelo-Silver, 2004), and iii) the social aspect of learning that takes place in a community of practitioners (Lave & Wenger, 1991). Students develop a sense of belonging to a larger global scientific community by accessing equipment at anytime from anywhere in the world a remote laboratory is available (Schauer et al., 2009). Their engagement with remote laboratories makes them contributors to understanding our global scientific community as each trial contributes to an ever growing database generated by users accessing each remote laboratory (Grober et al., 2007).

Some Challenges of Using Remote Laboratories

Even though the equipment is automated, it is recommended that students observe what happens during the lab rather than leave instructions in queue (Grober et al., 2007). This creates challenges for classroom teachers when a class wants to conduct trials with equipment and have to wait to access the single piece of equipment. It is also recommended that whenever possible, a hands-on laboratory should be conducted in class (Hatherly et al., 2009). Even though remote laboratories provide access to equipment, it is pre-determined for students, and as a result they are unable to work through the design and implementation phases associated with inquiry activities.

RECOMMENDATIONS

Technologies provide some opportunities for students to become active in lessons as teachers incorporate interactive demonstrations where students can make predictions and ask higher order questions with the results available instantaneously (Perkins et al., 2006). Additionally, students are able (within limits) to generate their own questions, determine how that data will be interpreted, as well as share and exchange ideas when they engage with remote laboratories (Schauer et al., 2009). For some teachers, the above noted benefits will require a pedagogical shift, specifically one that incorporates more inquiry-based opportunities for students. This pedagogical shift will require access to professional development focusing on ICT use, as identified by Uluyol and Sahin (2016) as one of the leading reasons teachers do not integrate technologies into the classroom. Specifically, the integration of technology in the overall learning environment will require professional development around teachers' TPACK. Furthermore, researchers (Badia, Meneses, & Sigales, 2013; Buckenmeyer, 2010; Hechter & Vermette, 2014; Qablan, Abuloum, & Al-Ruz, 2009; Ward & Parr, 2010) have reiterated the call for ongoing teacher professional development, specific in form and focus, while housed under the umbrella of a TPACK framework. The focus should be on contextualizing technology integration (Ruggiero & Mong, 2015); utilizing technology, introducing technology to students, and considering educational purposes (Zimlich, 2015); providing opportunities to learn from peers how to integrate technology (Lin & Lee, 2015); and building confidence in and highlighting benefits of using ICT rather than focusing on skills (Ward & Parr, 2010).

FUTURE RESEARCH DIRECTIONS

The survey of current literature in the face of available technologies for physics, and the broader STEM education, raise questions and point to opportunities for future research. The consistent findings of technologies focusing on presentation rather than inquiry presents a major concern around the integration of technology in the classroom.

Future studies should focus on the potential of a well-supported TPACK framework to assist teachers in their technological choices, with the aim to engage and increase student confidence through inquiry activities, while at the same time bolstering 21st century skills of collaboration by participating in a global community of emerging STEM-ists. To this end, professional supports are necessary to assist teachers as they re-envision the curriculum in terms of technological integration, with increased use of simulations and remote laboratories.

CONCLUSION

The affordances offered by ICT benefit science learning. Potential benefits of integrating digital literacy in science education have been documented, including increasing student motivation, expanding the pedagogical resources available to science teachers (Al-Alwani, 2005), promoting cognitive development, and highlighting relevance via relating science to students' real-life experiences (Webb, 2005). Digital literacy is important for science learning as it i) assists students to learn more effectively with the range of ICT-enabled affordances that have the capacity to motivate and enable better understanding of science concepts, and ii) lessens the working memory's cognitive load, while learning science that is ICT-based (Ng, 2011). DeCoito's (2014a) study of pre-service teachers' digital competencies cited a number of potential challenges to implementing digital literacies in the classroom, including the need for teacher training in order to effectively utilize digital literacies and achieve success in terms of teaching and learning science. Similar findings have been reported in studies focusing on digital scientific timelines to teach about the nature of science (DeCoito, 2014b), and digital games to enhance scientific and technological literacy while teaching STEM concepts (DeCoito & Richardson, 2016). The aforementioned findings are corroborated by Dipetta and Woloshyn (2009), and Brush and Bannon (1998), who maintain that without appropriate training, teacher educators and teachers will continue to use technology as a supplement to instruction rather than a tool to facilitate instruction. Additional challenges include the fact that the time taken for technological training infringes on time for learning content, and concern that an emphasis on technology may distract students and teachers from teaching and learning (Dipetta & Woloshyn, 2009). These challenges are warranted, as despite the affordances offered by ICT, the development of digital literacy will require a focus on technology for initial periods of time.

There is no doubt that simulations and remote laboratories support teaching and learning physics concepts. In both cases, simulations and remote laboratories provide opportunities for students to engage with otherwise abstract concepts and manipulate the factors that influence phenomena in physics. Both technologies also support student understanding of the application of mathematics in learning physics as students are able to engage with the phenomenon without becoming mired in the mathematical calculations. To promote the development of inquiry skills, it is essential that these digital tools not be used in isolation. Jewett (2015) has found in practice that the hands-on exploration is also an essential piece necessary to motivate students and spark questions about the underlying workings of physical phenomena.

Despite the affordances offered by digital literacies, the observations of Songer, Lee, and Kam (2002) from over a decade ago still persist as reasons why teachers do not use digital resources available through the Internet. Some reasons include: low levels of science content or computer knowledge; large class sizes; student access to technology at home; and unreliable technology and Internet connectivity. In addition to the aforementioned reasons, a lack of professional development initiatives focusing on effective integration of digital literacies in science education remains the greatest obstacle to technological integration in practice. There is a dire need for teacher professional development (in-service and pre-service) that focus on the nature of the pedagogical principles associated with particular digital literacies. Furthermore, in order to adopt a TPACK framework, teacher induction programs will need to be more innovative in terms of providing various opportunities for pre-service teachers to become immersed in digital literacies (DeCoito, in press) thus developing digital competencies, as well as self-efficacy. The provision and use of digital literacies have the potential to assist pre-service teachers in helping students develop 21st century skills, enhance teachers' and students' scientific and technological literacy, and improve attitudes toward teaching and learning science.

REFERENCES

Al-Alwani, A. (2005). *Barriers to Integrating Information Technology in Saudi Arabia Science Education* (Doctoral dissertation). The University of Kansas.

Andrews, T. M., Leonard, M. J., Colgrove, C. A., & Kalinowski, S. T. (2011). Active learning not associated with student learning in a random sample of college biology courses. *CBE Life Sciences Education, 10*(4), 394–405. doi:10.1187/cbe.11-07-0061 PMID:22135373

Annetta, L. A., Frazier, W. M., Folta, E., Holmes, S. Y., Lamb, R., & Cheng, M. T. (2013). Science teacher efficacy and extrinsic factors toward professional development using video games in a design-based research model: The next generation of STEM learning. *Journal of Science Education and Technology, 22*(1), 47–61. doi:10.1007/s10956-012-9375-y

Badia, A., Meneses, J., & Sigales, C. (2013). Teachers Perceptions of Factors Affecting the Educational Use of ICT in Technology-Rich Classrooms. *Electronic Journal of Research in Educational Psychology, 11*(3), 787–808. doi:10.14204/ejrep.31.13053

Ball, D. L., & Cohen, D. K. (1999). Developing practice, developing practitioners: Toward a practice-based theory of professional education. In G. Sykes & L. Darling-Hammond (Eds.), *Teaching as the learning profession: Handbook of policy and practice* (pp. 3–32). San Francisco: Jossey Bass.

Banchi, H., & Bell, R. (2008). The Many Levels of Inquiry. *Science and Children, 46*(2), 26–29.

Barron, B., & Darling-Hammond, L. (2008). *Teaching for meaningful learning: A review of research on inquiry-based and cooperative learning*. George Lucas Educational Foundation.

Bell, R., Smetana, L., & Binns, I. (2005). Simplifying inquiry instruction. *Science Teacher (Normal, Ill.), 72*(7), 30–34.

Belshaw, D. (2011). *The Essential Elements of Digital Literacies*. Retrieved from http://dougbelshaw.com/ebooks/digilit/

Bennett, F. (2002). The future of computer technology in K-12 education. *Phi Delta Kappan, 83*(8), 621–626. doi:10.1177/003172170208300811

Bøe, M. V., & Henriksen, E. K. (2013). Love it or leave it. Norwegian students motivations and expectations for post-compulsory physics. *Science Education, 97*(4), 550–573. doi:10.1002/sce.21068

Bolstad, T., Gilbert, J., McDowall, S., Bull, A., Boyd, S., & Hipkins, R. (2012). *Supporting future-oriented learning and teaching. A New Zealand perspective*. New Zealand Ministry of Education.

Brush, T., & Bannon, S. (1998). Characteristics of technology leaders: A survey of school administration in the United States. *International Studies in Educational Administration, 26*(2), 47–56.

Buckenmeyer, J. A. (2010). Beyond computers in the classroom: Factors related to technology adoption to enhance teaching and learning. *Contemporary Issues in Education Research, 3*(4), 27–35. doi:10.19030/cier.v3i4.194

Bybee, R. W., Taylor, J. A., Gardner, A., Van Scotter, P., Powell, J. C., Westbrook, A., & Landes, N. (2006). *The BSCS 5E Instructional Model: Origins, Effectiveness, and Applications*. Colorado Springs, CO: BSCS.

Cai, , Lu, B., Fan, Z., Indhumathi, C., Lim, K. T., Chan, C. W., & Li, L. et al. (2006). Bio-edutainment: Learning life science through X gaming. *Computers & Graphics, 30*(1), 3–9. doi:10.1016/j.cag.2005.10.003

Carlone, H. (2004). The cultural production of science in reform-based physics: Girls access, participation, and resistance. *Journal of Research in Science Teaching, 41*(4), 392–414. doi:10.1002/tea.20006

Carr, D., & Bossomaier, T. (2011). Relativity in a rock field: A study of physics learning with a computer game. *Australasian Journal of Educational Technology, 27*(6), 1042–1067. doi:10.14742/ajet.928

Chen, C.-H., & Wang, K.-C., & Lin, Y.-H. (2015). The Comparison of Solitary and Collaborative Modes of Game-based Learning on Students' Science Learning and Motivation. *Journal of Educational Technology & Society, 18*(2), 237–248.

Cornell University. (2009). *Digital Literacy*. Retrieved from https://digitalliteracy.cornell.edu/

Cox, T. (2008). Learning Styles and Students' Attitudes toward the Use of Technology in Higher and Adult Education Classes. *Institute for Learning Styles Journal, 1*, 1–13.

Czerniak, C. M. (1990, April). *A study of self-efficacy, anxiety, and science knowledge in pre-service elementary teachers.* Paper presented at the annual meeting of the National Association of Research in Science Teaching, Atlanta, GA.

Czerniak, C. M., & Haney, J. J. (1998). The effect of collaborative concept mapping on elementary preservice teachers anxiety, efficacy, and achievement in physical science. *Journal of Science Teacher Education, 9*(4), 303–320. doi:10.1023/A:1009431400397

Dagiéne, V. (2011). The development of ICT competency in pre-service teacher education. *International Journal of Digital Literacy and Digital Competence, 2*(2), 1–10. doi:10.4018/jdldc.2011040101

De la Torre, L., Sanchez, J., Dormido, S., Sanchez, J. P., Yuste, M., & Carreras, C. (2011). Two web-based laboratories of the FisL@bs network: Hookes and Snells laws. *European Journal of Physics*, *32*(2), 571–584. doi:10.1088/0143-0807/32/2/027

DeCoito, I. (2014a, May). *Engaging Learning Communities through Digital Literacies in Teacher Education.* Paper presented at the annual meeting of the Canadian Society for the Study of Education, St. Catharines, ON.

DeCoito, I. (2014b). Teaching about the nature of science through digital scientific timelines. In P. D. Morrell & K. Popejoy (Eds.), *A few of our favorite things: Teaching ideas for K-12 science methods instructors*. Sense Publishers.

DeCoito, I. (in press). Addressing Digital Competencies, Curriculum Development, and Instructional Design in Science Teacher Education. In *Encyclopedia of Information Science and Technology.* Hershey, PA: IGI Global.

DeCoito, I., & Richardson, T. (2016). Focusing on Integrated STEM Concepts in a Digital Game. In M. Urban & D. Falvo (Eds.), *Improving K-12 STEM Education* (pp. 1–23). Hershey, PA: IGI Global.

DiPetta, T., & Woloshyn, V. E. (2009). Exploring teacher-candidates' experiences, beliefs and to technology as an instructional learning tool following instruction in a "technology rich" classroom. *Journal of Educational Technology*, *6*(1), 54–64.

diSessa, A. A. (1993). Toward an epistemology of physics. *Cognition and Instruction*, *10*(2-3), 105–225. doi:10.1080/07370008.1985.9649008

diSessa, A. A. (2006). A history of conceptual change research: threads and fault lines. In R. K. Sawyer (Ed.), *The Cambridge Handbook of the Learning Sciences* (pp. 265–281). New York: Cambridge University Press.

Dominguez, A., Saena-de-Navarrete, J., de-Marcos, L., Fernandez-Sanz, L., Pages, C., & Martinez-Herraiz, J.-J. (2013). Gamifying learning experiences: Practical implications and outcomes. *Computers & Education*, *63*, 380–392. doi:10.1016/j.compedu.2012.12.020

Duit, R., Niedderer, H., & Schecker, H. (2007). Teaching physics. In S. K. Abell & N. G. Lederman (Eds.), *Handbook of Research in Science Education* (pp. 599–629). Mahwah, NJ: Erlbaum.

Finkelstein, N. D., Perkins, K. K., Adams, W., Kohl, P., & Podolefsky, N. (2005). Can Computer Simulations Replace Real Equipment in Undergraduate Laboratories? *PERC Proceedings 2004*. doi:10.1063/1.2084711

Forbes, C. T. (2009). *Preservice elementary teachers' development of pedagogical design capacity for inquiry–an activity-theoretical perspective* (Unpublished Dissertation). University of Michigan.

Gray, L., Thomas, N., & Lewis, L. (2010). *Teachers' Use of Educational Technology in U.S. Public Schools: 2009 (NCES 2010-040)*. Washington, DC: National Center for Education Statistics, Institute of Education Sciences, U.S. Department of Education.

Grober, S., Vetter, M., Eckert, B., & Jodl, H. J. (2007). Experimenting from a distance - remote controlled laboratory (RCL). *European Journal of Physics*, *28*(3), 127–141. doi:10.1088/0143-0807/28/3/S12

Hansson, L., Hansson, O., Juter, K., & Redfors, A. (2015). Reality-Theoretical Models-Mathematics: A ternary perspective on physics lessons in upper-secondary school. *Science & Education*, *24*(5-6), 615–644. doi:10.1007/s11191-015-9750-1

Hatherly, P. A., Jordan, S. E., & Cayless, A. (2009). Interactive screen experiments: Innovative virtual laboratories for distance learners. *European Journal of Physics*, *30*(4), 751–762. doi:10.1088/0143-0807/30/4/008

Hechter, R. P., & Vermette, L. A. (2014). Tech-savvy science education? Understanding teacher pedagogical practices for integrating technology in K-12 classrooms. *Journal of Computers in Mathematics and Science Teaching*, *33*(1), 27–47.

Higgins, S., Xiao, Z., & Katsipataki, M. (2012). *The Impact of Digital Technology on Learning: A Summary for the Education Endowment Foundation*. London: EEF.

Hmelo-Silver, C. E. (2004). Problem-based learning: What and how do students learn? *Educational Psychology Review*, *16*(3), 235–266. doi:10.1023/B:EDPR.0000034022.16470.f3

Hofer, M., & Grandgenett, N. (2012). TPACK development in teacher education: A longitudinal study of preservice teachers in a secondary M.A.Ed. program. *Journal of Research on Technology in Education*, *45*(1), 83–106. doi:10.1080/15391523.2012.10782598

Jewett, J. W. Jr. (2015). Bringing (century-old) technology into the classroom. Part I: Teaching mechanics and thermodynamics with antiques. *The Physics Teacher*, *53*(7), 428–431. doi:10.1119/1.4931013

Johnston, J., & Ahtee, M. (2006). Comparing primary student teachers attitudes, subject knowledge and pedagogical content knowledge needs in a physics activity. *Teaching and Teacher Education*, *22*(4), 503–512. doi:10.1016/j.tate.2005.11.015

Kapucu, S. (2014). Salient beliefs of pre-service primary school teachers underlying an attitude "liking or disliking physics". *Science Education International, 25*(4), 437–458.

Karam, R. (2015). Introduction of the thematic issue on the interplay of physics and mathematics. *Science & Education, 24*(5-6), 487–494. doi:10.1007/s11191-015-9763-9

Koehler, M. J., & Mishra, P. (2009). What is technological pedagogical content knowledge? *Contemporary Issues in Technology & Teacher Education, 9*(1), 60–70.

Koehler, M. J., Mishra, P., & Yahya, K. (2007). Tracing the development of teacher knowledge in a design seminar: Integrating content, pedagogy, and technology. *Computers & Education, 49*(3), 740–762. doi:10.1016/j.compedu.2005.11.012

Korpershoek, H., Kuyper, H., & van der Werf, M. P. C. (2015). Differences in students school motivation: A latent class modelling approach. *Social Psychology of Education, 18*(1), 137–163. doi:10.1007/s11218-014-9274-6

Lave, J., & Wenger, E. (1991). *Situated learning. Legitimate peripheral participation.* Cambridge, UK: Cambridge University Press. doi:10.1017/CBO9780511815355

Lee, Y.-J. (2011). Utilizing formative assessments to guide student learning in an interactive learning environment. *Journal of Educational Technology Systems, 39*(3), 245–260. doi:10.2190/ET.39.3.c

Lemke, J. (2004). *Why study digital game worlds? Notes toward a basic research agenda for learning technologies.* Retrieved from http://www-personal.umich.edu/~jaylemke/games.htm

Lin, M. H., & Lee, J. Y. (2015). Data-driven learning: Changing the teaching of grammar in EFL classes. *ELT Journal, 69*(3), 264–274. doi:10.1093/elt/ccv010

Livingstone, S. (2012). Critical reflections on the benefits of ICT in education. *Oxford Review of Education, 38*(1), 9–24. doi:10.1080/03054985.2011.577938

Markic, S., Eilks, I., & Valanides, N. (2008). Developing a tool to evaluate differences in beliefs about science teaching and learning among freshman science student teachers from different science teaching domains: A case study. *Eurasia Journal of Mathematics, Science & Technology Education, 4*(2), 109–120.

Marshall, J. A., & Young, E. S. (2006). Preservice teachers theory development in physical and simulated environments. *Journal of Research in Science Teaching, 43*(9), 907–937. doi:10.1002/tea.20124

Martin-Hansen, L. (2002). Defining Inquiry: Exploring the many types of inquiry in the science classroom. *Science Teacher (Normal, Ill.)*, *69*(2), 34–37.

Maziewski, A., Dobrogowski, W., & Zablotskii, V. (2007). GloLab: Creating a global Internet-accessible laboratory. *Physics Education*, *42*(1), 72–75. doi:10.1088/0031-9120/42/1/009

McConnell, W. J., & Dickerson, D. (2013). Making connections between the digital and physical world. *Science and Children*, *51*(4), 38–43. doi:10.2505/4/sc13_051_04_38

Milton, P. (2015). *Shifting Minds 3.0: Redefining the Learning Landscape in Canada*. C21 Canada.

Mishra, P., & Kohler, M. J. (2006). Technological pedagogical content knowledge: A new framework for teacher knowledge. *Teachers College Record*, *108*(6), 1017–1054. doi:10.1111/j.1467-9620.2006.00684.x

Muehrer, R., Jenson, J., Friedberg, J., & Husain, N. (2012). Challenges and opportunities: Using a science-based video game in secondary school settings. *Cultural Studies of Science Education*, *7*(4), 783–805. doi:10.1007/s11422-012-9409-z

National Research Council (NRC). (2000). *How people learn. Brain, mind, experience, and school*. Washington, DC: The National Academies Press.

Natural Science and Engineering Research Council of Canada. (2010). *Women in Science and Engineering in Canada*. Ottawa, Canada: Author.

Ng, W. (2011). Why digital literacy is important for science teaching and learning. *Teaching Science*, *57*(4), 26–32.

Nielsen, W., Miller, A. K., & Hoban, G. (2015). Science teachers response to the digital education revolution. *Journal of Science Education and Technology*, *24*(4), 417–431. doi:10.1007/s10956-014-9527-3

Nordine, J. (2011). Motivating calculus-based kinematics instruction with Super Mario Bros. *The Physics Teacher*, *49*(6), 380–382. doi:10.1119/1.3628271

NRC. (2011). *A Framework for K-12 Science Education: Practices, Crosscutting Concepts, and Core Ideas*. Washington, DC: The National Academies Press.

Partnership for 21st Century Skills. (2009). *Twenty-first Century Skills Map*. National Science Teachers Association.

Pask, C. (2003). Mathematics and the science of analogies. *American Journal of Physics*, *71*(6), 526–534. doi:10.1119/1.1545761

Perkins, D. (2009). *Making learning whole: How seven principles of teaching can transform education*. San Francisco, CA: Jossey-Bass.

Perkins, K., Adams, W., Dubson, M., Finkelstein, N., Reid, S., Wieman, C., & LeMaster, R. (2006). PhET: Interactive simulations for teaching and learning physics. *The Physics Teacher*, *44*(1), 18–23. doi:10.1119/1.2150754

Psycharis, S. (2013). Exploring the effects of the computational experiment approach to the epistemic beliefs, the motivation, the use of modeling indicators and conceptual understanding in three different computational learning environments. *Journal of Education and Training Studies*, *1*(1), 69–87. doi:10.11114/jets.v1i1.32

Qablan, A. M., Abuloum, A., & Al-Ruz, J. A. (2009). Effective integration of ICT in Jordanian schools: An analysis of pedagogical and contextual impediments in the science classroom. *Journal of Science Education and Technology*, *18*(3), 291–300. doi:10.1007/s10956-009-9151-9

Quitadamo, I., Kurtz, M., Cornell, C., Griffith, L., Hancock, J., & Egbert, B. (2011). Critical-thinking grudge match: Biology vs. chemistry - examining factors that affect thinking skill in non-majors science. *Journal of College Science Teaching*, *40*(3), 19–25.

Redish, E., & Kuo, E. (2015). Language of physics, language of math: Disciplinary culture and dynamic epistemology. *Science & Education*, *24*(5-6), 561–590. doi:10.1007/s11191-015-9749-7

Redish, E. F., Saul, J. M., & Steinberg, R. N. (1998). Student expectations in introductory physics. *American Journal of Physics*, *66*(3), 212–224. doi:10.1119/1.18847

Richardson, T., & DeCoito, I. (2017, January). *Teachers and technology in STEM education – Present practice and future directions.* Paper presented at the annual meeting of the Association for Science Teacher Education, Des Moines, IA.

Rondon, S., Sassi, F., & deAndrade, C. (2013). Computer game-based and traditional learning method: A comparison regarding students' knowledge retention. *BMC Medical Education*, *13*(30), 1–8. PMID:23442203

Ruggiero, D., & Mong, C. J. (2015). The teacher technology integration experience: Practice and reflection in the classroom. *Journal of Information Technology Education Research*, *14*, 161–178.

Sadler, P. M., Sonnert, G., Hazari, Z., & Tai, R. (2012). Stability and volatility of STEM career interest in high school: A gender study. *Science Education*, *96*(3), 411–427. doi:10.1002/sce.21007

Sasseville, B. (2004). Integrating Information and Communication Technology in classroom: A comparative discourse analysis. *Canadian Journal of Learning and Technology*, *30*(2), 5–27.

Schauer, F., Ožvoldová, M., & Lustig, F. (2009). Integrated e-Learning - New Strategy of Cognition of Real World in Teaching Physics. *World Innovations in Engineering Education and Research*, *9*, 119–135.

Seiter, E. (2008). Practicing at home: Computers, pianos, and cultural capital. In T. McPherson (Ed.), *Digital youth, innovations, and the unexpected* (pp. 27–52). Cambridge, MA: MIT Press.

Settlage, J. (2007). Demythologizing science teacher education: Conquering the false ideal of open inquiry. *Journal of Science Teacher Education*, *18*(4), 461–467. doi:10.1007/s10972-007-9060-9

Sherin, B. (2006). Common sense clarified: The role of intuitive knowledge in physics problem solving. *Journal of Research in Science Teaching*, *43*(6), 535–555. doi:10.1002/tea.20136

Shulman, L. (1986). Those who understand: Knowledge growth in teaching. *Educational Researcher*, *15*(2), 4–14. doi:10.3102/0013189X015002004

Shulman, L. S. (1987). Knowledge and teaching: Foundations of the new reform. *Harvard Educational Review*, *57*(1), 1–22. doi:10.17763/haer.57.1.j463w79r56455411

Singer, S., Hilton, M., & Schweingruber, H. (Eds.). (2005). *America's Lab Report: Investigations in high school science. Committee on High School Science Laboratories: Role and Vision, Board on Science Education, Center for Education, Division of Behavioral and Social Sciences and Education, National Research Council.* Washington, DC: The National Academies Press.

Skaalvik, E. M., & Skaalvik, S. (2008). Teacher self-efficacy: conceptual analysis and relations with teacher burnout and perceived school context. In R. Craven, H. W. Marsh, & D. McInerney (Eds.), *Self-processes, learning, and enabling human potential* (pp. 223–247). Information Age Publishing.

Songer, N., Lee, H.-S., & Kam, R. (2002). Technology-rich inquiry science in urban classrooms: What are the barriers to inquiry pedagogy? *Journal of Research in Science Teaching*, *39*(2), 128–150. doi:10.1002/tea.10013

Steinberg, R. (2000). Computers in teaching science: To simulate or not to simulate? *American Journal of Physics*, *68*(S1S37), S37–S41. doi:10.1119/1.19517

Swackhamer, L., Koellner, K., Basile, C., & Kimbrough, D. (2009, Spring). Increasing the self-efficacy of in-service teachers through content knowledge. *Teacher Education Quarterly,* 63-78.

Taber, K. S. (2008). Conceptual Resources for Learning Science: Issues of Transience and Grain-Size in Cognition and Cognitive Structure. *International Journal of Science Education, 30*(8), 1027–1053. doi:10.1080/09500690701485082

Trumper, R. (2006). Factors affecting junior high school students interest in physics. *Journal of Science Education and Technology, 15*(1), 47–58. doi:10.1007/s10956-006-0355-6

Uluyol, Ç., & Sahin, S. (2016). Elementary school teachers ICT use in the classroom and their motivators for using ICT. *British Journal of Educational Technology, 47*(1), 65–75. doi:10.1111/bjet.12220

Ward, L., & Parr, J. M. (2010). Revisiting and Reframing Use: Implications for the Integration of ICT. *Computers & Education, 54*(1), 113–122. doi:10.1016/j.compedu.2009.07.011

Warsh, D. (2007). *Knowledge and the Wealth of Nations: A story of economic discovery.* W.W. Norton & Company.

Webb, M. (2005). Affordances of ICT in science learning: Implications for an integrated pedagogy. *International Journal of Science Education, 27*(6), 705–735. doi:10.1080/09500690500038520

Wegener, M., McIntyre, T. J., McGrath, D., Savage, C. M., & Williamson, W. (2012). Developing a virtual physics world. *Australasian Journal of Educational Technology, 28*(3), 504–521. doi:10.14742/ajet.847

Wright, W. (2007, March). *Will Wright: Spore, birth of a game* [Video file]. Retrieved from https://www.ted.com/talks/will_wright_makes_toys_that_make_worlds

Zhang, Z., & Martinovic, D. (2008). ICT in teacher education: Examining needs, expectations and attitudes. *The Canadian Journal of Learning and Technology, 34*(2), 149–166.

Zimlich, S. L. (2015). Using technology in gifted and talented education classrooms: The teachers' perspective. *Journal of Information Technology Education: Innovations in Practice, 14*, 101–124.

KEY TERMS AND DEFINITIONS

Digital Literacy: A literacy that encompasses a person's competence, usage, and their ability to manipulate digital platforms and media to accomplish a task or reach a goal. This extends beyond basic use to include creativity and innovative abilities.

Digital Video Game (DVG): Any game played on a digital platform. There are a multitude of games available but can be generally classified based on the level of control the user has in the gaming environment.

Digital Web Based Tools: Technological means accessed through the Web to enhance teaching and learning.

Information and Communication Technologies (ICT): The collection of digital technologies (such as computers, mobile devices, radios) that are used to share information between parties.

Phenomenological Primitives (p-prims): Fragmentary knowledge, including abstract and fundamental principles, constructed by individuals to make sense of the world.

Remote Labs: Physical labs designed to conduct a specific investigation, which are accessed through the Internet.

Simulations: A non-contextual presentation of information that allows the user to focus on a single phenomenon in order to understand factors that impact its behaviour.

Technological, Pedagogical, and Content Knowledge (TPACK): A framework used to highlight the intersection of the three areas of knowledge, and their relationship and impact on teaching and learning in classrooms.

Twenty-First Century Learning Skills: Skills that include creativity and innovation, critical thinking and problem solving, communication, collaboration, information literacy, media literacy, ICT literacy, flexibility and adaptability, initiative and self-direction, social and cross cultural skills, productivity and accountability, leadership and responsibility.

Web 2.0 Technologies: Web technology providing user generated web pages and the interaction between users rather than first Web iterations that were generally static.

Chapter 8
Using ICT in STEM Education: A Help or a Hindrance to Student Learning?

Jean-François Hérold
Aix-Marseille Université, France

Jacques Ginestié
Aix Marseille University, France

ABSTRACT

The aim of this paper is to examine the use of digital technologies in Science, Technology, Engineering, and Mathematics Education (STEM). We will discuss the effectiveness of the teaching-learning process in terms of the elements that could possibly promote learning with the use of Information and Communication Technologies (ICT) in STEM education. This will be done, first, by taking a learner-centred approach to an activity that students carried out using ICT to performing a task set by the teacher (cognitive engagement in the task, motivation, nature of knowledge built). The aim was to understand how ICT could be a cognitive aid for the student. Second, a teacher-centred perspective to the development of prescribed tasks (form, knowledge carried by the task) was used to identify how ICT can be adapted to aid student learning in STEM education.

INTRODUCTION

Huge financial support has, for several years, been provided by states, institutions, and communities in the deployment of information and communication technology (ICT) for education (OCDE, 2015; Pedro, 2012). These massive investments in ICT aim, on the one hand, to prepare children for the future evolutions in technologies

DOI: 10.4018/978-1-5225-2525-7.ch008

and society, specifically in our contemporary context of globalisation and the development of a knowledge society. On the other hand, there is the idea that due to a massive increase in their potential, digital technologies offer increasingly important possibilities for supporting student learning, assisting teachers in their mission, or even replacing the teachers. Because digital technologies symbolise progress, and thus the necessity to prepare students for the future, these policies have a dual aim: to learning with ICT and to learn about ICT.

Science, Technology, Engineering, and Mathematics (STEM) frequently deal with all new technologies, including, specifically, digital technologies. It is easy to see this from many stages of this long tradition and people are likely to remember the different approaches in the 1970s with tools such as Lego-Logo, Fishertechnik, or any others. At the same time, there has been considerable pedagogical innovation that includes using a problem solving approach, a project based approach, and, more recently, an investigation based approach. Although some of these research findings show that these pedagogical innovations are somewhat effective from the point of view of student learning, Connor, Karmokar and Whittigton (2015) consider that the most significant element of all these approaches is that of a student who is active in his or her learning. So, for these authors, problem-based learning is a subset of inquiry-based learning, and inquiry-based learning is a subset of active learning. If some research suggest the relative efficiency of inquiry-based learning (Lazonder & Harmser, 2016), these authors also raise the question of what type of guidance is adequate in order to promote students' learning in such an approach. For Kirschner, Sweller and Clark (2006), inquiry-based instruction induces a heavy cognitive load for the learner by imposing a search for a problem-space that may then be an obstacle to learning. Therefore, for these authors, it is necessary to provide the learners with specific guidance to help them to learner efficiently in processing the information of the problem-space and, thus to enable them to learn by reducing the cognitive load imposed by the different tasks.

Many combinations of ICT and pedagogical innovations have marked the development of STEM in recent years. However, their effectiveness has never really been studied and it impossible to say which one impacts the other. This hierarchy is mostly due to the advantages of ICT and the development of a number of applications. In many cases, these applications are presented (by authors, editors, producers, etc.) as the best way for facilitating learning. A number of studies published in scientific journals pose this efficiency as evidence that is based on the satisfaction of actors, teachers, and students. Such studies respond to the question of the attractiveness of new pedagogical organisations. The students can be satisfied by ICT, but this interest decreases at the same time as their use increases in their environment. This satisfaction says nothing about the learning efficiency. A lot of questions are still open: what and how they learn? If they learn, whether this is due to: ICT, the pedagogical organisation, the interactions between teacher and students and others?

All the above questions are relevant for assessing the policies around the massive deployment of digital technologies, as well for understanding the efficiency of the pedagogical organisations (with or without ICT), and for appreciating the cognitive development of students. These three levels cannot be reduced to a simple satisfaction evaluation and the results can differ from one to another. In other words, people suppose that ICT can support and enhance learning. For example, in the presentation of the UNESCO project entitled "Enhancing teacher education for bridging the quality gap in Sub-Saharan Africa" [UNESCO, 2014] it is clearly written that the use of ICT enhances student learning and improves the quality of instruction. Nevertheless, analyses show that there is no evidence of this, and thus the role of the digital technologies in facilitating learning remains to be demonstrated (Michko, 2007). The present chapter presents some elements of the debate around ICT from the point of view of the efficiency of learning.

BACKGROUND

Impact of ICT on Learning and Efficiency

The question of the ICT integration efficiency is central and widely documented. Usually research results provide evidences of efficiency either in certain domains of learning or with certain kinds of students. For example, some studies show the positive role of ICT for promoting the understanding of scientific ideas but they observe no differences between students using or not using ICT for the high level of understanding (Hogarth, Bennett, Lubben, Campbell, & Robinson, 2006). In other studies, the impact of ICT seems to be more limited in biology than in the physical sciences (Onuoha-Cajetan, 2007). For Higgins, Xiao and Katsipataki (2012), "the research evidence over the last 40 years about the impact of computer and digital technologies on learning consistently identifies positive benefits" (p. 3). They also point out that "the variety of digital technologies, the diversity of contexts and settings]…[make it difficult to identify clear and specific implications for educational practice in schools" (p. 15). The meta-analysis conducted by Belland, Walker, Kim and Lefler (2016) shows that computer-based scaffolding has a consistently positive effect on cognitive outcomes across various contexts of use, scaffolding characteristics, or levels of assessment; they also found that there is no difference in effect sizes among scaffolding with fading, adding, fading and adding, or no fading or adding.

Some studies that focus on a specific tool show that using technology as an instructional tool effectively enhances students' learning (Gulek & Demirtas,

2005) with the use of a laptop having an impact on their achievement, especially in mathematics. In their review of research on mobile apps for science learning, Zydney and Warner (2016) show that students' learning of basic knowledge is enhanced. Other studies have reached the same conclusion but in different domains of knowledge (Ardac and Sezen's (2002) research on the effectiveness of computer-based chemistry instruction; Aydin and Cagiltay's (2012) study of a microware engineering course). The second-order meta-analysis conducted by Tamim, Bernard, Borokhovski, Abrami, and Schmid (2011), that included 25 meta-analyses (from 1993 to 2007), mainly focused on students' achievement or performance and concluded that "the debate about technology's role in education has not been fully solved" (p. 5).

Moreover, if digital equipment in schools has improved significantly, especially in wealthy countries, some reports show that implementation in classrooms does not necessarily follow the same dynamics (Punie, Zinnbauer, & Cabrera, 2006). Thus, if the use of ICT as a teaching and learning tool is highly encouraged by the educational authorities, there are still some difficulties for teachers in their pedagogical implementation (lack of training courses, little appropriate software for learning activities). Thus, if we consider the outcomes of TIMSS 2007 (Trends in International Mathematics and Science Study is a series of international assessments of the mathematics and science knowledge of students around the world) in science classes, 60% of EU students were taught by teachers who had never used a computer for studying natural phenomena in the context of simulations, and 51% of the teachers had never asked the students to use a computer to perform procedures or scientific experiments (Eurydice, 2011). However, changes of teachers' practices mean that teachers must know how to use ICT for pedagogical activities (technological skills and knowledge about the use of ICT), must also be able to discuss how ICT can be used to support learning tasks, be able to create teaching plans that include the use of different ICT, be able to discuss the characteristics of ICT, and be able to describe how the use of a specific ICT can support students' learning (UNESCO, 2011). In order to promote students' learning with the use of ICT, Almerich, Orellana, Suarez-Rodriguez and Diaz-Garcia (2016) point out that it is necessary to study the relation between the use of ICT in class by teachers, their own competences with ICT, and their attitudes in class. For these authors, "teachers are the key element to introduce ICT into educational practice" (p. 111). This is why it is necessary to study the connection between teachers' practices, their use of ICT, and the effects on students' performance. This will enable us to discern when there is no effect of the use of ICT on student achievement. This null effect may occur as a result of the combination of the positive effects of ICT use and the negative effects of inadequate teaching practices, as the hypothesised by Falk, Mang, and Woessmann (2015) in their discussion paper about 2011 the TIMSS test.

To sum up, as Mayer (2010) says, many people claim that ICT can improve the effectiveness of the teaching-learning process but these assertions are not really supported by research outcomes. In many cases the use of ICT in education focuses primarily on technology and to a much lesser extent on learning. In this context, in order to use ICT effectively with the objective to enable all students to learn, it is necessary to understand what can hinder learning and what can promote learning in a teaching-learning situation based-on ICT as the main resource. Both of the research questions in this study — in STEM education, how do students learn with ICT and how is possible to promote learning with ICT — become crucial for understanding the part played by ICT and the part played by the pedagogical approach of the teacher. This question is also asked in the working paper of Kärkkäinen and Vincent-Lancrin (2013) in their case study of the "HP Calalyst Initiative": "how can technology-supported learning help to move beyond content delivery and truly enhance STEM education so that students develop a broad mix of skills?" (p. 11). From this perspective, the analysis of the teaching situation in the classroom allows us to understand the nature of the effectiveness of the teaching-learning process.

Teaching situation with ICT can be considered as a system constituted by a student (among other students in the classroom), the teacher, the knowledge to learn and ICT organisation. These interactions cannot be understood as simple cause-effect relationships. One possibility is then to analyse the activity of these actors in real time in a teaching situation in which ICT is instrumental and, in particular, to analyse the learning undertaken by learners, or not, during their activity (Ginestié, 2013; Hérold, 2014). Thus, analysing the activity of the student and identifying the elements of the teaching situation that interfere with learning and those that promote learning, will enable a better understand of how to use ICT in the classroom (Hérold, 2014).

We will therefore discuss the effectiveness of the teaching-learning process in terms of the elements that could potentially promote learning with the use of ICT in STEM education. This will be done, first, by taking a learner-centred approach to an activity that students carried out using ICT to performing a task set by the teacher (cognitive engagement in the task, motivation, nature of knowledge built). The aim is to understand how ICT could be a cognitive aid for the student. Second, a teacher-centred perspective to the development of prescribed tasks (form, knowledge carried by the task) is used to identify how ICT can be adapted to aid student learning in STEM education.

Learning With ICT in Class

Digital technologies are becoming increasingly common in STEM's teachings and their use takes on different forms. Many applications have been used in the classroom, such as those used in STEM education: Computer-Aided Experimentation (CAEx),

Computer-Aided Design (CAD), Computer-Aided Design and Manufacturing (CADM), Computer-based Mock-up (CBM), Computer-Based Modelling (CBM), simulators, serious games, etc. These applications have led to enormous development in educational distance learning, such as Open Distance Learning (ODL), e-learning, Massive Open Online Courses (MOOC), Corporate Open Online Courses (COOC), Open Online Training For All (OOTFA), Small Private Online Classes (SPOC), Courses at Scale (CaS), and so on. This abundance trivialises new forms of access to knowledge and suggests that, with a connection, anyone can succeed in successfully training online. Thus, in STEM courses the student uses the computer to perform multiple and various prescribed tasks that may entail either some information research on the Internet or the use of specific software, many of which are industrial products. We can then make two observations: (1) in STEM courses the prescribed tasks that use ICT are very different from each other and therefore do not involve the same kind of knowledge; and (2) ICT use in these teaching-learning situations is highly diverse in nature and can range from the simple "exerciser" to the highly sophisticated software environment. The use of industrial software in STEM courses is a special case because industrial software is not inherently learning software.

Mayer (2010) asserts that, "learning with technology involves learning situations in which the instructional experience is created with the aid of a physical device, such as a computer or the Internet" (p.186). From our point of view, this means being able to define learning that can be promoted by such ICT. This requires, first, identifying the different types of ICT used in class for learning STEM knowledge, and second, determining the learning that can be promoted for each ICT identified. For the latter, it is also necessary to define the limits of the tools used in classrooms from the perspective of student learning.

ICT Used in Class

It is necessary to remember that in STEM education, not all ICT used in teaching-learning situations are intended to assist in learning, but can be software, devices, or technical systems that were developed for general industrial use. Few of these have been modified and adapted for learning applications. This reference to the real tools used by professionals is important because it legitimates the aims of learning and all the processes that the teaching of learning brings into play in STEM education. These technologies are introduced as tools from various points of view: (1) as objects of knowledge to be studied and (2) as resources to solve problems. The process of teaching-learning is a complex process based-on a permanent dialogue between individuation and socialisation, between knowledge the use of tools, and knowledge provided by the use of these tools.

Graesser, Chipman and King (2008) provide some examples of such possible devices to promote learning: computer-based training (tool with lessons, tests, and possible feedbacks), online multimedia lessons (lessons with pictures, photos, and videos), hypermedia (Web pages with links), interactive simulations (the student has a control over the tool and she/he can set parameters, run a simulation, pause to analyse intermediate results, and so on), intelligent tutoring systems (that take into account the learner's responses and adapt the course accordingly), animated pedagogical agents (use of virtual pedagogical agents to help the student understand and learn a lesson), virtual environments (highly realistic virtual environments by implementing learner avatars), and serious games (games whose objective is to provide student with the possibility to acquire knowledge).

De Vries (2001) proposes a further categorization of ICT depending on the function performed by the device: presenting information (course materials, tutorials, etc.), suggesting exercises ("drills"), teaching (intelligent tutorials), captivating the attention and motivation of the student (educational games), providing a space for exploration (hypermedia), providing an environment for discovery (interactive simulation, micro-worlds), and providing a space for exchange (computer supported collaborative learning).

The use of ICT in class also corresponds to the educational use of industrial and/or business applications, especially in STEM education (spreadsheets, 3D-CAD or "three-dimensional Computer Aided Design", engineering tools, model-based design tools such as Matlab© and Simulink©, Rapid Application Development tools, Integrated Development Environments).

Some Limitations

Researchers have studied a number of these devices. For example, in Mayer (2005), the possibilities offered by five different learning environments and their limitations are discussed: animated pedagogical agents (Moreno, 2004); virtual environments (Cobb & Fraser, 2004); games, simulations, and micro-worlds (Rieber, 2004); hypermedia (Dillon & Jobst, 2004), and e-courses (Clark, 2004). These authors showed that some positive effects on learning can indeed be demonstrated in the use of these devices in educational situations in regard to factors such as student motivation, commitment to the prescribed task, pleasure, interactivity, and making educational content more attractive. However, the results also show some limitations. For example, for simulations, the fact that this type of learning is very difficult for students who are novices, or for micro-worlds, the fact that students must be able to know how to use a micro-world and, above all, want to use it, and for games, the problem of integration of the teaching content into the game (Rieber, 2004). Another example refers to hypermedia and the fact that the ability of the student to explore information depends mainly on the engagement of the student (Dillon & Jobst, 2004). However, according to Tricot (2007), sometimes processing a document from a hypermedia may have such a cognitive cost, due to the extraneous cognitive load (Sweller & van Merriënboer, 2005), that the learner has a great difficulty in completing the prescribed task, which in turn limits his/her involvement in the task.

Thus, while many studies highlight the gains of using ICT in class, the results of other studies suggest the need to relativize such gains. As Mayer (2010) states, "many strong claims are made for the potential of new technologies to transform education and training around the world, but few of the claims have been substantiated by research evidence or even tested in rigorous scientific research" (p. 182). Therefore, we need research that show what works with technologies and what does not. For example, Amadieu & Tricot (2014) show that:

- If ICT can be motivating for learners, it is nevertheless important to consider the learning activities carried out with ICT because an activity can be motivational with a given tool but that may not be the case with another activity;
- The educational use of ICT should take into account the knowledge and skills of learners who will have to use a particular tool and the knowledge needed for proper use of the tool itself;
- If ICT enables many interactions with pedagogical content, and thus makes the learner more active, this learner activity must have a function for the learning of the student. This implies that the student has the knowledge to carry out this activity and that the proposed activity will contribute to learning;
- The vast majority of ICT used in class does not provide immediate feedback that is powerful enough to be useful to the student;
- Using ICT in class is not the same as using ICT at home. Learning in school imposes specific tasks (the "prescribed tasks" that are "educational tasks") as carriers of specific knowledge (knowledge to learn that is defined by the teacher and refers to the prescribed task, and the knowledge necessary to use ICT).

In conclusion, we can make two conclusions. First, the majority of educational software is designed by researchers who are also developers of the tool. Therefore, they have a technology-centred approach that also partially takes into account the characteristics of the learner; to use a quote by Duchâteau (2001), in many cases, the developers of the educational software are essentially interested in "how do that?" and too little in "why do that?". Second, in STEM education, as noted earlier, teachers use many ICT tools that are mainly developed for general industrial use. These tools are not learning tools; they were designed for use by specialists in a particular subject area. Furthermore, the HMI (Human Machine Interface), the help provided by the software company, the ergonomics of the tool, and its use, are designed for experts in a particular subject area and not for learners of this area. From our perspective, this implies the need for serious consideration of the characteristics of the tool by the teacher in order to provide students with a teaching situation that, if it relies on this kind of tool, is truly effective for student learning (Hérold, 2014). This then requires an *a priori* analysis of the prescribed task which such tool in order to determine all the knowledge that are truly involved in the activity to be performed by the student in class, and not only knowledge to teach.

A LEARNER-CENTRED APPROACH

Understanding How a Student Learns

Multiple research studies have attempted to answer the question of how people learn at school. Each scientific discipline has a range of theories of learning. It would be difficult here, in just a few lines, to address all the responses to the question of how people learn at school. However, we have at our disposal different syntheses of research findings on learning at school and on how to implement a learning environment that would be relatively efficient from the standpoint of student learning. For example, there is that of Halpern, Graesser and Hakel (2007), or the one offered on the Carnegie Mellon University website. These syntheses highlight the need for a learner-centred approach in order to be able to define a truly effective learning environment. For example, it is important to take into account the student's prior knowledge. In a situational context that is unfamiliar to the student, he/she often uses prior knowledge that is inappropriate ("inadequate prior knowledge") and knowledge learned in previous situations. When knowledge is used inappropriately it can impede learning and it also consequences in the number and type of errors in complex tasks (Anthony, 1996; Bastien, 1997; Bastien & Bastien-Toniazzo, 2004; Baviskar, Hartle, & Whitney, 2009; Cooper, Tindall-Ford, Chanderler, & Sweller, 2001; Hérold, 2014; Schneider & Stern, 2010). However, when knowledge is robust, meaningfully organised, and corresponds to the situation, this prior knowledge can help learning (see Séjourné (2002) — drawing on an example of hypermedia in physics, referred to as "Labdoc Sound and Vibration", the design of this hypermedia software was based on hypotheses such as these). Another example is about the choice and the organisation of the learning materials. Several principles refer to the learning material organisation effects: coherence effect (the fact that it is necessary to design the learning material so that the main points are prominent and remove irrelevant materials), forms of narrative for the learning material (stories and example cases must be singled out), and using multiple and varied examples for the understanding of an abstract concept (Halpern et al., 2007).

There are also some principles that are more specific to multimedia learning, such as the contiguity effects (ideas should be presented contiguously in space and time in the multimedia environment), dual code effects (information is remembered well when it is delivered in verbal and pictorial modes), and the coherence effect (important to remove irrelevant material from the learning material).

This last point is important in the case of the use of ICT in class, especially if ICT is a tool developed for an industrial use. Indeed, most of the time, the HMI of ICT includes details that do not help students to learn the key elements. This can be explained by John Sweller's Cognitive Load Theory. Sweller's cognitive load theory (Sweller, 2010; Sweller, van Merriënboer, & Paas, 1998) is a psychological theory that takes into account the phenomena of cognitive overload in learning activities. The starting point of Sweller's research work is to ask what are the elements of a learning situation that enable student to learn, or what are the elements that result in the student's failure to learn? Through this approach Sweller seeks to identify the elements of cognitive overload in the teaching-learning activities, especially with regard to the learning materials used by teachers (Sweller, 2010). Cognitive load theory makes a distinction between three forms of cognitive load (Sweller & van Merriënboer, 2005): intrinsic cognitive load, which is the one linked to the cognitive resources that are attributed to memorising information in the working memory (cognitive load associated with cognitive resources mobilised by the complexity of learning content in the treatment and implementation of the prescribed task); essential cognitive load, which is the one linked to the construction and automation of knowledge (essential cognitive load dependent on the characteristics of the knowledge to be learned — the less knowledge in memory that the student has about the learning situation, the more important is the essential cognitive load); and extraneous cognitive load, which is the one that is linked to the resources allocated by the cognitive system to activities that do not directly contribute to do main activity, such as solving a problem (extraneous cognitive load is due to the cognitive processing required by the learning material, it corresponds to the cognitive processing of all information that is not directly related to learning the knowledge covered by the learning situation). A high extraneous cognitive load will limit the possibility of using resources linked to essential cognitive load, thus limiting the acquisition and automation of knowledge and, as a result, rendering them difficult to recall (see an example of research results in Dunleavy, Dede, and Mitchell (2009) on augmented reality simulations in science courses).

Therefore, in order to facilitate the learning of a given knowledge, it is necessary to minimise the extraneous cognitive load (organise and adapt the information to be processed by the student, provide adaptive helps, offer suitable guidance, etc.), and limit the intrinsic cognitive load by analysing the prescribed task to determine the quantity of elements of knowledge to be processed and their levels of interaction for the implementation of the prescribed task (Sweller, 2010; Sweller & van Merriënboer, 2005).

Finally, it is also important to take into account the role of emotion and motivation in learning. Indeed, learning results are in the dynamic interaction of cognitions with emotion and motivation (Schneider & Stern, 2010). It is now clear that learning is not only function of cognition (knowledge, skills, cognitive processes involved in learning), but also that learning involves motivation and emotion. Without taking into account motivation and emotion in learning, it is impossible to properly understand what is happening in class during the teaching-learning process (Yun Dai & Sternberg, 2004). The three dimensions of learning, that is, cognition, emotion, and motivation, are not only involved in the issue of whether or not students learn, but are also closely linked. Indeed, it is now accepted that the student's emotions in class have some influence on her/his performance in learning. Even if, for now, few studies provide sufficiently strong results on this issue mainly due to methodological problems and also due to the fact that a such term as emotion does not yet have a clear definition — as Linnenbrink (2006) notes, there is "considerable overlap in the way in which the authors define affect, moods and emotions" (p. 308). Nevertheless, it was demonstrated that emotion moderates cognitive processes by driving selective processes during the mental representation of the situation (the focus on the relevant information of the situation in order to elaborate a goal), and also affects decision-making during problem solving activities.

On the other hand, research on motivation, conducted by Bandura, shows that the perception of a student on the value of a learning activity and the perception that she/he can have of her/his competence to perform this activity and to control it, are the main sources of motivation (Bandura, 1993, 2007). The value given by a student to a learning activity in class can been of different kinds. The value can be associated with: the usefulness of the knowledge learn via the learning activity (in order to progress, to learn more about a specific subject, to be able to do something knew); the pleasure of discovering new knowledge in order to be able to control it (in STEM education, for example, this can be knowing how to manage a new material, or how to improve a sophisticated measurement tool); the opportunity for the student to have a teacher's recognition (a good score on the assessment, good answer to an oral question in class); showing other students that the student is better than they are (notion of performance, to show that her/his is able to do something that the others cannot do). The perception that the student can have on her/his ability to be competent to perform the learning activity (Bandura's "self-efficacy") is mainly due to a belief system about her/his ability to perform the proposed activity, systems that is built, more or less, by social learning (Bandura, 1993) in various activities (not only learning activities), during which the student receives social feedback concerning her/his performances (Bandura, 1993).

From this we can see that motivation is hardly linked with affect — feelings such as pleasure. But is it linked to emotion? The difficulty here is in the definition of the terms: we can distinguish moods and emotions based on intensity or on duration, we can also discuss how to incorporate affect (emotion?) into models of motivation, or how to integrate these three areas of cognition, emotion, and motivation (Linnenbrink, 2006). Nevertheless, when we discuss the positive effects of learning in school situations with ICT in STEM education, it is necessary to know when people say that the positive effect of ICT on learning is the motivation of the student to do the prescribed tasks, whether this positive effect is occurring because of affect or whether it is really from motivation, as discussed above. This is why we must be very careful when we talk about a positive effect of the use of ICT in class, especially in respect of motivation. Indeed, this positive effect is more commonly associated with pleasure than with motivation (within the meaning given by Bandura).

Learning by Instruction: A Complex Activity for Students

Learning by instruction is a complex activity that is difficult to analyse. Providing an answer to the question of how student learn in class is not simple. To help to answer this question it is necessary to understand how a teaching-learning situation can be effective and why.

In a few words, we can say that in order to be effective a teaching-learning situation must be motivating for the student who agrees to engage in the prescribed tasks and makes the necessary effort to learn the knowledge covered by these tasks. The teaching-learning situation must take into account the achievements of students in terms of prior knowledge that can be more or less familiar; the teaching-learning situation must be compatible with the possibilities of processing and storage of the cognitive system of the student at the time; and, finally, the teaching-learning situation must take account the effect of emotion and motivation on the learning of the student.

INSTRUCTION USING ICT IN CLASS

The Teaching-Learning Situation: A Complex System

In class, the teaching-learning situation corresponds to a system of interactions between the activity of the teacher and the activity of the student (Ginestié & Tricot,

2013). Being interested in both the teacher's activity and the student's activity entails trying to understand the functioning of the system constituted by the school situation, and the teaching-learning processes that govern it (Ginestié & Tricot, 2013).

The school situation is a system constituted by different elements. The first refers to the knowledge to teach, knowledge chosen by the teacher for a specific situation. From a learner-centred approach, the heart of the system is, of course, the student with her/his cognitive profile, her/his emotive profile, and her/his motivation profile. This student is not alone in the classroom and is interacting with other students. In STEM education the student uses mostly uses ICT for learning activities. Thus, a second element of the system is the tool used by the student (which can be a simple sheet of paper or a sophisticated computer-based system). At the same time, there is a dual aim: to understand the possibilities of all the tools each has at his/her disposal for acting increasingly efficiently; and an aim to understanding the limits of each tool, not only as a limit to his/her own possibility of action but also as a limit to his/her own understanding. To achieve this dual aim, the development of procedural schemes (how to do) must be aligned with the development of semiotic schemes (why we are doing it and why we are doing it like this). The pedagogical situations have to make sense for the pupils, not only because these are the rules of the school, but also, and mainly, because they make sense to the social activities outside of the school. Finally, as Brown, Collins, and Duguid (1989) note, knowledge is "being in part a product of the activity, context, and culture in which it is developed" (p. 32). Here the concepts of "context" and "culture" could be indicated by the term "environment".

So, from this point of view, the teaching-learning situation is akin to a system of different elements (a student, some knowledge to teach, a teacher, other students, a tool (which could be ICT), and an environment) that are all interrelated. This is why, in order to be able to understand how a teaching-learning situation could be effective, or not, it is necessary to take into account, not only the different elements of the teaching-learning situation, but also the interrelations that could exist between these different elements. To understand how a student learns in class it is then necessary to qualify and to quantify these different interrelations.

However, in accepting that there is a real challenge to understanding the interactions of the system and the corresponding teaching-learning situation by studying the activity of the different actors (teacher and the student) within the system, we must admit that the study of these interactions is particularly difficult (Ginestié & Tricot, 2013). These difficulties are, first, theoretical (theories, concepts, models), and, second, methodological: studying the interactions of the system requires having to handle many variables, each of a different nature. Among the possible methodologies there is the possibility to neutralise some elements of the system: we can neutralize the teacher, the other students, and we study the interactions between a student and the tool whilst taking into account the interrelations with the knowledge for learning and the environment. Alternatively, we can neutralise the other students and the tool, and so on. It is subsequently necessary to define some suitable research methods. Middleton (2008) discusses some methods and techniques for researching technology education: classroom case studies; process tracing method, which is an observational method of data elicitation; comparative analysis; capturing knowledge and activity through, on the one hand observations of activities that are recorded and, on another hand, artefacts of resources used in these activities. Obviously, these methods and research techniques can be used in the different areas of STEM education.

Task and Activity

For a student in a teaching-learning situation, learning mainly consists in doing learning tasks prescribed by the teacher. If we want to understand why a student learns or does not learn in class, then it is necessary to analyse the task prescribed by the teacher (which defines what to do) and to analyse the activity generated by the student in performing this task (what the student really does to perform the prescribed task). The prescribed task must be meaningful in terms of the knowledge involved in the teaching situation and should not describe the actions that the student must perform to achieve the prescribed task. The prescribed task is defined by the teacher in terms of expected results and also in terms of the resources and the constraints of the task.

How the prescribed task is performed depends on the student (her/his knowledge at the time she/he must perform the required task, her/his emotion, her/his motivation to do the prescribed task) and the activity of the student. This activity has observable components (actions, traces of the activity that could be written traces or digital traces in the case of the use of ICT, and sometimes the student's verbalisations), and unobservable components (the cognitive activity of the student and the effect of emotion and motivation on this cognitive activity). The analysis of the student's activity will therefore allow us to deduce the activated cognitive processes and the mobilised knowledge used to carry out the prescribed task. However, when the teacher prescribes a task, she/he identifies a goal but not necessarily how to achieve this goal. When the student notes the prescribed task, her/his cognitive system will build a mental representation of the goal, a mental representation based on present knowledge in memory and made accessible by the characteristics of the situation. The student's cognitive interpretation of the prescribed task conveys the nature of the activity that will enable the student to perform the prescribed task. If this interpretation of the prescribed task leads the student's cognitive system to develop a different goal to the one set by the prescribed task (due to a lack of knowledge or inadequate knowledge), then the prescribed task becomes an effective task, one that corresponds to the student's activity but is different from the prescribed task. Interpreting this effective task through the student's activity is potentially an important source of information about the student's level of learning.

Very often, the prescribed task supports a lot of implicit knowledge for students; while for the teacher, due to his/her expertise, these implicit knowledge is tacit. In many cases it is this tacit knowledge that impede student learning. For example, in the case of a paper-and-pencil activity in geometry, the teacher may ask the student to draw a triangle (but the teacher does not specify how to do the drawing), only indicating the length of the sides of the triangle. In reading the prescribed task the student will therefore interpret the goal (here, the triangle plot). Yet, often what is observed is that the student uses her/his ruler to draw the first side of the triangle (which generally corresponds to the first side length given by the wording) and then draws the second side of the triangle with her/his ruler, then, in many cases, finding her/himself in a deadlock to draw the last side of the triangle (in according to the orientation the student has given to the first two sides of the triangle). In this example, the tacit knowledge is a skill that matches the drawing of a triangle with a compass. If this skill is not a "strong" knowledge (Ritter, Anderson, Koedinger, & Corbett, 2007) for the student, then this knowledge is not activated by the student's cognitive system at the time of the interpretation of the wording. The student uses her/his ruler to perform the drawing because side lengths are mentioned in the wording. Therefore, the skill, trace a given length segment with a ruler, is activated as more accessible in memory. However, if the exercise is done with an IT tool such as *Geogebra©*, the tacit knowledge will not be at all the same. In that case, the tacit knowledge will be knowledge that enables the student to correctly use the proper procedure in the software: first, select the icon in order to draw a "Segment with Given Length", state the length in the pop-up window, select the icon "Circle with Centre through Point", choose the item "Circle with Centre and Radius" in the sub-menu, and so on. Thus, with the same knowledge, in this example drawing a triangle, as a paper-and pencil activity or as a task that is prescribed using an IT tool, the student's activity will not be at all the same. So, the knowledge that the student is able to mobilise to achieve the prescribed task will not be the same for both situations. Hence the need for the teacher to carefully analyse the activity that will be done by the student to achieve the prescribed task, and thus to highlight the knowledge that the student must mobilise to carry out the prescribed task, especially with an IT tool that imposes its own procedures.

In technology education, we observe some difficulties when a student must measure the intensity of an electric current with a multimeter; in 99% of cases, she/he uses the voltmeter function and not the ammeter and the effective "know-how" she/he has built is for measuring voltage. This "know-how" has become a "strong" knowledge in the memory and is therefore easily accessible. This is why most of the students are in deadlock for the task of measuring current in class. With virtual labs that simulate electronic circuits, the problem would not be the same, for example, in order to scope a signal, the student merely has to point to a test point with the mouse. However in reality, to scope a signal it is necessary to use two points for the measure: the test point and the ground. The tacit knowledge in this last activity is the fact that it is necessary to consider two points in order to make the measure, something that is not always the case with certain simulators.

FUTURE RESEARCH DIRECTIONS

In order to understand what the student really learns in class, it is necessary to understand what the teacher really prescribes as student tasks, especially in terms of tacit knowledge. The activity of the student performing the set task could then be analysed to identify how the student interpreted the task, and what the student really did according to her/his knowledge in memory, whilst also taking account of the possible effect of emotion and motivation on her/his activity. In so doing, it is certain that the problem-solving approaches and project-based learning are likely to provide richer information about the nature of the students' activities, and thereby provide a better understanding of the basic learning processes of students. Indeed, learning based on project activity means that students become active in a process that is characterised by recurrent cycles of analysis, synthesis, action, and reflection (Hérold & Ginestié, 2011). The same applies to problem-solving activity: the student has to elaborate a goal and if the problem is too complex, she or he has to identify sub-goals from the interpretation of the task. In these two cases, this requires, from our point of view, defining prescribed tasks with the sub-goals clearly identified.

What Must Teachers Take into Account When They Use ICT in Class in Order to Improve Students' Learning?

The use of these sophisticated tools is not so easy that theirs interfaces let foresee. They suppose a high level of abstraction with a great mastery of the possibilities and their users need to anticipate the solution and the result of each action carried out by the subject. When ICT is used in a teaching-learning situation it is necessary to analyse that it could be done with this ICT, but not only. As we have already noted, in order to understand how a student learns in class we must also take into account the activity of the student in carrying out the prescribed task with ICT rather than only describing the ICT and its different possibilities for action. The activity of a student depends of the nature of the prescribed task in terms of the knowledge involved, the prior knowledge of the student at this moment, and also the effects of emotion and/or motivation on the student's activity.

We have previously established that one of the barriers to learning with ICT was due to the fact that when the cognitive processing (knowledge, cognitive processes) required to carry out the prescribed task with ICT generates a cognitive load that exceeds the cognitive ability of the student, then the student does not learn. In many cases teachers try to overcome this problem by providing strong procedural guidance for using these tools. For example, for a demonstration task in geometry using a geometry software such as *Geogebra*©, there will be a learning aid for students to help them to trace the figure, but only if the students have previously learned how to use the software (reducing extraneous cognitive load) and are able to use the adequate prior knowledge to trace the figure, are they then able to make a useful mental representation of the goal in order to solve the problem. On the other hand, the teacher must guide the students by carrying out a demonstration (reduction of the intrinsic cognitive load), by direct interaction (dialogue), or by using a learning document — the software cannot do everything.

Another example in technology education is an object design task with a 3D printing system that has the potential to print the object, thus reducing the intrinsic load for the student who is then able to compare the numerical model, developed via the 3D-CAD, with the printed object. By encouraging the process of researching solutions, the 3D printing system acts to support the evaluation of a possible solution, which is always difficult to represent at the beginning of the design process. Nevertheless, the use of 3D-CAD, by the nature of its HMI, can impose extraneous cognitive load such that the student will find her/himself in great difficulty and unable to use the possibilities offered by the software.

The teacher must take these limitations into account in the implementation of her/his teaching when she/he chooses to use ICT in class, especially in the case of a tool that was developed for industrial use. For this, teachers can rely on a number of principles provided by research. For example, they can provide assistance in planning (establishment of sub-goals to achieve the solution to the problem, selection of relevant information to achieve each of these sub-goals); provide a visual representation of the solution (assistance with the development of an adequate mental representation of the sub-goal); and provide immediate feedback to the student on her/his actions and her or his answers (Hérold (2008, 2012) provides an example of a digital tool for which these principles have been implemented using the example of an exerciser in mathematics, and Belland et al. (2016) take a similar approach to computer-based scaffolding in STEM education that demonstrates the possibility for interventions that extend the students' capabilities as they engage with the central problem in problem-centred instructional approaches, as well as the conditions for the effectiveness of this intervention with relevant activities from STEM education).

Mayer (2010) presents numerous principles to promote learning with technology. First, the student must be able to select the pertinent information in the numerical environment in order to build a mental representation using her or his prior knowledge in memory. For this, the relevant information for the prescribed task should be easily accessible, whether this is for a written document or for ICT. This should enable the student to engage in appropriate cognitive processing without cognitive overload. This is why, in order to promote learning, teachers must be careful not to overload the treatment of relevant information for the prescribed task. It is possible to do this by analysing the prescribed task in terms of relevant information for the mental representation of the task, taking account of the prior knowledge of the student. Second, teachers must monitor irrelevant information that is often present on the HMI of ICT used in class, especially in STEM education. For Mayer (2010), "instruction is the manipulation of the learner's environment in order to promote learning" and "learning with technology is caused by instructional methods and not by instructional media" (p. 197).

CONCLUSION

If people want to use digital technology to promote students' learning it is necessary that the design and implementation of the teaching-learning situation that uses ICT meet certain conditions. The most important is the fact that it is necessary to fully understand how students learn in class, and not to think about general pedagogical methods that more or less work. The characteristics of the teaching situation must be taken into account, especially if the teacher chooses to use ICT for the prescribed task.

Furthermore, the prescribed task must be analysed with respect to the activity that the students will really carry out and not only in terms of the knowledge and skills that the teacher assumes to be necessary in order to carry out the prescribed task.

REFERENCES

Almerich, G., Orellana, N., Suarez-Rodriguez, J., & Diaz-Garcia, I. (2016). Teacher 's information and communication technology competences: A structural approach. *Computers & Education*, *100*, 110–125. doi:10.1016/j.compedu.2016.05.002

Anthony, G. (1996). When mathematics students fail to use appropriate learning strategies. *Mathematics Education Research Journal*, *8*(1), 23–37. doi:10.1007/BF03217287

Ardac, D., & Sezen, A. H. (2002). Effectiveness of computer-based chemistry instruction in enhancing the learning of content and variable control under guided versus unguided conditions. *Journal of Science Education and Technology, 11*(1), 39–48. doi:10.1023/A:1013995314094

Aydin, E., & Cagiltay, N. (2012). A new RF and microwave engineering course enriched with advanced technologies. *Computer Applications in Engineering Education, 20*(4), 634–645. doi:10.1002/cae.20432

Bandura, A. (1993). Perceived self-efficacy in cognitive development and functioning. *Educational Psychologist, 28*(2), 117–148. doi:10.1207/s15326985ep2802_3

Bandura, A. (2007). *L'auto-efficacité*. Bruxelles, Belgium: de Boeck.

Bastien, C. (1997). *Les connaissances de l'enfant à l'adulte*. Paris, France: Armand Colin.

Bastien, C., & Bastien-Toniazzo, M. (2004). *Apprendre à l'école*. Paris, France: Armand Colin.

Baviskar, S. N., Hartle, R. T., & Whitney, T. (2009). Essential criteria to characterize constructivist teaching: Derived from a review of the literature and applied to five constructivist-teaching method articles. *International Journal of Science Education, 31*(4), 541–550. doi:10.1080/09500690701731121

Belland, B. R., Walker, A. E., Kim, N. J., & Lefler, M. O. (2016). Synthetizing results from empirical research on computer-based scaffolding in STEM education: A meta-analysis. *Review of Educational Research, 20*(10), 1–36.

Brown, J. S., Collins, A., & Duguid, P. (1989). Situated Cognition and the Culture of Learning. *Educational Researcher, 18*(1), 32–42. doi:10.3102/0013189X018001032

Clark, R. C. (2004). Multimedia learning in e-courses. In R. E. Mayer (Ed.), *The Cambridge Handbook of Multimedia Learning* (pp. 589–616). New-York, NY: Cambridge University Press.

Cobb, S., & Fraser, D. S. (2004). Multimedia learning in virtual reality. In R. E. Mayer (Ed.), *The Cambridge Handbook of Multimedia Learning* (pp. 525–548). New-York, NY: Cambridge University Press.

Connor, A. M., Karmokar, S., & Whittington, C. (2015). From STEM to STEAM: Strategies for enhancing engineering and technology education. *International Journal of Engineering Pedagogies, 5*(2), 37–47. doi:10.3991/ijep.v5i2.4458

Cooper, G., Tindall-Ford, S., Chanderler, P., & Sweller, J. (2001). Learning by imaging. *Journal of Experimental Psychology: Human Perception and Performance: Applied*, *7*(1), 68–82. PMID:11577621

De Vries, E. (2001). Les logiciels dapprentissage: Panoplie ou éventail? *Revue Française de Pédagogie*, *137*(1), 105–116. doi:10.3406/rfp.2001.2851

Dillon, A., & Jobst, J. (2004). Multimedia learning with hypermedia. In R. E. Mayer (Ed.), *The Cambridge Handbook of Multimedia Learning* (pp. 1–16). New-York, NY: Cambridge University Press.

Duchâteau, C. (2001). Synthèse du 5° colloque Hypermédias et Apprentissage. In *Proceedings of 5° colloque Hypermédias et Apprentissage* (pp. 376-382). Grenoble, France: INRP.

Dunleavy, M., Dede, C., & Mitchell, R. (2009). Affordances and limitations of immersive participatory augmented reality simulations for teaching and learning. *Journal of Science Education and Technology*, *18*(1), 7–22. doi:10.1007/s10956-008-9119-1

Eurydice. (2011). *Chiffres clés de l'utilisation des TIC pour l'apprentissage et l'innovation à l'école en Europe*. Bruxelles, Belgium: Commission Européenne. Retrieved from: http://eacea.ec.europa.eu/education/eurydice/documents/key_data_series/129FR_HI.pdf

Falk, O., Mang, C., & Woessmann, L. (2015). *Virtually no effect? Different uses of classroom computers and their effect on student achievement*. Bonn, Germany: IZA.

Ginestié, J. (2013, 3-6 december). *Giving meaning to the use of tools: some elements of discussion about ICT in the French curriculum of technology Éducation.* Paper presented at the 27th PATT Conference, Christchurch, New-Zealand.

Ginestié, J., & Tricot, A. (2013). Activité d'élèves, activité d'enseignants en éducation scientifique et technologique. *Recherches en Didactique des Sciences et Technologie*, *8*, 9–22.

Graesser, A. C., Chipman, P., & King, B. G. (2008). Computer-Mediated Technologies. In M. J. Spector, D. Merrill, & J. van Merrienboer (Eds.), *Handbook of Research on Educational Communications and Technology* (pp. 211–224). New-York, NY: Erlbaum.

Gulek, J. C., & Demirtas, H. (2005). Learning with technology: The impact of laptop use on student achievement. *The Journal of Technology, Learning, and Assessment*, *3*(2), 4–38.

Halpern, D. F., Graesser, A. C., & Hakel, M. (2007). 25 learning principles to guide pedagogy and the design of learning environments. Washington, WA: Association for Psychological Science Taskforce and Lifelong Learning Work and Home. Retrieved from http://www.psy.memphis.edu/learning/whatweknow/index.html

Hérold, J.-F. (2008). Conception et réalisation d'un outil informatisé d'aide à l'apprentissage: une approche cognitive de l'apprenant. In M. Sidir, G.-L. Baron, & E. Bruillard (Eds.), *Communication et apprentissage instrumentés en réseau* (pp. 265–276). Paris, France: Hermès Lavoisier.

Hérold, J.-F. (2012). Analyse cognitive de l'activité de l'élève pour une personnalisation d'un environnement numérique d'apprentissage. *STICEF, 19*, 285–307.

Hérold, J.-F. (2014). *Learning with an IT tool: analysis of the student's activity.* Paper presented at the Interaccion Conference, Tenerife, Spain. doi:10.1145/2662253.2662303

Hérold, J.-F., & Ginestié, J. (2011). Help with solving technological problems in project activities. *International Journal of Technology and Design Education, 21*(1), 55–70. doi:10.1007/s10798-009-9106-8

Higgins, S., Xiao, Z., & Katsipataki, M. (2012). *The impact of digital technology on learning: A summary for the education endowment foundation.* Durham, UK: Education Endowment Foundation and Durham University.

Hogarth, S., Bennett, J., Lubben, F., Campbell, B., & Robinson, A. (2006). *ICT in science teaching*. London, UK: EPPI-Centre.

Kärkkäinen, K., & Vincent-Lancrin, S. (2013). Sparking Innovation in STEM Education with Technology and Collaboration: A Case Study of the HP Catalyst Initiative. *OECD Education Working Papers, 91*. 10.1787/5k480sj9k442-en

Kirschner, P. A., Sweller, J., & Clark, R. E. (2006). Why minimal guidance during instruction does not work: An analysis of the failure of constructivist, discovery, problem-based, experiential, and inquiry-based teaching. *Educational Psychologist, 41*(2), 75–86. doi:10.1207/s15326985ep4102_1

Lazonder, A. W., & Harmsen, R. (2016). Meta-Analysis of Inquiry-Based Learning Effects of Guidance. *Review of Educational Research.*

Linnenbrink, E. A. (2006). Emotion research in education: Theoretical and methodological perspectives on the integration of affect, motivation and cognition. *Educational Psychology Review, 18*(4), 307–314. doi:10.1007/s10648-006-9028-x

Mayer, R. E. (2005). *Cambridge Handbook of Multimedia Learning*. New-York, NY: Cambridge University Press. doi:10.1017/CBO9780511816819

Mayer, R. E. (2010). Learning with technology. In H. Dumont, D. Istance, & D. Benavides (Eds.), *The nature of learning: using research to inspire practice* (pp. 173–198). New-York, NY: OECD. doi:10.1787/9789264086487-10-en

Michko, G. M. (2007). *A Meta-analysis of the Effects of Teaching and Learning with Technology on Student Outcomes in Undergraduate Engineering Education*. Houston, TX: University of Houston.

Middleton, H. (2008). *Researching technology education. Methods and techniques*. Rotterdam, The Netherlands: Sense Publishers.

Moreno, R. (2004). Multimedia learning with animated pedagogical agents. In R. E. Mayer (Ed.), *The Cambridge Handbook of Multimedia Learning* (pp. 507–524). New-York, NY: Cambridge University Press.

OCDE. (2015). *Students, Computers and Learning: Making the Connection*. New-York, NY: OCDE Publishing. doi:10.1787/9789264239555-en

Onuoha-Cajetan. (2007). *Meta-analysis of the effectiveness of computer-based laboratory versus traditional hands-on laboratory in college and pre-college science instructions* (Unpublished doctoral dissertation). Minneapolis Capella University, Minneapolis, MN.

Pedro, F. (2012). Trusting the unknown: the effect of technology use in education. In S. Dutta & B. Bilbao-Osorio (Eds.), *Global Information Technology Report. Living in a Hyperconnected World* (pp. 135–148). Geneva, Switzerland: SRO-Kunding.

Punie, Y., Zinnbauer, D., & Cabrera, M. (2006). *A Review of the Impact of ICT on Learning*. Bruxelles, Belgium: Joint Research Centre.

Rieber, L. P. (2004). Multimedia learning in games, simulations and microworlds. In R. E. Mayer (Ed.), *The Cambridge Handbook of Multimedia Learning* (pp. 549–568). New-York, NY: Cambridge University Press.

Ritter, S., Anderson, J. R., Koedinger, K. R., & Corbett, A. (2007). Cognitive Tutor: Applied research in mathematics education. *Psychonomic Bulletin & Review*, *14*(2), 249–255. doi:10.3758/BF03194060 PMID:17694909

Schneider, M., & Stern, E. (2010). The cognitive perspective on learning: Ten cornerstone findings. In OCDE (Ed.), The nature of learning: Using research to inspire practice (pp. 69-90). Paris, France: OCDE.

Séjourné, A. (2002). Concepts of hypermedia in physics: The importance of resources from the point of view of learning. *Educational Media International, 39*(3-4), 317–329. doi:10.1080/09523980210166035

Sweller, J. (2010). Element interactivity and intrinsic, extraneous and germane cognitive load. *Educational Psychology Review, 22*(2), 123–138. doi:10.1007/s10648-010-9128-5

Sweller, J., & van Merriënboer, J. (2005). Cognitive load theory and complex learning: Recent developments and future direction. *Educational Psychology Review, 17*(2), 147–177. doi:10.1007/s10648-005-3951-0

Sweller, J., van Merriënboer, J., & Paas, F. (1998). Cognitive architecture and instructional design. *Educational Psychology Review, 10*(3), 251–296. doi:10.1023/A:1022193728205

Tamim, R. M., Bernard, R. M., Borokhovski, E., Abrami, P. C., & Schmid, R. F. (2011). What forty years of research says about the impact of technology on learning a second-order meta-analysis and validation study. *Review of Educational Research, 81*(1), 4–28. doi:10.3102/0034654310393361

Tricot, A. (2007). Apprentissages et documents numériques. Paris, France: Belin.

UNESCO. (2011). *ICT competency framework for teachers.* Paris: UNESCO. Retrieved from: http://iite.unesco.org/pics/publications/en/files/3214694.pdf

UNESCO. (2014). *L'éducation pour tous en Afrique subsaharienne en 2010. Rapport pour la réunion du Groupe de Haut Niveau.* Retrieved from: http://www.unesco.org/new/fr/cfit

Yun Dai, D., & Sternberg, R. J. (2004). Beyond constructivism: toward an integrated understanding of intellectual functioning and development. In D. Yun Dai & R. J. Sternberg (Eds.), *Integrative perspectives on intellectual functioning and development* (pp. 3–38). London, UK: Lawrence Erlbaum Associates.

Zydney, J. M., & Warner, M. (2016). Mobile apps for science learning: Review of research. *Computers & Education, 94*, 1–17. doi:10.1016/j.compedu.2015.11.001

KEY TERMS AND DEFINITIONS

Activity: Action that the subject does to achieve the task, with two dimensions: an observable dimension (verbalizations, actions, written and/or numerical traces etc.) and an unobservable dimension (cognitive processes, emotions).

Cognitive Activity: Components of a student's activity as cognitive processes and knowledge used to carry out a task.

Digital Technology: Term used to describe the use of digital resources to teach and to learn in school activities in order to analyse complex situations, find and use information, solve problems, understand concepts.

Efficiency: Qualifier used to mean that the teaching-learning process must allow all students to learn.

Learner-Centred Approach: A pedagogical approach based on proposing prescribed tasks, meaningful to the student and the teacher's promoting the student's learning by helping him or her to discover the solutions of the problems posed by the prescribed task, allowing the student to build new knowledge which involves for the teacher, the understanding difficulties of the student and why he or her could be in difficulty.

Task: This is what to be done; a task is defined by a goal to be attained and the conditions for achieving that goal.

Teaching-Learning Process: Process that takes place during any teaching sequence and that involves different entities in interaction: a student among others students, a teacher, some knowledge to learn, a learning tool, a context.

Chapter 9
STEM Teaching and Learning via Technology-Enhanced Inquiry

Michael L. Connell
University of Houston – Downtown, USA

Sergei Abramovich
State University of New York at Potsdam, USA

ABSTRACT

This chapter aims to address several limitations of Technological Pedagogical Content Knowledge (TPACK) – a theoretical model used in the application of technology when teaching STEM disciplines. To this end, a supplement to TPACK drawn from the Action on Objects (AO) framework (Connell, 2001) is suggested. To illustrate the value of the proposed enhancement of TPACK, an example integrating science, technology, and mathematics is provided. The Texas College and Career Readiness Standards are used to demonstrate the relationship between the proposed theoretical modification of the leading model and the current teaching practice involving such scientific activities as measuring, record keeping, analyzing, conjecturing and evaluating. Additional suggestions and applications of the TPACK/AO model are provided.

INTRODUCTION

STEM Education has undergone changes regarding the manner in which core foundational understandings of content are best developed (Breiner, Harkness, Johnson, & Koehler, 2012; National Research Council, 2003). These changes have contributed to a growing need for revision of existing pedagogy (Hulleman

DOI: 10.4018/978-1-5225-2525-7.ch009

& Harackiewicz, 2009; Labov, Reid, & Yamamoto, 2010). The need for such pedagogical changes in many cases have been both enabled and exacerbated by the application of increasingly powerful educational technology in the K-16 schools (e.g., Abramovich & Cho, 2009; Bodzin, Fu, Bressler, & Vallera, 2015; Hall & Chamblee, 2013; Jang, 2012; Lyons & Tredwell, 2015; So & Ching, 2011; Valanides & Angeli, 2008; Winkel, 2013; Yu & Yu, 2002).

If we take a view of pedagogy as an intersection of student need, teacher ability, and content requirements, the addition of educational technology into the STEM classroom creates new problematics which need to be addressed. In particular, when the teacher is unsure how to apply technology, the content being taught can be diluted or misinterpreted. This is a natural consequence resulting from an increased focus on the development of procedural skills needed to use the technology itself, something that leads to decreased time available for the content. In these cases, students often develop either a broad superficial understanding of a few independent content areas or a small set of procedures only useful for a select set of problems.

Attempts to address such problems may lead to an adoption of a teacher replacement model which incorporates an integrated learning system (ILS) to bypass teachers completely and standardize content delivery. Such efforts to "teacher-proof" content lead to the creation of pre-programmed instructional modules which fall back on a behavioral view of the content (Williams, 1999; Martens, Daly Begeny, & VanDerHeyden, 2011) where each learning objective devolves to an isolated item to be memorized.

Once this viewpoint is adopted, notions of efficiency come into play, in particular, "faster is better" and become the instructional focus. Within this orientation the ILS is designed to take advantage of the computer's speed in enabling a student to "produce" the end product as quickly as possible using immediate feedback to simply posed memory recall items. This "faster is better" belief is then applied using the tremendous speed of the computer, making it possible for the student to know within milliseconds whether or not the memory prompt was addressed correctly. In fairness, it should be noted that such on-the-spot responses may be appropriate for learning in STEM content domains that are very restricted and require a high degree of memorization. For example, development of initial terminology could be well addressed by such methods.

However, outside of such special cases, the immediate feedback produced by implementing this model of instruction does not develop higher order thinking skills. Although the student knows whether or not the answer provided was right or wrong nearly instantaneously, the reason for this answer is not addressed. At a foundational level, the computer in this case does not provide sufficient time for the student to think about the implications of the problem or to perform basic self-confirmations. To put it bluntly, the computer's speed has done away with the time necessary for the student to think and internalize.

BACKGROUND

If a student is to internalize and construct meanings from experiences as required in STEM education, then there must be time to reflect upon the experiences and how they connect with the students' existing knowledge base. When students know within a split-second if a given answer is correct there is no need for further reflection concerning their beliefs, intuitions, prior experiences, relation to other units of knowledge, and so forth. Effective STEM teaching, with its focus on critical analysis and problem solving, should not be reinforcing student beliefs that all that counts is the correct answer. Yet this viewpoint is even planned for and presented literally millions of times each day in numerous Computer Assisted Instruction (CAI) programs. Despite this, immediate feedback is the approach taken in many general computer assisted instruction models. In other words, the computer, rather than enabling problem solving, just makes it "easy" for students to solve problems, replacing their thinking skills by purely artificial intelligence of a computer (Abramovich, 2016).

In addition, programmed learning as typically implemented in ILS shells breaks down concepts into easily memorized bits and pieces to facilitate evaluation and presentation. From the efficiency perspective described earlier, such fragmentation of curricular elements is viewed positively. It contributes to a high rate of correct responses, enables the ILS to ascertain right/wrong immediately, and make decisions regarding optimal path of presentation quickly and efficiently (Harnisch & Connell, 1991). However, when an integrated set of understandings leading to effective application of varied concepts, as the current emphasis in STEM education requires, this fragmentation of the curriculum can have disastrous results.

This is unfortunate, for when educational technology is utilized effectively learners in STEM fields can personally experience the same types of data and tools for exploration and analysis as those used by practicing mathematicians, engineers, and scientists. For example, digital fabrication is not only essential for engineering applications such as 3-D printing but has proven to be a powerful pedagogical tool (Abramovich, 2011; Abramovich & Connell, 2015). In particular, the process of digital fabrication of different geometric shapes formed by the combinations of graphs of the basic functions from the secondary mathematics curriculum – polynomial, exponential, circular – requires deep conceptual understanding of two-variable inequalities that define those shapes. An educational paradigm of digital fabrication

(Gershenfeld, 2005; Bull, Knezek, & Gibson, 2009; Walter-Herrmann & Büching, 2013) has grown out of ubiquitous movement of the 1980s to introduce a pedagogy of student-computer interaction into the schools as a way of improving teaching and learning of mathematics and science. Conceptually, digital fabrication makes it possible "to explore how to represent a functional description of a system in a physical form and likewise to which extent a functional description of a physical system can be abstracted" (Dittert & Krannich, 2013, p. 173). As Nake (2013) noted, digital fabrication is a space where abstract algorithms that guide computational tools and specific productions of art (e.g., a computer aided design) meet. For example, in mathematics, fractal curves, bifurcation diagrams, and images of different kinds of attractors in dynamical systems (Gleick, 1987; Leonov & Kuznetsov, 2013; Boeing, 2016) or even more elementary 2-D graphs, all described by abstract mathematical models, are examples of digital fabrication.

The problem space created through technology enabled explorations surrounding the digital fabrication process provides learners with problems that are rich in both mathematical and engineering sophistication. Furthermore, these problem situations can then be used by the instructor as a mechanism to develop links between procedural skills of using technology in constructing graphs of functions, the factual graphs knowledge related to identifying the loci necessary for the fabrications to be created, and the conceptual knowledge required for the fabrication of the specific parts of these graphs and their intersections (Abramovich & Connell, 2015).

Developing such links requires robust understanding of algebraic inequalities and an ongoing mindful interaction between the human and the computer. Furthermore, these problem situations can then be a mechanism to develop links between *procedural* skills of using technology to graph functions, identification of the *factual* knowledge needed to identify the loci necessary, and the *conceptual* knowledge required to fabricate the specific parts of these graphs and their intersections. When used in such a fashion, educational information technology allows for the dynamic scaffolding of difficult-to-understand abstract concepts. Jonassen (2000) promotes the idea of "using selected computer applications as cognitive tools for engaging and enhancing multiple forms of thinking in learners", which he calls Mindtools.

Traditional computer learning applications assess the effects of computer technologies on the learner where the learner has no input into the process. Mindtools assess the effects of learning with computer technologies when learners enter into an intellectual partnership with the computer (Jonassen, 2000, p. 4).

In today's classroom, application of educational technologies has the potential to increase classroom teachers' ability to guide students' search for the same class of reproducible patterns in nature by displaying real-time data in visual forms that are more easily understood by learners at increasingly younger ages. As this chapter shall illustrate, with the addition of the technology enhanced object there is tremendous leveling of the playing field in terms of scientific and mathematical thinking. Educational technology now routinely supports the Action on Objects approach (Connell, 2001) as a tool to generate meaningful information by the students' *actions*, to capture the essence of the activity by allowing the students to *organize* their work in powerful structures, and to create formal *records of action* that may be shared or used in later problem-solving endeavors and shared globally via the internet.

BRIDGING PCK AND TPACK

Technological Pedagogical Content Knowledge (TPACK, sometimes abbreviated as TPCK) is a theoretical framework that has enjoyed widespread applications as it applies to the integration of technology in the teaching and learning process (Angeli & Valanides, 2009; Mishra & Koehler, 2006; Niess, 2005). Historically, TPACK can trace its origin to early thoughts about teaching and learning emerging at the confluence of Shulman's (1986) notion of pedagogical content knowledge (PCK) and that of Maddux's (1984) Type I/Type II technology applications. The concept of PCK was motivated by observation that "questions about the *content* of the lessons taught, the questions asked, and the explanations offered" (Shulman, 1986, p. 8, italics in the original) are unjustifiably missing from research on teaching. Likewise, the Type I/Type II concept was prompted by a cogitation that because "so many educators seem to be simply taking it for granted that microcomputers are desirable, ... such uncritical and unrealistic optimism is doomed from the start and [without good research] could result in a backlash reaction against educational computing" (Maddux, 1984, p. 36). Thus, at the confluence of the calls for subject matter knowledge and the appropriate use of technology, the TPACK concept was developed.

According to Shulman (1986), the genre of PCK may "include, for the most regularly taught topics in one's subject area, the most useful forms of representation of those ideas, the most powerful analogies, illustrations, examples, explanations, and demonstrations - in a word, the ways of representing and formulating the subject that make it comprehensible to others" (p. 9). PCK in this interpretation also includes an understanding of what makes the learning of specific topics easy or difficult. In particular, application of PCK provides a way of addressing awareness and understanding of the conceptions and preconceptions that students of different ages and backgrounds bring with them to the learning of those most frequently taught topics and lessons.

When these ideas were first discussed, they were commonly represented by a Venn diagram with content in one circle and pedagogy in another circle with PCK as the intersection of the two circles. As such, it was often used to explain/contrast a great scholar who can't teach (high content knowledge and low pedagogical knowledge), an overconfident teacher who audaciously avoids any discourse in the classroom (high pedagogical knowledge and low content knowledge), and a teacher-scholar (high content knowledge and high pedagogical knowledge and the ability to integrate the two) who teaches at the deep structure level (Shulman, 2005). This initial conceptualization of PCK quickly spread as it made sense, was intuitive and it was quite easy to recognize how each of the domains overlap representing new integrated knowledge domains (Graham, 2011).

With the addition of technology, the logical thing to do, given the initial Venn diagram, was to simply add another circle (Figure 1). When this was done the most widespread representations of TPACK would be similar to Figure 1.

Figure 1. TPACK
Reproduced by permission of the publisher, © 2012 by tpack.org

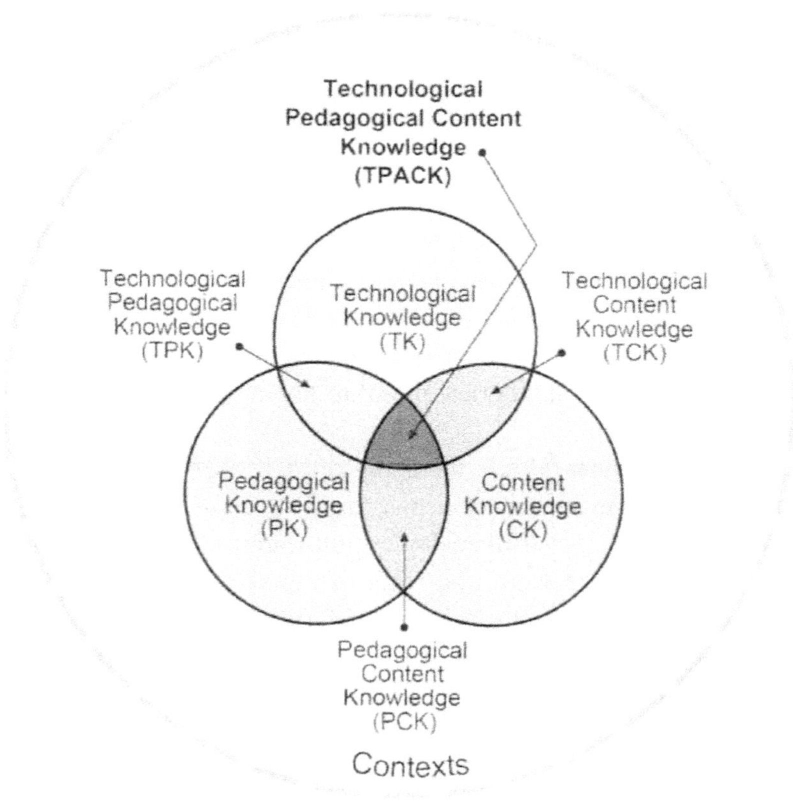

At the early stages, both TPACK and the instructional approaches which were generated by the application of TPACK represented a seemingly adequate descriptive set of theoretical frameworks for implementing, discussing, and researching technology in teaching and learning for all ages. In the years that followed, however, the proliferation of many competing TPACK constructs hints at broader concerns as to whether TPACK is in fact a completely established theoretical framework (Graham, 2011).

Some noted weaknesses of TPACK are inherited from Shulman's original diagrammatic representation, years of subsequent discussion and refinement of PCK, and "fuzzy" definitions (Gess-Newsome, 2002; Graham, 2011). Considering the intersections of the three circles shown in Figure 1, TPACK appears to address everything, but by doing so it creates a host of possible boundary conditions (i.e., TPK, TCK, and TPCK). There have also been many unsuccessful attempts to define, identify, or distinguish these boundary condition constructs (e.g., Angeli & Valanides, 2009; Archambault & Barnet, 2010; Graham et al., 2009; Yurdakul et al., 2012), leading many of these same researchers to question the balance between complexity and parsimony in the entire model. In some part, due to TPACK's comprehensive nature, it appears to lack a prescriptive capacity to help clarify how to best implement technology to increase the effectiveness of teaching and learning (Archambault & Barnett, 2010; Graham, 2011).

To further complicate the increasingly "fuzzy" nature of the TPCK framework, technological knowledge itself adds another layer of fuzziness as the underlying technology is not universally defined. A typical solution to solve the technology definition issue was to create the TPACK constructs, where each attempted to isolate a specific technology (i.e., ePCK, ICT-TPACK). This resulted in a focus on instructional technology and reductionist models where most of the teaching and learning decisions were treated in isolation as described in the ILS example from the introduction.

In concrete terms, the TPCK theoretical framework only truly existed at the intersection of three isolated teacher-based concerns (i.e., TPK, TCK, and TPCK). But, to provide a useful theoretical framework the simultaneous consideration of all three domains within learner-centered environment must be present.

TCK AND ACTION ON OBJECTS

To avoid some of the "fuzzy" problems which had arisen in the application of TPACK, such as the lack of definition in terms of underlying technologies and varying definitions for similar terms, the Action on Objects model (Connell, 2001) was adapted to provide specific guidance in the practical application of TCK. This adaptation provided needed specificity for the TCK approach which had been lacking in the early TPACK formulation. In practice, however, this use is best viewed as a supplement to the TPACK approach and not as a replacement for it.

An example of this adaptation would be initial student *action* upon a *physical object*, such as a pile of counters, used to develop conceptual understanding of the basic operations of arithmetic. A digital picture, or sketch, could then be created using any of a wide variety of applications to record the actual counters and in turn serve as a *graphic object* for further actions by the student. Technologies such as spreadsheets, databases, graphing modules, etc., in addition to serving as further representations of the problem space, have embedded computational and representational power which create natural entry points for more abstracted *student actions* with technology as outlined by the TCK framework. The addition of such entry points enabled by this use of Action on Objects approach allows TCK to obtain a degree of specificity lacking in prior formulations.

To see what this would look like in actual practice consider a student using web-based educational technology to explore the periodic table of the elements. This task can be approached using three levels of technology-enabled objects. First, look at a very simple case where a technologically enabled *Object* simply presents static information - often in the form of facts to be memorized - to a student. An example of such a *Low Level Interaction* may be found at Periodic Table (n. d.) and should be intimately familiar to most readers - the Periodic Table itself (Figure 2).

Figure 2. The Periodic Table: An example of a low-level interaction

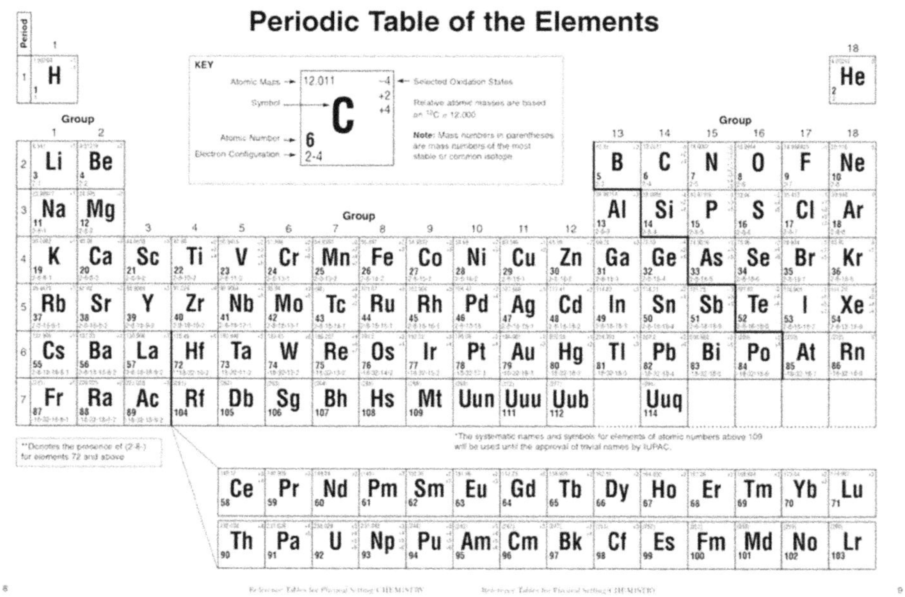

In this example, technology simply presents this object as information - as a webpage. Although it is excellent to have this informational chart, the level of interaction is low and the resulting understanding is primarily of promoting student awareness of existing information. However, in this example one does not take advantage of the true power of technology, and, apart from saving paper and ink, such use of technology does not have any advantage over a simple printed page or a wall chart. Nonetheless, it is quite surprising how often students encounter this type of model in technology use. In such cases, students are presented with a text or information display which is presented in a static form which cannot be directly interacted with.

There is at least one time, however, when such a static representation works very well and that is when the students themselves create it. For example, PowerPoint could be used to record the results of a web-quest exploring elements and their related chemical properties. Such an element web-quest would allow students to explore a variety of information sources, practice classifying, note-taking, editing, and selecting digital photographs representing these ideas. In this case, although the final product is a static representation, the process leading to its creation is a highly interactive and meaningful activity.

In a more powerful Medium Level Interaction, the student can act directly upon the technological object itself. In this case, the understandings which emerge are created by the student who acts upon the technologically enabled object (whose properties were both programmed and presented in a form allowing for easy manipulation by the student). In this simple situation, the student chooses to press certain keys to try for a particular result, and the device reacts. When a student uses a calculator, for example, this is typically the level of interaction they experience. This is true whether using a traditional calculator or a computer-based calculator such as that shown in Figure 3.

Along this line, a very rich set of useful calculators can be found online at (Calculator Soup, n. d.), where a very impressive listing of potential tools has been provided. Among other tools, they include calculators for loan, mortgage, time value of money, algebra, trigonometry, fractions, physics, statistics, time and date, and conversions. Since many of the pages offer sample approaches and/or possibly applicable equations, they can be quite helpful to the student trying to understand their calculations.

Such tools can be extremely powerful in allowing students to explore parts of the universe that would otherwise be unavailable to them. For example, using the interactive tool at http://htwins.net/scale/, it is possible to explore the relative sizes of objects in the universe ranging from the smallest units of space time to the largest cosmic structures (Figure 4).

Figure 3. A computer-based "calculator"

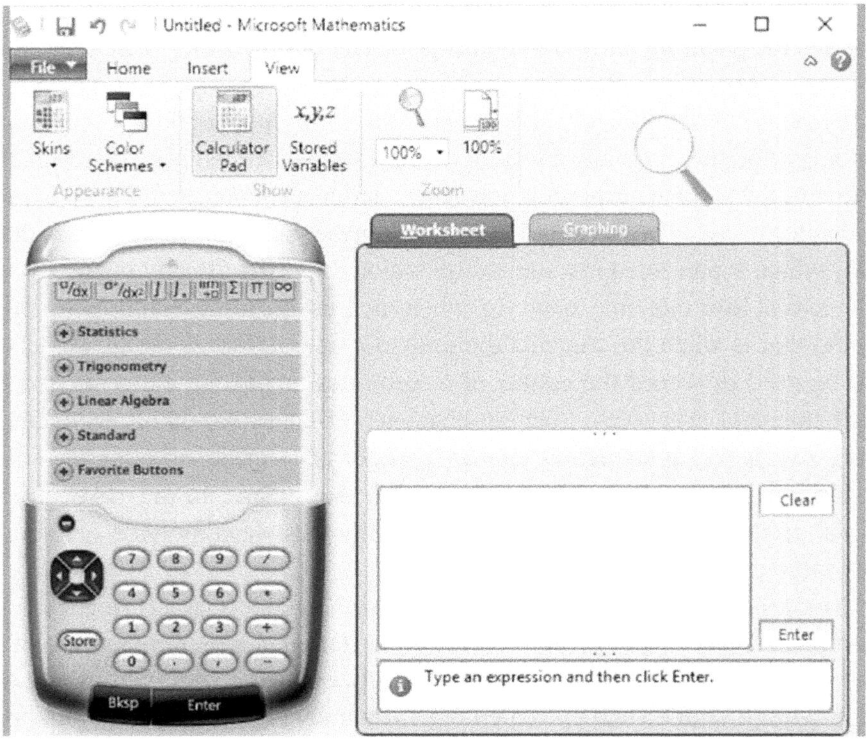

Figure 4. The scale of the universe

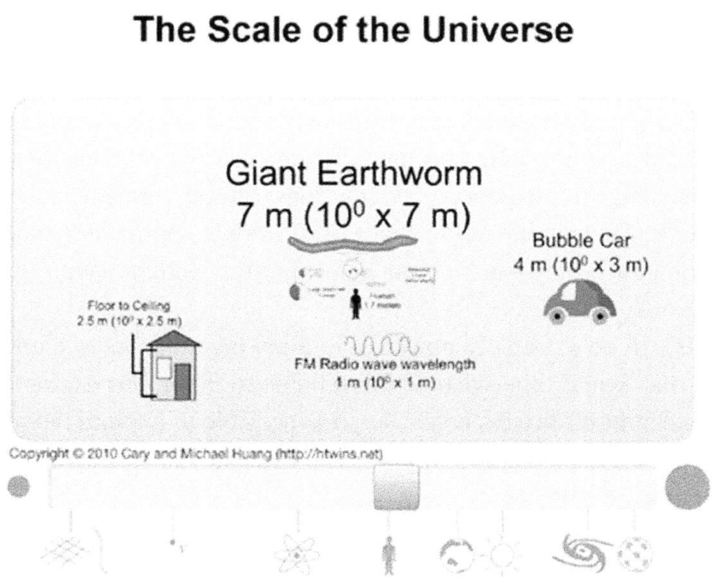

Going back to our focus on the Periodic Table, a slightly more interactive example illustrating *Medium Level Interaction* may be found at (It's Elemental, n. d.). Figure 5 shows the information presented when the student uses this object to explore the element Lithium.

In this example, the object does more than just provide a static image. In addition to the traditional Periodic Table, the object links to additional information which is then presented to the student. This information includes the origin of the element name, how it should be pronounced, common uses, etc. This ability of the components of the environment to interact with the student provides for a tremendous leverage in their utility and power over simple static representations and can even provide alternate representations to the student.

An example of a High Level Interaction, found at http://www.ptable.com/, is shown in Figures 6 - 8, demonstrating one path through the information potentially presented to the student in their further exploration of Lithium.

Although similar in form to the *Low Level Interaction* shown earlier, this more powerful example does more than just provide static information. If, for example, the student hovers their pointer over the *Metals* label the representation changes, highlighting those elements sharing this property (Figure 7) which is then presented to the student.

Should the student have chosen to select directly the LI icon, an external website – in this case a wiki – is provided.

As this example shows, in a High Level Interaction, the direction of the interaction still originates with an action being performed by a student, generally in response to a problem situation or problem solving goal. As was the case with Medium Level Interactions, the student performs actions of their choice directly upon the object. The object may, depending upon the supporting programming or context, link to other

Figure 5. The Periodic Table: An example of a medium level interaction

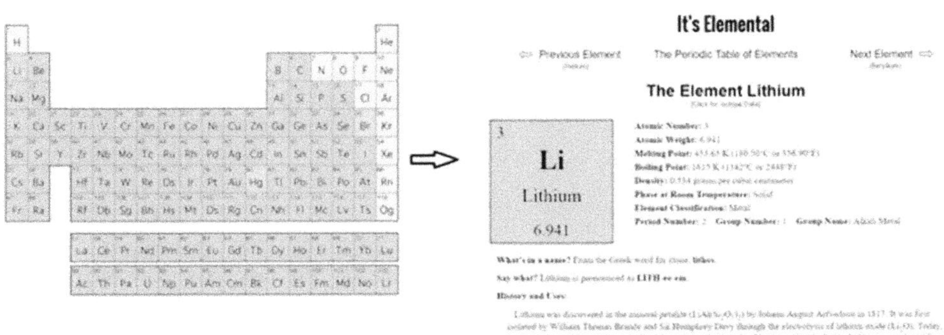

Figure 6. The Periodic Table: An example of a high level interaction – part one

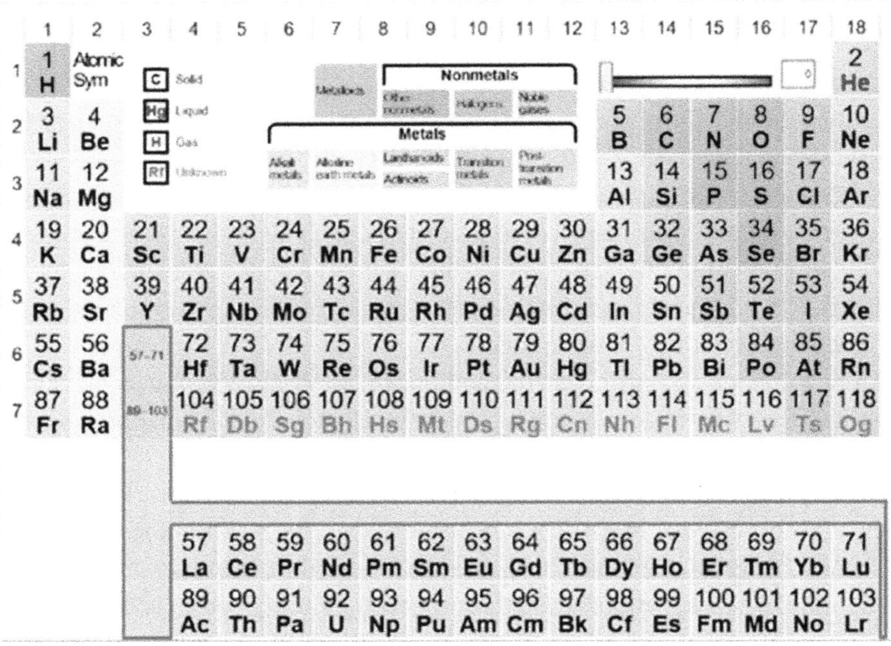

Figure 7. The Periodic Table: An example of a high level interaction – part two

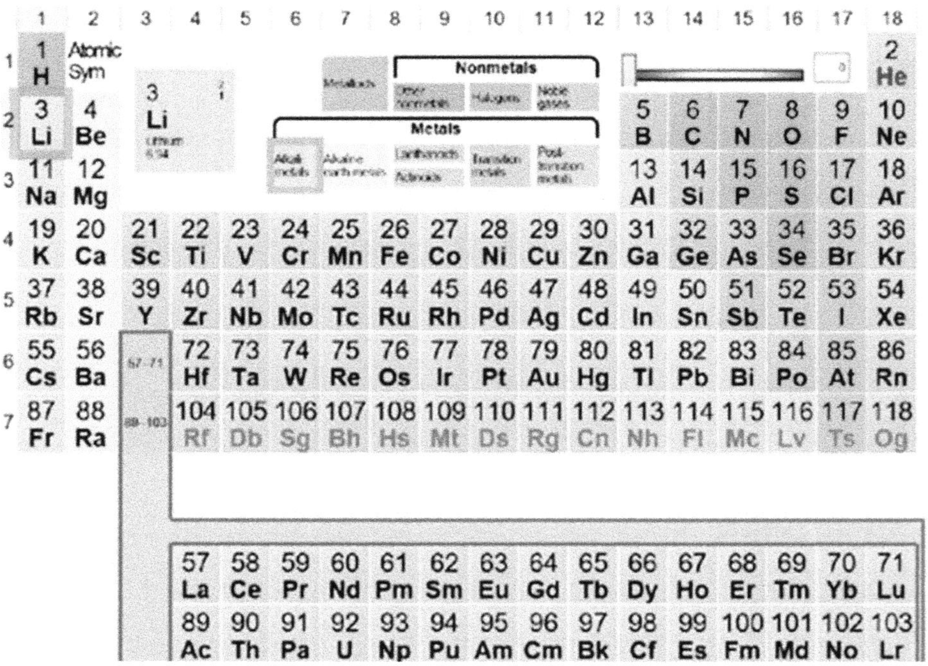

Figure 8. The Periodic Table: An example of a High level interaction – part three

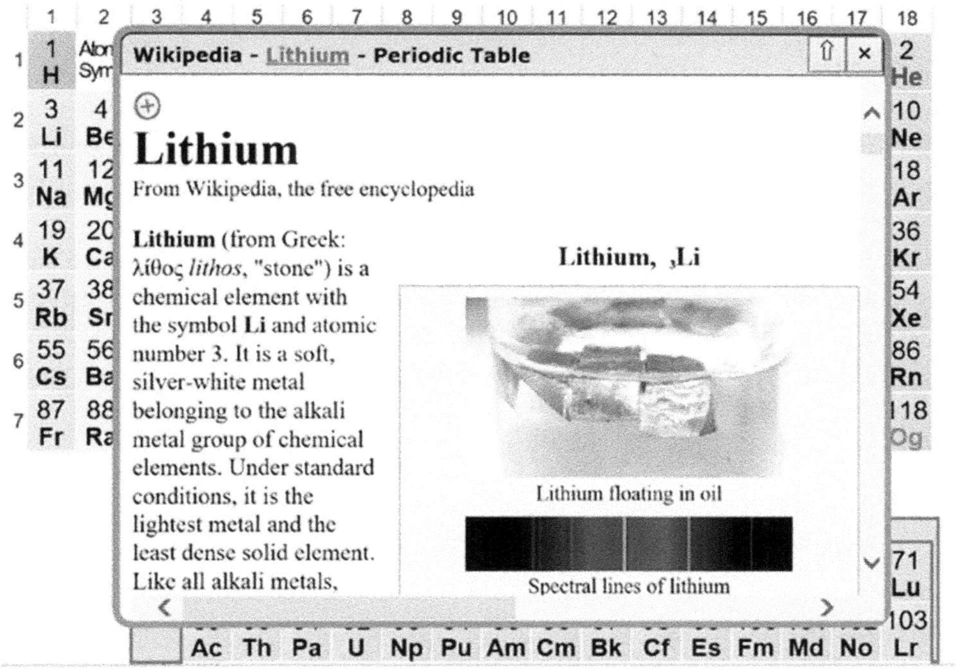

conceptually related objects for additional processing or representational purposes. In other cases, the student may interact with intermediary objects to create specific representations and tools of their own design and choice.

These objects represent a new way of thinking, discussing, and building scientific understanding in an inquiry-oriented environment (Schank & Kozma, 2002). Furthermore, such use of multiple representations in combination supports higher comprehension of phenomena (Kozma, Russell, Jones, Marx, & Davis, 1996; Ainsworth, 2006). Studies suggest that student's generation and manipulation of this type of object lead to higher comprehension levels when compared to viewing teacher-generated representations including moment-by-moment sequences of events, causality, and measures of causal and dynamic reasoning (Gobert & Clement, 1999). In technology enabled activities, students produce and share representations while teachers serve as models and guides to interactive peer collaboration - including using various representations in their own teaching and discourse (Lemke, 2000; Schank & Kozma, 2002).

Such "mindful engagement" enables students to build conceptual knowledge (Salomon, Perkins, & Globerson, 1991), and provides "instantaneous visual information" as they perform experiments (Kuech & Lunetta, 2002) allowing for modifying hypotheses and evaluating results (Friedler, Nachmias, & Linn, 1990; Romanenko, Neverov, & Nikitina, 2012) via the multiple representations such as tables, graphs, simulations, and animations. Digital technologies, which, in the words of Vygotsky (1978), have potential using "first order symbols ... directly denoting objects of actions ... [as means of acquiring] second order symbolism" (p. 115), have been suggested as tools providing the real-time link between a concrete experience and its symbolic representation (Abramovich, 2005, 2014; Kaput, 1992; Mokros & Tinker, 1987).

In looking at the psychological construction of meaning in the various STEM fields, meaning is best made by students' performance with actual actions upon actual objects. In order for this to occur, the object must become real in the minds of the students. It must become instantiated. It must possess well-defined attributes in one's mind so that the eventual symbol generated by the experience has well-defined properties.

AN ILLUSTRATION

Burning the Candle[1]

The activity requires students to use a variety of technologically supported tools to investigate various factors surrounding the burn rate of candles. In particular, unused birthday candles of various lengths and diameters are provided together with lighters, timers of various sorts, computers/tablets with supporting software installed and the Internet access provided.

The activity itself is particularly appropriate for a STEM exploration as it requires careful measurement, record keeping, analysis, hypothesis creation and evaluation. From a technology perspective, multiple tools are easily available and appropriate. This enables the discussion of which tools were chosen, why, and how this choice will impact future actions. It should be noted that this reflects one of a huge number of potential trajectories which any given class of students might take.

From a content perspective, there is large number of areas represented from mathematics, science and their mutual overlap. Appendix includes the list of Texas College Readiness Standards (http://www.thecb.state.tx.us/collegereadiness/crs.pdf), a context in which the first author prepares teachers, that shows some of the major areas addressed.

When the candle activity is presented in objective form, it is often stated as: With your partner determine the burn rate of unused birthday candles of different lengths and diameters. As students gather data and plan their analysis, it is quite common for multiple student-posed questions to be presented. Among these are:

"Is the burn rate constant over the length of the candle?", "How long would it take to totally consume a candle of *x* length?", and "What causes the change in burn rate that one observed?" These questions, and many others, are naturally generated as the students explore the problem setting.

Framing the activity to require a partner ensures that there will be opportunity for discussion and collaboration. It also could result in multiple replications of the activity as different pairs in the class participate. Both of these aspects reflect conscious teacher guidance. The teacher's role is often most present in the framing of the activity and the selection of what tools are available to the students. In accordance with the TCK framework such decisions are the subject of careful consideration and based upon the needs and experiences of the specific group of students. Tasks must also be selected, as was done in this example, which lend themselves to multiple approaches coupled with the possibility of using technological tools of various types. These tools, whether digital or physical, must be both available and familiar to the students. By selecting the overarching question to be explored and providing a rich and varied tool set for the students to select, the learning demands of the content can be thoroughly addressed.

As the students explore the initial teacher-posed problem of determining the burn rate and their own self-generated problems they must choose between these wide variety of tools and applications. Since these will vary from classroom to classroom the principles of Action on Object enhanced TCK are used to guide the decision-making process and subsequent actions and recordings.

Given the teacher-posed question of determining the burn rate, students quickly recognize the need to measure elapsed time and candle length. They might also recognize that properties of the candle itself (which is a type of Object) should be carefully examined. For example, a typical birthday candle, as shown in Figure 9, is not uniform in its diameter. The initial portion of the candle is shaped like a cone. This means that the diameter is not constant over the candle length which comes to influence the initial burn rates for the candle. It is not uncommon when a class explores this activity for some groups to make note of this, while other groups miss it completely. The explanatory power of this observation, or lack thereof, can itself be the subject of additional inquiry using either the same tools or a more sophisticated set.

As definitions are developed and explored (the "burn rate", for example), the students begin the process of identifying what the knowns and unknowns are, the relationships between them, and begin to create a plan for exploration. At this point, they generally recognize the need to record their measurements of length and time. Most groups will choose to standardize the length of time between the measurements. Not every group will select the same time interval, however, which later lends itself to further discussion of exploration possibilities. It is worth noting

Figure 9. Typical birthday candle showing the conical taper

that in this approach towards instruction the goal is not to speedily arrive at a single answer in the shortest time possible. Rather, a primary instructional goal is to allow students to explore, compare, and contrast a variety of approaches towards the same problem. Generalizations can then be made and finally a formal record of findings is produced. The parallels between this approach and the canonic forms of scientific reasoning are not accidental.

It is important to recognize that not every construction of meaning is of equal worth. There are truths and relationships contained within the activity which must be uncovered if the fuller learning objectives are to be met. Good teaching which leads to bad content understanding clearly violates the underlying premises of TCK, TPACK and basic pedagogy.

Let us assume that the students have access to a computer with a spreadsheet program. This is a very natural digital tool for students to use, particularly, if they have had earlier experience with the tool itself (which is highly recommended). As an aside, it is possible for a variety of tools to be used in the data collection. For example, one group might use a ruler and stopwatch while another group may choose to create a digital recording of the candle as it burns in front of a grid graduated in millimeters to be used for later analysis. The choice of method is, again, recorded and made available for discussion at the conclusion of the activity.

Assuming that a pair of students have decided to measure time in seconds, length in millimeters, and use a consistent time interval of 30 seconds, it is entirely possible that the spreadsheet would be used to generate a record of activity like that shown in Figure 10. Provided that sufficient precision and care were taken the method of measurement is not relevant at this point although it may become a discussion item later.

Figure 10. A record for candle burning

Time (sec.)	Length, *l* (mm)
0	
30	
60	
90	

Now that this record of action is in place, the power of the technology comes to play. The initial record of action can now be the Object of further actions on the part of the student. The nature of precisely what these actions are is dictated by the form the data was recorded in and the tools available within the spreadsheet itself. Typical beginning actions might include constructing a graph, examining the table of results, or posting the findings to a classroom website where they can be shared and discussed with other groups. Figure 11 shows a typical student-generated worksheet that addresses this choice of actions.

Depending upon the experience and sophistication of the students, more advanced actions might include creating a regression equation, examining correlations between recorded variables, looking for statistically significant outliers among the classroom records, etc. Figure 12 shows some of these more advanced analyses. As was mentioned in the introduction, this example serves to illustrate how today's classroom technology routinely supports the Action on Objects approach (Connell, 2001) as a tool to record the information generated by the students to capture the essence of the activity. The approach allows the students to organize their work in powerful structures and to create formal records of action that may be used in later problem solving endeavors and even shared globally via the Internet.

FUTURE RESEARCH DIRECTIONS

When thinking about future research directions, one has to keep in mind that educational technologies provide "mindful engagement" allowing students to build conceptual knowledge (Salomon et al., 1991; Sheppard, Macatangay, Colby, & Sullivan, 2009; Turkle, 2008), "instantaneous visual information" as student perform

Figure 11. Spreadsheet record showing basic graph and table views

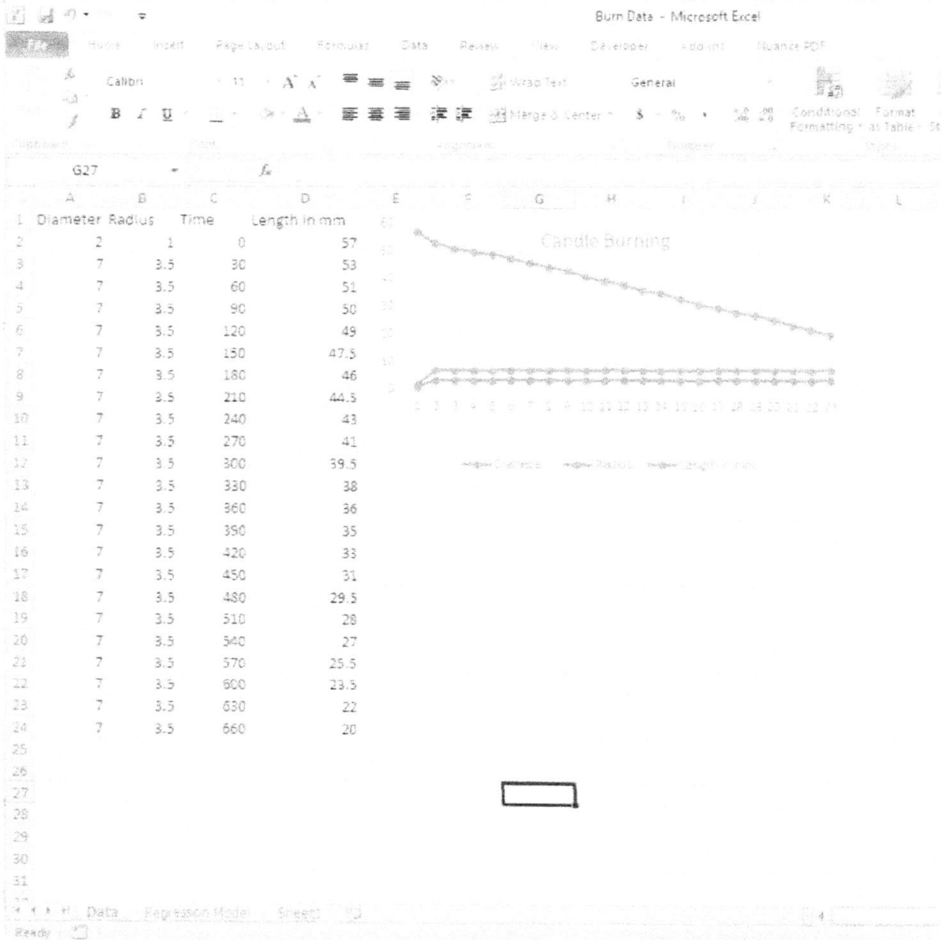

Diameter	Radius	Time	Length in mm
2	1	0	57
7	3.5	30	53
7	3.5	60	51
7	3.5	90	50
7	3.5	120	49
7	3.5	150	47.5
7	3.5	180	46
7	3.5	210	44.5
7	3.5	240	43
7	3.5	270	41
7	3.5	300	39.5
7	3.5	330	38
7	3.5	360	36
7	3.5	390	35
7	3.5	420	33
7	3.5	450	31
7	3.5	480	29.5
7	3.5	510	28
7	3.5	540	27
7	3.5	570	25.5
7	3.5	600	23.5
7	3.5	630	22
7	3.5	660	20

experiments (Kuech & Lunetta, 2002). In doing so, the students modify hypotheses, and evaluate results (Cox & Brna, 2016) via the multiple representations such as tables, graphs, simulations, and animations. Digital technologies have been suggested as tools providing the real-time link between a concrete experience and its symbolic representation emphasizing the primordial nature of the first order symbols over the second order symbolism (Vygotsky, 1978).

These phenomena include events that are unobservable in the classroom due to their size, duration, or location and the accompanying multimedia databases (Furness, Winn, & Yu, 1997). Static images, animations, simulations, virtual reality, and distributed learning environments (Bodzin & Park, 2002) can be used to transition from passive learning to interactive learning (Yair, Mintz, & Litvak, 2001).

Figure 12. Spreadsheet record showing advanced analysis made possible by technology

SUMMARY OUTPUT

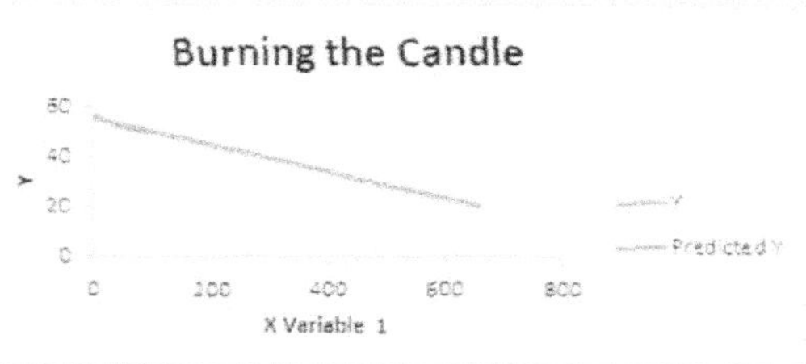

Regression Statistics	
Multiple R	0.999
R Square	0.998
Adjusted R Square	0.997
Standard Error	0.547
Observations	23.000

ANOVA

	df	*SS*	*MS*	*F*	*Significance F*
Regression	1.00	2572.51	2572.51	8585.61	0.00
Residual	21.00	6.29	0.30		
Total	22.00	2578.80			

	Coefficients	*Standard Error*	*t Stat*	*P-value*	*Lower 95%*	*Upper 95%*	*Lower 95.0%*	*Upper 95.0%*
Intercept	55.36	0.22	250.49	0.00	54.90	55.82	54.90	55.82
X Variable 1	-0.05	0.00	-92.66	0.00	-0.05	-0.05	-0.05	-0.05

RESIDUAL OUTPUT

Observation	*Predicted Y*	*Residuals*
1.00	55.36	1.64
2.00	53.77	-0.77
3.00	52.18	-1.18
4.00	50.58	-0.58
5.00	48.99	0.01
6.00	47.39	0.11
7.00	45.80	0.20
8.00	44.20	0.30
9.00	42.61	0.39
10.00	41.01	-0.01
11.00	39.42	0.08
12.00	37.83	0.17
13.00	36.23	-0.23
14.00	34.64	0.36
15.00	33.04	-0.04
16.00	31.45	-0.45
17.00	29.85	-0.35

The computer, with its object-oriented interfaces and tools geared specifically to enable the user to perform specific actions upon specific objects, lends itself perfectly to an action upon object model of learning and instruction. Some of these include Step-by-Step Development of Mental Activities (Galperin & Talizyna, 1979; Leontiev, 1981), Action Reification (Sfard, 1994), and Action on Objects (Connell, 2001) to name a few. This linkage between suggestions from learning theories and the object classes engendered by the technology is far too priceless to allow it going to waste.

In looking at the psychological construction of the meaning of a STEM-related concept, we observe that meaning is best made by students' performance with actual actions upon actual objects. In order for this to occur, the object must become real in the minds of the students. It must become instantiated. It must possess well-defined attributes so that the eventual symbol generated by the experience has well-defined properties. And it is here that a conflict seems to occur between our educational preferences (experimental background and theoretical prescriptions) and the nature of the object.

Therefore, a particular item of interest of future research has to do with the nature of the action and the nature of the object in the ongoing construction of STEM-related concepts. Whitehead, Griffin, and Sherburne (1979) argue quite convincingly that it is through the increasing sophistication and power of abstraction enabled through symbol systems based upon more primitive objects that we are able to leverage forward our thinking. There is a long history regarding the nature of the foundational object upon which one acts and the importance of this foundational object in the generation of later conceptualizations.

CONCLUSION

The range of problems which are currently being addressed by computer applications in STEM education has steadily increased with uses in many cases limited only by the imagination of those who would be users. Concurrently with this growth in general technological application, there has been an increasing naive perspective that the use of technology, in and of itself, will solve many of our educational problems merely by being made available in the classroom. Unless we ask the difficult questions concerning at what point technology enters into educational efforts and evaluate the affects upon desired educational values, we run a great risk of misapplication which could seriously damage the educational enterprise.

For example, as Clark and Meyer (2016) have noted "...the human cognitive system is limited and, when it comes to instruction, less is often more... often students learn more content when less glitz is presented." Furthermore, the very features which contribute to a computer package suitable for a skilled user - speed, accuracy, flexibility, etc. – can make it unsuitable for a student attempting to construct a conceptual understanding of the topic. Careful application of some variant of the TPACK framework, as supplemented by the Action on Object approaches outlined in this chapter, has the potential to provide sufficient structure and guidance for teachers to effectively utilize technology in the STEM classroom. Regardless of which approach is eventually adopted, however, it is clear that the computer must be more than a black-box which always gives the right answer or provides the optimal path.

The box itself must be transparent and its methods understandable to the student. The adoption of such a methodology, where the goals, values and assumptions of education - including a careful examination of the role of the teacher - are carefully considered the adoption of technology in STEM education can live up to its promise as a powerful educational tool.

REFERENCES

Abramovich, S. (2005). Early algebra with graphics software as a Type II application of technology. In C. D. Maddux & D. L. Johnson (Eds.), Classroom integration of Type II uses of technology in education (pp. 21-33). New York, NY: The Haworth Press.

Abramovich, S. (2011). Technology as a context for secondary pre-teachers' use of algebraic inequalities in a capstone course. In *Proceedings of Society for Information Technology & Teacher Education International Conference 2011*. Chesapeake, VA: Association for the Advancement of Computing in Education.

Abramovich, S. (2014). On the duality of transition from visual to symbolic in the teaching of school mathematics. *Open Mathematical Education Notes*, *4*, 9–18.

Abramovich, S. (2016). Do computers enable problem solving or just make it "easy"? *Computer Tools in Education*, *2*, 45–54.

Abramovich, S., & Cho, E. K. (2009). Mathematics, computers, and young children as a research-oriented learning environment for a teacher candidate. *Asia Pacific Education Review*, *10*(2), 247–259. doi:10.1007/s12564-009-9015-2

Abramovich, S., & Connell, M. L. (2015). Digital fabrication and hidden inequalities: Connecting procedural, factual and conceptual knowledge. *International Journal of Technology in Teaching and Learning*, *11*(2), 76–89.

Ainsworth, S. (2006). DefT: A conceptual framework for considering learning with multiple representations. *Learning and Instruction*, *16*(3), 183–198. doi:10.1016/j.learninstruc.2006.03.001

Angeli, C., & Valanides, N. (2009). Epistemological and methodological issues for the conceptualization, development, and assessment of ICT–TPCK: Advances in technological pedagogical content knowledge (TPCK). *Computers & Education*, *52*(1), 154–168. doi:10.1016/j.compedu.2008.07.006

Archambault, L. M., & Barnett, J. H. (2010). Revisiting technological pedagogical content knowledge: Exploring the TPACK framework. *Computers & Education*, *55*(4), 1656–1662. doi:10.1016/j.compedu.2010.07.009

Bodzin, A., & Park, J. C. (2002). Using a non-restrictive web-based forum to promote reflective discourse with preservice science teachers. *Contemporary Issues in Technology & Teacher Education*, *2*(3). Retrieved from http://www.citejournal.org/vol2/iss3/science/article1.cfm

Bodzin, A. M., Fu, Q., Bressler, D., & Vallera, F. L. (2015). Examining the enactment of Web GIS on students geospatial thinking and reasoning and tectonics understanding. *Computers in the Schools*, *32*(1), 63–81. doi:10.1080/07380569.2015.1004278

Boeing, G. (2016). Visual Analysis of Nonlinear Dynamical Systems: Chaos, Fractals, Self-Similarity and the Limits of Prediction. *Systems*, *4*(4), 37. doi:10.3390/systems4040037

Breiner, J. M., Harkness, S. S., Johnson, C. C., & Koehler, C. M. (2012). What is STEM? A discussion about conceptions of STEM in education and partnerships. *School Science and Mathematics*, *112*(1), 3–11. doi:10.1111/j.1949-8594.2011.00109.x

Bull, G., Knezek, G., & Gibson, D. (2009). Editorial: A rationale for incorporating engineering education into the teacher education curriculum. *Contemporary Issues in Technology & Teacher Education*, *9*(3), 222–225. Retrieved from http://www.editlib.org/p/32411

Calculator Soup, L. (n.d.). *Calculator Index*. Retrieved January 10, 2017, from http://www.calculatorsoup.com/calculators/

Clark, R. C., & Mayer, R. E. (2016). *E-learning and the science of instruction: Proven guidelines for consumers and designers of multimedia learning*. Hoboken, NJ: John Wiley & Sons. doi:10.1002/9781119239086

Connell, M. L. (2001). Actions upon objects: A metaphor for technology enhanced mathematics instruction. In D. Tooke & N. Henderson (Eds.), *Using information technology in mathematics* (pp. 143–171). Binghamton, NY: Haworth Press.

Cox, R., & Brna, P. (2016). Twenty years on: Reflections on Supporting the use of external representations in problem solving.... *International Journal of Artificial Intelligence in Education*, *26*(1), 193–204. doi:10.1007/s40593-015-0054-z

Dittert, N., & Krannich, D. (2013). Digital fabrication in educational contexts – ideas for a constructionist workshop setting. In J. Walter-Herrmann & C. Büching (Eds.), *FabLab: Of machines, makers and inventors* (pp. 173–180). Bielefeld, Germany: Transcript Verlag. doi:10.14361/transcript.9783839423820.173

Friedler, Y., Nachmias, R., & Linn, M. C. (1990). Learning scientific reasoning skills in microcomputer based laboratories. *Journal of Research in Science Teaching*, *27*(2), 173–192. doi:10.1002/tea.3660270208

Furness, T. A., III, Winn, W., & Yu, R. (1997). The impact of three dimensional immersive virtual environments on modern pedagogy: Global change, VR and learning. In Proceedings of Workshops in Seattle, Washington, and Loughborough, England in May and June 1997.

Galperin, P., & Talizyna, T. (1979). Contemporary condition of the SSDMA theory. *Vestnik MGU. Psychology (Savannah, Ga.)*, *14*(4), 54–63.

Gershenfeld, N. (2005). *Fab: The coming revolution on your desktop – from personal computers to personal fabrication*. New York, NY: Basic Books.

Gess-Newsome, J. (1999). Pedagogical content knowledge: An introduction and orientation. In J. Gess-Newsome & N. Lederman (Eds.), *Examining pedagogical content knowledge: The constructs and its implication for science education* (pp. 3–17). Dordrecht, The Netherlands: Kluwer Academic Publishers.

Gleick, J. (1987). *Chaos: Making a new science*. Elmsford, NY: Pergamon Press.

Gobert, J. D., & Clement, J. J. (1999). Effects of student-generated diagrams versus student-generated summaries on conceptual understanding of causal and dynamic knowledge in plate tectonics. *Journal of Research in Science Teaching*, *36*(1), 39–53. doi:10.1002/(SICI)1098-2736(199901)36:1<39::AID-TEA4>3.0.CO;2-I

Graham, C. R. (2011). Theoretical considerations for understanding technological pedagogical content knowledge (TPACK). *Computers & Education*, *57*(3), 1953–1960. doi:10.1016/j.compedu.2011.04.010

Graham, C. R., Burgoyne, N., Cantrell, P., Smith, L., St. Clair, L., & Harris, R. (2009). Measuring the TPACK confidence of inservice science teachers. *TechTrends*, *53*(5), 70–79. doi:10.1007/s11528-009-0328-0

Hall, J., & Chamblee, G. (2013). Teaching algebra and geometry with GeoGebra: Preparing pre-service teachers for middle grades/secondary mathematics curriculum. *Computers in the Schools*, *30*(1-2), 12–29. doi:10.1080/07380569.2013.764276

Harnisch, D. L., & Connell, M. L. (1991). *An Introduction to educational information technology* (3rd ed.). Kawasaki, Japan: NEC Technical College.

Hulleman, C., & Harackiewicz, J. M. (2009). Promoting interest and performance in high school science classes. *Science, 326*(5958), 1410–1412. doi:10.1126/science.1177067 PMID:19965759

It's Elemental. (n.d.). Retrieved January 10, 2017, from http://education.jlab.org/itselemental/

Jang, S.-J. (2012). Developing a technology-team teaching model to enhance the TPACK of pre-service science teachers. In S. Abramovich (Ed.), *Computers in education* (Vol. 2, pp. 71–90). New York, NY: Nova Science Publishers.

Jonassen, D. (2000). *Mindtools for schools: Engaging critical thinking with technology*. Columbus, OH: Merrill.

Kaput, J. J. (1992). Technology and mathematics education. In D. A. Grouws (Ed.), *Handbook of research on mathematics teaching and learning* (pp. 515–556). New York, NY: Macmillan.

Kozma, R. B., Russell, J., Jones, T., Marx, N., & Davis, J. (1996). The use of multiple, linked representations to facilitate science understanding. In *NATO Symposium on International Perspectives on the Psychological Foundations of Technology-Based Learning Environments, Crete, Greece, Jul 1992, and at the 5th EARLI Conference, Aix-en-Provence, France, Sep 1993.* Mahwah, NJ: Lawrence Erlbaum Associates.

Kuech, R. K., & Lunetta, V. N. (2002). Using digital technologies in the science classroom to promote conceptual understanding. *Journal of Computers in Mathematics and Science Teaching, 21*(2), 103–126.

Labov, J. B., Reid, A. H., & Yamamoto, K. R. (2010). Integrated biology and undergraduate science education: A new biology education for the twenty-first century? *CBE Life Sciences Education, 9*(1), 10–16. doi:10.1187/cbe.09-12-0092 PMID:20194802

Lemke, J. L. (2000). Across the scales of time: Artifacts, activities, and meanings in ecosocial systems. *Mind, Culture, and Activity, 7*(4), 273–290. doi:10.1207/S15327884MCA0704_03

Leonov G. A., & Kuznetsov N. V. (2013). Hidden attractors in dynamical systems. From hidden oscillations in Hilbert–Kolmogorov, Aizerman, and Kalman problems to hidden chaotic attractor in Chua circuits. *International Journal of Bifurcation and Chaos, 23*. DOI: 10.1142/S0218127413300024

Leontiev, A. N. (1981). The problem of activity in psychology. In J. V. Wertsch (Ed.), *The concept of activity in Soviet psychology* (pp. 37–72). Armonk, NY: M. E. Sharpe.

Lyons, C. D., & Tredwell, C. T. (2015). Steps to implementing technology in inclusive early childhood programs. *Computers in the Schools*, *32*(2), 152–166. doi:10.1080/07380569.2015.1038976

Maddux, C. D. (1984). Educational microcomputing: The need for research. *Computers in the Schools*, *1*(1), 35–41. doi:10.1300/J025v01n01_04

Martens, B. K., Daly, E. J., Begeny, J. C., & VanDerHeyden, A. (2011). Behavioral approaches to education. In W. W. Fisher, C. C. Piazza, & H. S. Roane (Eds.), *Handbook of applied behavior analysis* (pp. 385–401). New York, NY: The Guilford Press.

Mishra, P., & Koehler, M. J. (2006). Technological pedagogical content knowledge: A framework for teacher knowledge. *Teachers College Record*, *108*(6), 1017–1054. doi:10.1111/j.1467-9620.2006.00684.x

Mokros, J. R., & Tinker, R. F. (1987). The impact of microcomputer-based labs on childrens ability to interpret graphs. *Journal of Research in Science Teaching*, *24*(4), 369–383. doi:10.1002/tea.3660240408

Nake, F. (2013). Considering algorithmics and aesthetics. In J. Walter-Herrmann & C. Büching (Eds.), *FabLab: Of machines, makers and inventors* (pp. 79–92). Bielefeld, Germany: Transcript Verlag. doi:10.14361/transcript.9783839423820.79

National Research Council. (2003). Improving undergraduate instruction in science, technology, engineering, and mathematics: Report of a workshop. Steering Committee on Criteria and Benchmarks for Increased Learning from Undergraduate STEM Instruction. In R. A. McCray, R. DeHaan, & J. A. Schuck (Eds.), Committee on Undergraduate Science Education, Center for Education, Division of Behavioral and Social Sciences and Education. Washington, DC: The National Academies Press.

Niess, M. L. (2005). Preparing teachers to teach science and mathematics with technology: Developing a technology pedagogical content knowledge. *Teaching and Teacher Education*, *21*(5), 509–523. doi:10.1016/j.tate.2005.03.006

Periodic Table. (n.d.). Retrieved January 10, 2017, from http://www.printablee.com/postpic/2015/10/periodic-table-with-oxidation-numbers_400959.png

Romanenko, V., Neverov, V., & Nikitina, G. (2012). Computer-assisted laboratory instruction at the tertiary level. In S. Abramovich (Ed.), *Computers in education* (Vol. 2, pp. 91–105). New York, NY: Nova Science Publishers.

Salomon, G., Perkins, D. N., & Globerson, T. (1991). Partners in cognition: Extending human intelligence with intelligent technologies. *Educational Researcher, 20*(3), 2–9. doi:10.3102/0013189X020003002

Schank, P., & Kozma, R. (2002). Learning chemistry through the use of a representation-based knowledge building environment. *Journal of Computers in Mathematics and Science Teaching, 21*(3), 253–279.

Sfard, A. (1994). Reification as the birth of metaphor. *For the Learning of Mathematics, 14*(1), 44–55.

Sheppard, S. D., Macatangay, K., Colby, A., & Sullivan, W. M. (2009). *Educating engineers*. San Francisco, CA: Jossey-Bass.

Shulman, L. S. (1986). Those who understand: Knowledge growth in teaching. *Educational Researcher, 15*(2), 4–14. doi:10.3102/0013189X015002004

Shulman, L. S. (2005). Signature pedagogies in the professions. *Daedalus, 134*(3), 52–59. doi:10.1162/0011526054622015

So, W.-M. W., & Ching, N.-Y. F. (2011). Pupil science learning in resource-based e-learning environments. *Journal of Computers in Mathematics and Science Teaching, 30*(2), 203–223.

The Scale of the Universe. (n.d.). Retrieved January 10, 2017, from http://htwins.net/scale/

Turkle, S. (Ed.). (2008). *Falling for science: Objects in mind*. Cambridge, MA: MIT Press.

Valanides, N., & Angeli, C. (2008). Learning and teaching about scientific models with a computer modeling tool. *Computers in Human Behavior, 24*(2), 220–233. doi:10.1016/j.chb.2007.01.005

Vygotsky, L. S. (1978). *Mind in society*. Cambridge, MA: MIT Press.

Walter-Herrmann, J., & Büching, C. (Eds.). (2013). *FabLab: Of machines, makers and inventors*. Bielefeld, Germany: Transcript Verlag. doi:10.14361/transcript.9783839423820

Whitehead, A. N., Griffin, D. R., & Sherburne, D. W. (1979). *Process and reality: An essay in cosmology*. New York, NY: Free Press.

Williams, R. L. (1999). The behavioral perspective in contemporary education. *Teacher Educator, 35*(2), 44–60. doi:10.1080/08878739909555225

Winkel, B. (2013). Computers have taken us to the brink in mathematics ... and we have balked. *Computers in the Schools, 30*(1-2), 148–171. doi:10.1080/07380569.2013.768940

Yair, Y., Mintz, R., & Litvak, S. (2001). 3D-Virtual reality in science education: An implication for astronomy teaching. *Journal of Computers in Mathematics and Science Teaching, 20*(3), 293–305.

Yu, F. Y., & Yu, H. J. J. (2002). Incorporating e-mail into the learning process: Its impact on student academic achievement and attitudes. *Computers & Education, 38*(1), 117–126. doi:10.1016/S0360-1315(01)00085-9

Yurdakul, I. K., Odabasi, H. F., Kilicer, K., Coklar, A. N., Birinci, G., & Kurt, A. A. (2012). The development, validity and reliability of TPACK-deep: A technological pedagogical content knowledge scale. *Computers & Education, 58*(3), 964–977. doi:10.1016/j.compedu.2011.10.012

KEY TERMS AND DEFINITIONS

Action on Objects: A learning approach which, in part, provides guidance in selecting instructional referents, the actions students are to perform, and the content one would expect to emerge. Originally developed for mathematics by Connell in the 1990's, it was later expanded to supplement TPACK by Slough and Connell in 2006.

Computer Assisted Instruction: An interactive instructional technique where a computer is used to present instructional material and monitor learning that takes place often using a combination of text, graphics, sound and video.

Digital Fabrication: A process that joins design with production through the use of 3D modeling software or computer-aided design.

Integrated Learning System (ILS): A hardware/software combination designed to deliver instructional content coupled with the ability to measure, monitor, and maintain the content utilizing an array of assessment and management tools.

Pedagogical Content Knowledge (PCK): Commonly used ways of representing and formulating a subject that make it comprehensible to others. In Shulman's original discussions, it is viewed as the overlap between a teacher's content knowledge and pedagogical knowledge.

Science, Technology, Engineering and Mathematics (STEM): A set of deeply connected subject areas both in the real world and in how students learn within each.

Technological Pedagogical Content Knowledge (TPACK): Sometimes abbreviated as TPCK, a theoretical framework expanding upon Shulman's PCK concept, that has enjoyed widespread applications as it applies to the integration of technology in the teaching and learning process.

ENDNOTE

1 See *Appendix* for a highly-scripted version of this activity to illustrate the scaffolding which might be necessary for a novice.

APPENDIX

Sample Activity Illustrating Proposed Chapter Approach: Burning the Candle

Activity Objective:

Determine the burn rate of unused birthday candles of different lengths and diameters.

Is the burn rate constant over the length of the candle? Explain changes in the burn rate for single candles and for candles of different sizes.

Texas College Readiness Standards

Mathematics

I. Numeric Reasoning: C. 1
II. Algebraic Reasoning: D. 1
III. Geometric Reasoning: A. 1
IV. Measurement Reasoning: A. 1; C. 1
VI. Statistical Reasoning: A.1; B. 1, 2, 4; C. 1, 2, 3
VII. Functions: A. 1; B. 1, 2; C. 2.
VIII. Problem Solving and Reasoning: A, B, C
IX. Communication and Representation: A, B, C
X. Connections: A, B

Science

I. Nature of Science: Scientific Ways of Learning and Thinking: A, B, C, D, E
II. Foundation Skills: Scientific Applications of Mathematics: A, B, D, E, F
III. Foundation Skills: Scientific Applications of Communication: A, C
IV. Cross-Disciplinary Themes: C, D, E

Cross-Disciplinary

I. Key Cognitive Skills: A, B, C, E
II. Foundational Skills: D, E

Materials:

- Birthday candles (a variety of sizes is preferred), lighter, Silly Putty, timer, ruler
- Access to a computer or tablet with Excel
- (Optional) Graph paper

Part 1

Engage/Explore:

At least 2 people are needed to conduct each experiment. One to start the timer the instant the other lights the candle; one to blow out the candle as soon as the timer reaches 30s; one to measure the remaining length of the candle in mm after each 30 burn.

1. Measure the candle's diameter and length from the base of its wick, in mm. Also, measure the height of the cone-shaped top from the base of its wick to where the candle becomes cylindrical. This measurement should be done in line with the cylindrical length.
2. Create a spreadsheet with at least two columns, Time (secs) and Length (mm). Include other columns as you see fit (see Figure 10).
3. Measure the candle from the base of the wick to the other end of the candle, in millimeters. For Time = 0 seconds, record this length in mm. Set the candle vertically into a wad of Silly Putty.
4. Set the timer to count down from 30 seconds to 0 seconds. Light the candle and start the timer.
5. As soon as the timer reaches 0 seconds, blow out the candle.
6. Remove the candle from the Silly Putty and measure its length.
7. Repeat and record the length for each 30 second burn.
8. Continue until the candle is too short to burn.

Explore:

Create a graph of Candle Length (y-axis) versus Burn Time (x-axis). Be sure to note down the original length and diameter of the candle; and the height of the conical tip of the candle. What other connections can you explore using the spreadsheet tools?

Explain:

Does the data form a straight line? Are there places where the graph changes? Are these consistent with changes in the shape of the candle? Account for any inconsistencies in timing and actual burn time.

Evaluate:

Check data, graphs and oral or written explanations.

Elaborate:

Compare your results with those from other groups. Consider the diameters of their candles. Did those with similar size candles produce similar graphs? Did those with different size candles produce different graphs? Can you explain the differences?

Chapter 10
Technology-Supported Inquiry in STEM Teacher Education:
From Old Challenges to New Possibilities

Marina Milner-Bolotin
University of British Columbia, Canada

ABSTRACT

The chapter describes the implementation of collaborative educational technologies in STEM teacher education to support teacher-candidates in acquiring inquiry-based teaching skills and positive attitudes about inquiry learning. The focus is on five different collaborative technology-enhanced pedagogies: (1) Peer Instruction, (2) collaborative design of conceptual questions with PeerWise, (3) data-driven STEM inquiry via using live data collection and analysis, (4) computer modeling-enhanced inquiry, and (5) collaborative reflection on peer teaching. Teacher-candidates experienced these pedagogical approaches first as learners, then reflected on them as future teachers, and lastly incorporated some of them during the practicum. As a result, teacher-candidates gained experience in promoting technology-enhanced inquiry in STEM education and began developing positive attitudes towards technology-enhanced inquiry-based STEM education.

INTRODUCTION

This chapter explores how modern collaborative educational technologies can be implemented in STEM teacher education in order to support teacher-candidates in (1) acquiring inquiry-based teaching skills, (2) forming positive attitudes about inquiry-based STEM education, and (3) building resiliency in the face of initial

DOI: 10.4018/978-1-5225-2525-7.ch010

failure of inquiry-based pedagogies (Milner-Bolotin, 2016b). The last two points are especially important, as there is ample research evidence that when instructors lack adequate support, they are likely to give up on implementing innovative research-based pedagogies in the face of early adoption failure (Lasry, Guillemette, & Mazur, 2014; Mazur, 1997a; Wieman & Perkins, 2005). Thus, inadequate support for K-12 and post-secondary STEM educators for reforming their teaching coupled with the limited opportunities for teacher collaboration result in an ever growing gap between research-informed and classroom-enacted educational practices (Cole & Knowles, 2000; Milner-Bolotin, 2014a). This explains why increased access to educational technologies does not necessarily guarantee a wider adoption of inquiry-based pedagogies and improved learning. In order to promote inquiry-based learning and active student engagement in STEM, K-12 and post-secondary instructors have to be supported in adopting new pedagogical approaches (Harris & Hofer, 2011; Shelton, 2015). This requires reimagining STEM teacher education, pedagogical preparation of post-secondary instructors, as well as in-service professional development (Lee & Tsai, 2010; Niess, 2005). If we fail to do so, many instructors will continue to ignore the mounting research evidence about how students learn and how STEM subjects can be taught more effectively (Bransford, Brown, & Cocking, 2002). Moreover, a large number of STEM instructors will inevitably revert to much safer but also less effective teacher-centered pedagogies they had experienced as students (Shelton, 2015; Wieman, 2012; Wieman & Perkins, 2005). This is one of the key reasons why the substantial governmental investments and half-a-century long education reform efforts in North America and in Europe aimed at promoting inquiry-based education have rarely brought the desired outcomes (Feder, 2010; Krajcik & Mun, 2014; National Research Council, 2012; Pollock, 2004).

The goals of STEM education, the role of inquiry, the tools available to modern students, as well as the population of students in K-12 STEM classrooms have changed dramatically over the last few decades (Cuban, 1990; National Research Council, 2013). In the 21st century we cannot afford to leave the majority of our population outside of STEM fields, as every student who disengages from STEM in K-12 closes multiple future opportunities for themselves inside and outside of STEM-related professions (Chachashvili-Bolotin, Milner-Bolotin, & Lissitsa, 2016; Let's Talk Science, 2013, 2015). In the current economic reality, STEM engagement should not be limited to a select few students, as it was common in North America during the post-Sputnik era, but should become an integral part of K-12 education for all (DeBoer, 1991; Let's Talk Science, 2012).

In this chapter we investigate how modern educational technologies can help to prepare the next generation of teachers who will be open to and capable of engaging their students in inquiry-based STEM learning (Jones & Leagon, 2014; Milner-Bolotin, 2016c). We focus on K-12 teacher education, as unlike their post-secondary colleagues, teachers in North America have to earn a teaching certification in order to be allowed to teach in the K-12 public school system. Teacher education is a perfect opportunity to expose future teachers to inquiry-based pedagogies and to available support networks. However, before discussing STEM teacher education that promotes technology-supported inquiry, it is important to highlight some relevant research on the topic.

BACKGROUND

Inquiry-Based STEM Education: Past, Present, and Future

A quick Google search for *inquiry-based STEM education* generates more than 1,210,000 results. There is little doubt that inquiry-based education has drawn ample attention from governments, educational administrators, and the community at large. Most of the documents produced by educational bodies in Canada, the United States, Europe and worldwide in the last decades emphasize inquiry, especially when it applies to STEM education (British Columbia Ministry of Education, 2015; BSCS, 2008; Crawford, 2007; Feder, 2010; R. R. Hake, 2007; National Research Council, 2013). Indeed, the term is often used broadly and indiscriminately. However, by labeling a "hands-on" classroom activity as inquiry-based, we cannot solve the problem of student growing disengagement from STEM fields and their declining performance on international assessments (Let's Talk Science, 2013; OECD, 2014). Inquiry-based STEM education is a complex, contextual and multi-faceted phenomenon that requires skillful facilitation and extensive subject and pedagogical knowledge from the teacher. Without contextualizing inquiry-based STEM learning, defining what it is and what it is not, we limit our ability (a) to evaluate if the teaching approaches implemented under the inquiry-based STEM education umbrella are indeed inquiry-based; (b) to understand how to support teachers in facilitating inquiry-based learning; and (c) to provide constructive feedback to teachers on the pedagogical effectiveness of these approaches and ways of enhancing them (Abd-El-Khalick et al., 2004; Eugenia Etkina et al., 2010).

The idea of learning by inquiry goes back thousands of years to the teachings of ancient Greeks. Socrates, Plato and Euclid used questioning and logic as a way to get to "true knowledge". Etymologically, the modern English verb *inquire* has its roots in the French *enquerre* and in the Medieval Latin *inquerere*, meaning to see into something, to find something out by asking and seeking information. However, inquiry as a pedagogical approach was not common in American schools until the early 20th century. John Dewey (1859-1953), an American philosopher, psychologist and educational reformer, brought inquiry-based learning and its role in a child's development into the spotlight (Dewey, 1938). Under the influence of the last wave of educational reforms in North America, 21st century scholars and educators keep redefining the spectrum of learning environments under the umbrella of inquiry-based STEM education (An, 2015; Eugenia Etkina et al., 2010; Eugenia Etkina, Planinši, & Vollmer, 2013; E. Etkina, Van Heuvelen, Brookes, & Mills, 2002; Hmelo-Silver, Duncan, & Chinn, 2012; Milner-Bolotin, 2012; National Research Council, 2012; Tan & Caleon, 2016). Still, they haven't yet reached an agreement regarding what inquiry-based STEM education is, and most importantly, what it is not. As Tan and Caleon (2016) write:

The current vision of science education in myriad educational contexts encourages students to learn through the process of science inquiry. Science inquiry has been used to promote conceptual learning and engage learners in an active process of meaning-making and investigation to understand the world around them. The science inquiry process typically involves asking questions and defining problems; constructing explanations and designing solutions; planning and carrying out investigations; analyzing and interpreting data; and engaging in argument from evidence. Despite the importance and provision of new directions and standards about science inquiry, ambiguities in conceptualizations of inquiry still exist. These conceptualizations may serve as barriers to students learning science. (p. 168)

While being fully aware of the challenges described above, in this chapter we start with the vision of inquiry-based STEM education found in the National Science Standards (National Research Council, 2012; Olson & Loucks-Horsley, 2000). Next we expand it, through emphasizing conceptual understanding and meaning making in STEM learning (Minner, Levy, & Century, 2010). Consequently, in this chapter, inquiry-based STEM learning is described as an environment where the learner:

1. **Engages** in answering meaningful STEM-oriented questions
2. **Relies** on gathered research evidence to answer these questions
3. **Formulates** testable conceptually-oriented explanations based on this evidence
4. **Designs** and conducts experiments (multiple iterations) to test these explanations
5. **Makes sense** of the results, connects them to prior knowledge, and refines the findings
6. **Communicates and justifies** research findings, and poses new questions.

This list highlights why creating inquiry-based STEM learning environments is so challenging. It is fraught with ambiguities both for the students and for the teachers designing these learning environments and evaluating student learning outcomes (Jonassen & Land, 2012). The more open these environments are, the more likely the teachers themselves might not know all the answers and will be required to engage in an authentic inquiry together with their students. This will require teachers to possess broad STEM knowledge, necessary inquiry skills, and confidence in their ability to figure things out as opposed to teaching from the textbook. Akin to scientific inquiry happening in research labs, inquiry-based learning relies on ongoing feedback, refinement and adjustment, which can be achieved through student collaboration, as well as continuous interactions between a student and a teacher. Therefore, inquiry-based teaching is a reflective and an iterative endeavour.

Pedagogically effective inquiry-based learning environments are still uncommon in K-12 classrooms. Thus, it is unlikely that current STEM teacher-candidates had much experience with them as K-12 students, and even less likely that they were encouraged to reflect on these environments as future teachers. Therefore, it is crucial that they experience inquiry in their STEM methods courses as learners and reflect on this process as future teachers before implementing inquiry-based lessons during the practicum and after graduation (Milner-Bolotin, Fisher, & MacDonald, 2013). In the next section, we delineate the conceptual theoretical framework incorporated in this chapter.

Theoretical Framework for Studying STEM Teacher Education

During the last half a century, education researchers have suggested multiple conceptual frameworks for investigating the development of teacher knowledge (Abbitt, 2011; Ball, Thames, & Phelps, 2008; Blömeke & Delaney, 2012; Corrigan, Gunstone, & Dillon, 2011; Hourigan & Donaghue, 2013; Koh & Divaharan, 2011). These frameworks can be used to study how future STEM teachers learn to teach by inquiry and acquire positive attitudes about inquiry-based STEM education. The original Pedagogical Content Knowledge (PCK) framework proposed by Shulman more than three decades ago (Shulman, 1986a, 1986b) will serve the backbone of the current study. Shulman emphasized that teachers should possess subject-specific content knowledge (CK), general pedagogical knowledge (PK), and pedagogical content knowledge (PCK), where PCK can be considered to be the overlap of the two. The PCK framework was later expanded to include the knowledge of educational technologies—technological knowledge (TK)—thus morphing into the Technological Pedagogical (and) Content Knowledge framework (TPACK) (Mishra & Koehler, 2006). The latter framework separates the content specific knowledge usually acquired by teacher-candidates during their B.Sc. degree (i.e., the knowledge of specific STEM disciplines, such as mathematics, physics or chemistry) from the knowledge of how these subjects can be taught in the K–12 context, and how technology can be used to enhance student learning (Figure 1).

In their conceptual framework for mathematics teacher competencies, Blömeke and Delaney (2012) referred to the TPACK aspect of teacher knowledge as cognitive abilities or professional knowledge. We will extend their theoretical framework to describe STEM teacher competencies (Figure 2). We will use TPACK in a greater sense, including not only the overlap of the three knowledge domains (CK, PK, and TK), but also each one of them. They also emphasized affective–motivational

Figure 1. Technological Pedagogical and Content Knowledge (TPACK) framework by Koehler and Mishra (2015)

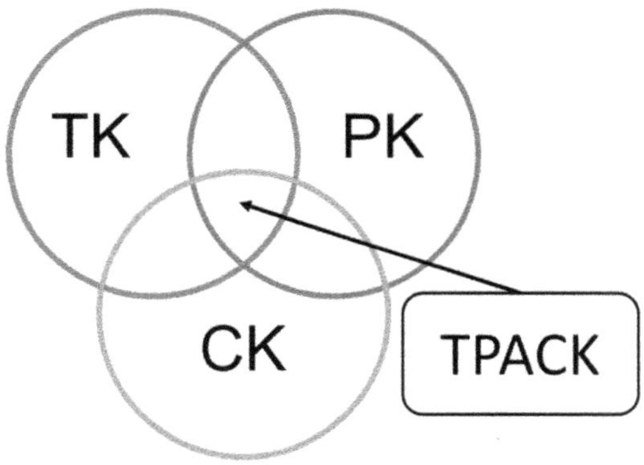

Figure 2. A big picture of STEM teacher competences, as a combination of cognitive abilities, affective characteristics and potential opportunities for TPACK growth as a result of collaboration with peers (Teacher Zone of Proximal Development or T-ZPD). Arrows indicate the interactions between different facets of teachers' professional competencies.

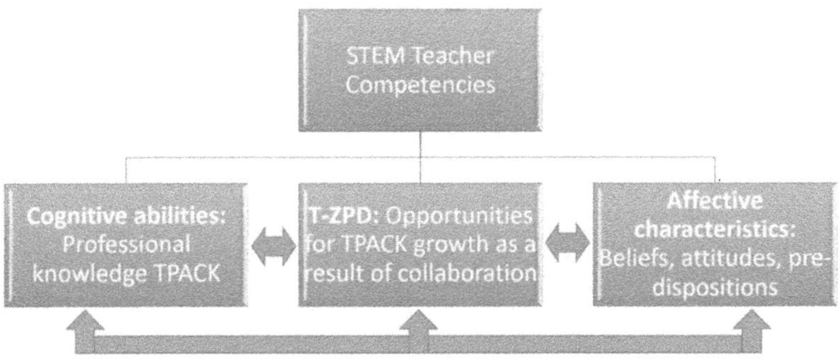

characteristics, such as professional beliefs, motivation, and self-regulation. This is relevant to all STEM fields, as teachers' attitudes and predispositions towards inquiry-based STEM learning are crucial for the successful implementation of these pedagogies.

This chapter builds on the original framework by Blömeke and Delaney (2012) by emphasizing that teachers' competencies, similarly to student knowledge, can be developed through collaboration with peers (Figure 2). Instead of looking at teachers' competencies as static, we consider their dynamic nature, asking: How can these competencies be developed if teachers are encouraged to collaborate with peers and more experienced teachers? Therefore, the theoretical framework adopted in this chapter focuses on the potential growth of STEM teachers' competencies as a result of this collaboration. This approach is an adaptation of the original Zone of Proximal Development concept suggested by Vygotsky, to the context of teaching and teacher-education (Vygotsky, 1978). We refer to it as the *Teacher Zone of Proximal Development* (T-ZPD). T-ZPD describes the gap between what a teacher has already mastered (the actual level of development, as expressed by their current TPACK) and what they can achieve when provided with opportunities to collaborate with peers and more experienced educators. T-ZPD represents the potential for the development of teachers' TPACK (Figure 3). The need to consider the dynamic nature of TPACK affected by teachers' interactions with peers, students, parents, administrators, and

Figure 3. Growth of teachers' TPACK due to collaboration with peers more experienced teachers. The light blue area in the left image shows the Teacher-Zone of Proximal Development (T-ZPD). As a result of collaboration all aspects of teachers' knowledge have grown, including their TPACK.

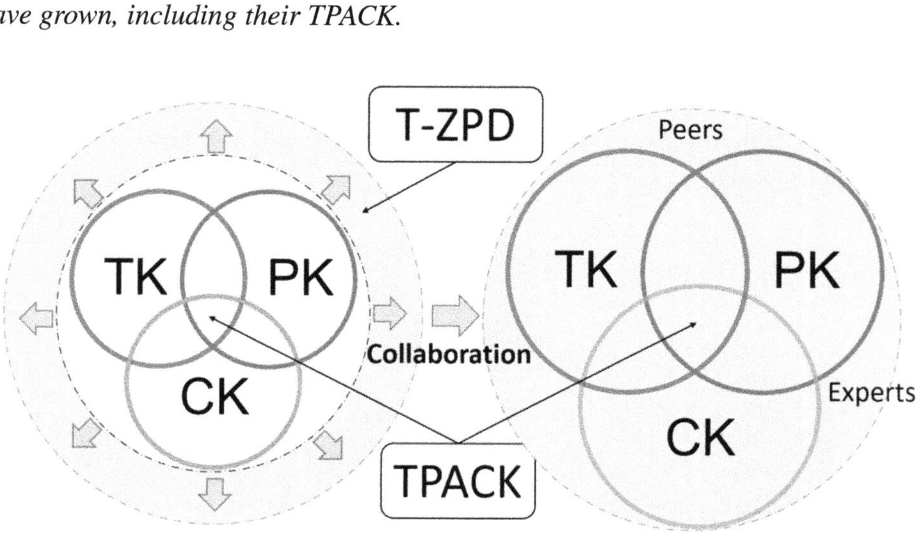

the society at large, reflects our belief in teaching as a highly professional endeavour. In order to keep up-to-date, teachers must always learn, update and question their knowledge, interact with others in the field, and continuously reflect on their practice. The view of ever-evolving mastery of teaching and the importance of a community in becoming and being an effective teacher is situated in Vygotsky's socio-cultural learning theory (Daniels, 2001; Vygotsky, 1978).

It is difficult to acquire inquiry-based teaching skills while working in isolation. It is more effective to master these skills in the process of apprenticeship and collaboration with peers. However, this collaboration should not be an isolated occurrence. This often happens when teachers attend one-time professional development events that are rarely followed up by long-term collaboration with peers focussed on the implementation of these pedagogies. These one-time professional development opportunities rarely bring sustained changes in teachers' practice (Luft & Hewson, 2014). In order to support teachers in moving towards inquiry-based STEM teaching, they have to have multiple opportunities to inquire about their own practice, adopt and adapt new pedagogies, collaborate with colleagues, and acquire the necessary TPACK(Burridge & Carpenter, 2013; Krajcik & Mun, 2014) (British Columbia Ministry of Education, 2015; Schmidt et al., 2011).

The following section describes five different collaborative technology-enhanced pedagogies that can be used in STEM teacher education in order to help teacher-candidates (1) acquire relevant inquiry-based teaching skills, (2) form positive attitudes about inquiry-based STEM education, and (3) build resiliency in the face of initial failure of inquiry-based teaching practices in their future classrooms (Milner-Bolotin, 2016b). All these pedagogies were used simultaneously in the STEM methods courses in order to provide teacher-candidates with ample opportunities for collaborative development of their teaching competencies and positive attitudes about inquiry-based STEM education, as well as the resiliency and the capacity needed to overcome initial obstacles in implementing inquiry in their own classrooms.

COLLABORATIVE INQUIRY IN STEM TEACHER EDUCATION

This section considers five different collaborative technology-enhanced pedagogies simultaneously implemented in STEM teacher education courses in order to prepare teacher-candidates for enacting inquiry-based teaching. These technologies include: Electronic response systems (implemented with Peer Instruction), PeerWise collaborative database, data acquisition tools, computer simulations and modeling software, and Collaborative Learning Annotation System (CLAS) (Milner-Bolotin, 2014b, 2016c) (Table 1). While there are many potential educational technologies relevant to teacher education, the choice of the five technologies mentioned above is deliberate. These technologies are all low cost or free; they are relevant to both K-12 and post-secondary classrooms; they promote collaboration and reflection; they help teacher-candidates acquire relevant TPACK (Koehler & Mishra, 2015); they help teacher-candidates develop positive attitudes about research-informed inquiry-based education; and lastly, they support teacher-candidates in cultivating resilience in the face of initial failure and the courage to persevere at inquiry-based

Table 1. Description and pedagogical goals of five different technology-enhanced learning environments used in STEM teacher education aimed at promoting inquiry-based teaching and learning

	Pedagogical description/technology	Pedagogical goals and how they are promoted
1. Peer Instruction	**Pedagogical description:** Technology is used to collect continuous student feedback and adjust teaching to address student conceptual difficulties via using multiple-choice conceptual questions. Peer Instruction relies on teacher's abilities to ask conceptual questions, identify student difficulties and support student inquiry as opposed to providing them with the answers. Despite the availability of technology and possibilities of using low-tech versions in small classes, Peer Instruction is not as widely used in K-12 as in post-secondary STEM classrooms. **Technology:** Electronic response systems, smart phones, personal computers, iPads, or a low tech version of QR cards used for individual student voting wherein a teacher uses a phone to collect data.	**The goals – to support teacher-candidates in:** • Developing their PCK • Reflecting on their STEM PCK and on the difference between learning by memorization and by understanding/questioning • Learning to identify student conceptual difficulties and to ask different levels of questions that promote inquiry and help build conceptual understanding (Bloom, 1956) **How these goals are promoted:** • By modeling effective conceptual questions during methods courses and inviting teacher-candidates to experience Peer Instruction as learners and teachers • By emphasizing the value of peer collaboration • By experiencing formative assessment and peer collaboration in practice as learners and reflecting on it as future teachers.
2. Peer Wise-based pedagogy	**Pedagogical description:** PeerWise database supports a community of learners who collaborate on designing multiple-choice questions. PeerWise and Peer Instruction both emphasize formative assessment using conceptual multiple-choice questions and peer collaboration. **Technology:** An online social platform for sharing, responding to, reflecting on, editing, and improving multiple-choice questions. It allows students to contribute their multiple-choice questions, respond to the questions asked by their peers, rate the quality of questions and explanations, and suggest improvements.	**The goals – to support teacher-candidates in:** • All of the above under Peer Instruction, plus • Designing meaningful conceptual STEM questions that open opportunities for inquiry • Learning to identify student conceptual difficulties and to ask different levels of questions that promote inquiry and help build conceptual understanding (Bloom, 1956) • Learning to give and receive constructive feedback. **How these goals are promoted:** • By designing, responding to and providing constructive feedback to conceptual questions • By thinking of meaningful distractors (wrong answers for multiple-questions) • By discussing these questions with peers • By learning to receive constructive feedback and to act on suggestions • By realizing that they still have a lot to learn.

continued on following page

Table 1. Continued

	Pedagogical description/technology	Pedagogical goals and how they are promoted
3. Real life data as an impetus for inquiry	**Pedagogical description:** Technology enables learners to explore "what-if" questions about the real world via collecting and analysing real-time data, conducting experiments and answering questions experimentally that could not have been answered before. The collected data can include temperature, pressure, position of objects, PH levels of different solutions, etc. This pedagogy invites students to ask and answer their own inquiry questions, thus engaging them in authentic STEM learning. **Technology:** Examples include Logger Pro sensors (Milner-Bolotin & Moll, 2008), various smart phone applications, etc.	**The goals – to support teacher-candidates in:** • Designing and facilitating inquiry-based STEM lessons both during their class discussions and in labs • Design inquiry-based labs • Learning to guide students in asking questions • Learning to explain real-life data and use it to test STEM hypotheses **How these goals are promoted:** • By helping teacher-candidates to experience data-driven inquiry • By making them comfortable with the use of various technologies for real-life data collection • By modeling investigations based on live data collection and analysis • By modeling the use of technology to bridge STEM to everyday life
4. Pedagogy that utilizes computer simulations & modeling	**Pedagogical description:** Various computer simulations and modeling tools are used to ask "what-if" questions about the real world or about more abstract phenomena. This pedagogy invites students to think more abstractly, while using authentic research tools. Learners can use these tools to ask and answer their own inquiry questions, thus engaging them in authentic STEM learning. **Technology:** Computer simulations such as PhET, data modeling and analysis tools, such as GeoGebra, programming tools, etc.	**The goals – to support teacher-candidates in:** • Inquiring about abstract (or not so abstract) phenomena through manipulating different parameters of the system (modeling, simulations) • Experiencing scientific inquiry first hand by asking and answering "what-if" questions • Acquiring STEM intuition, by learning to connect various representations of phenomena with the implications of these representations (e.g., the meaning of formulae), such as mathematical descriptions with scientific outcomes. • Linking theory to practice (virtual versus real labs) **How these goals are promoted:** • By helping teacher-candidates to experience data-driven inquiry and ways of scientific inquiry
5. Collaborative Learning Annotation System (CLAS)	**Pedagogical description:** This pedagogy emphasizes continuous reflection. Teacher-candidates are invited not only to practice their teaching in a supportive environment, but also to reflect on it and improve it as a result. **Technology:** CLAS is a media player that allows recording, sharing, and commenting on videos. Videos uploaded to CLAS are stored on a secure server at UBC and allow control over who has access to them. CLAS is also mobile compatible, allowing students to record videos and make comments directly from their mobile devices.	**The goals – to support teacher-candidates in:** • Becoming a reflective practitioner • Becoming a teacher who is open to peer collaboration via learning from others and supporting them • Becoming a confident and open-minded teacher who is open to suggestions in order to improve their practice. **How these goals are promoted:** • By learning to reflect on their teaching, provide constructive feedback, accept feedback and act on it • By practicing their teaching in a friendly environment • By collaborating with peers in a safe environment • By comparing multiple iterations of different lessons and reflecting on their pedagogical progress • By building self-confidence in their teaching abilities.

teaching. As Winston Churchill once said "Success is not final, failure is not fatal: it is the courage to continue that counts". This is the key message we would like to convey to future teachers: teaching is a life-long process of learning and personal growth, filled both with failures and successes. The second message is that reflection and peer collaboration are valuable tools for supporting teachers when they work on modifying their pedagogy and improving their practice.

These technologies were implemented in our STEM methods courses during a year-long secondary STEM Teacher Education Program (http://teach.educ.ubc.ca/). About 20 teacher-candidates took part in these 13-week 3 credit courses (Milner-Bolotin, Egersdorfer, & Vinayagam, 2016). The next sections outline how these courses supported the development of teacher-candidates' TPACK, positive attitudes about inquiry-based education, and confidence in facilitating inquiry-based learning in the future.

Peer Instruction: Learning STEM by Questioning and Collaborating

Peer Instruction pedagogy was originally proposed by a Harvard Physics Professor – Eric Mazur almost 30 years ago (Mazur, 1997b, 1997c). Peer Instruction is a form of live formative assessment that is continuously used by an instructor to support student learning and tailor instruction to students' needs. In the last decades Peer Instruction has become a very common and widely researched interactive engagement pedagogy in post-secondary STEM education (Fraser et al., 2014; Richard R. Hake, 1998; Lasry, Guillemette, Dugdale, et al., 2014; Wieman, 2012). Peer Instruction uses flashcards, electronic response systems (clickers), or mobile devices (virtual clickers) to engage students in meaningful learning by asking them carefully crafted conceptual multiple-choice questions that target common student difficulties and misconceptions (Lasry, 2008; Lasry, Mazur, & Watkins, 2008). The students are given a chance to think about these questions and provide individual responses. When Peer Instruction is used with clickers, the histogram of student responses is shared with the entire class. Then the students are invited to discuss their answers with their peers. This small group collaborative exercise is followed up by an all-class discussion led by the instructor (Milner-Bolotin, 2015a). With the proliferation of new technologies and free or low-cost versions of virtual clickers, such as plickers (Amy, 2016), Peer Instruction is becoming more and more popular in the K-12 educational system. We have written extensively about the use of Peer Instruction in STEM teacher education and will refer the readers to these papers (Lasry, 2008; Lasry, Watkins, Mazur, & Ibrahim, 2013; Milner-Bolotin, 2016b). Here we focus on how Peer Instruction implemented in STEM teacher education courses can support future teachers in promoting inquiry-based pedagogies in their own practice.

Peer Instruction relies on *conceptual* questions that ask learners to analyze phenomena without relying on formulae and plug-and-chug calculations (Beatty, Gerace, Leonard, & Dufresne, 2006). Figure 4 shows an example of such a question in the context of physics. The question challenges students to analyze the implications of Newton's second law in the case of a weight-pulley system. This conceptual question can inspire a student-led independent inquiry where such a system is set up and its acceleration is investigated using Logger Pro live data collection system (Milner-Bolotin, Kotlicki, & Rieger, 2007).

In order to design pedagogically effective conceptual questions that can be used in Peer Instruction, teachers have to recognize potential student difficulties and common misconceptions, and offer effective ways of helping students to inquire about the phenomenon. When teacher-candidates experience Peer Instruction as learners, they realize the power of this pedagogical approach for STEM education. They also become aware of the value of STEM education research for their own practice. This is one of the reasons why noted physics educator Lillian McDermott and her colleagues called on improving physics teacher preparation through incorporating educational research into teacher education practice (McDermott, Heron, Shaffer, & Stetzer, 2006).

Figure 4. An example of a conceptual multiple-choice physics question challenging students to analyze the implications of Newton's second law. This question was retrieved from Mathematics and Science Teaching and Learning through Technology online database (Milner-Bolotin, 2016a).

BLOCKS AND PULLEY

Two blocks are connected via a pulley. The blocks are initially at rest as block m_1 is attached to a wall. If string A breaks, what will the accelerations of the blocks be? (Assume no friction and strings don't stretch)

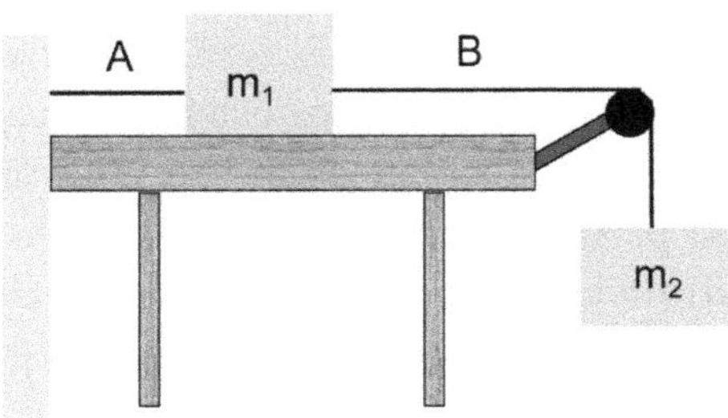

A. $a_1 = a_2 = 0$

B. $a_1 = a_2 = g$

C. $a_1 = a_2 = \frac{m_2 g}{(m_1 + m_2)}$

D. $a_1 = a_2 = \frac{m_1 g}{(m_1 + m_2)}$

E. $a_1 = a_2 = \frac{m_2}{(m_1 + m_2)}$

Peer Instruction is especially appropriate to STEM teacher education. Through the use of well-crafted conceptual questions teacher-candidates have an opportunity to realize that STEM mastery requires higher order thinking skills and an ability to think creatively and conceptually. As one teacher-candidate expressed it: "Peer Instruction gets you away from thinking physics is a math class. It gets you to really understand the science." After earning a Bachelor degree in the relevant subject, many teacher-candidates assume that they have mastered the content knowledge they will be teaching and they have little to learn. Yet, often this mastery is rather superficial: teacher-candidates often encounter the same conceptual difficulties as secondary students (Milner-Bolotin et al., 2013). Teacher-candidates often might know the answer to a question, but they do not know why the answer holds. They might have learned some facts and relationships, but they might not have had an opportunity to inquire why these relationships hold true, which is the key element in inquiry-based teaching and learning. Inquiry is driven by curiosity and doubt, which means STEM teachers should learn not to take things for granted, but keep asking questions. Lastly, while experiencing Peer Instruction as learners, future teachers learn to listen to others, to ask questions, to articulate their thinking, and understand where somebody else might have difficulty. Implementing Peer Instruction in teacher education courses is a powerful opportunity for reflection and for deepening teacher-candidates' TPACK, as well as helping them gain confidence in explaining complex phenomena conceptually without always relying on formulae and factual knowledge.

To support teacher-candidates in developing necessary questioning skills and enhancing their TPACK, we incorporated PeerWise, an online collaborative multiple-choice questions' database (Denny, 2016).

PeerWise: Learning to Ask STEM Questions Through Collaboration With Peers

PeerWise is a free online collaborative tool. It supports learners in designing and sharing their multiple-choice questions, commenting and responding to peers' questions, and eventually creating a shared resource of conceptual questions, explanations, solutions and suggestions (Milner-Bolotin, 2014b).

In our STEM methods courses teacher-candidates used PeerWise weekly as part of the PeerWise cycle sequence (Table 2). Teacher-candidates were asked on a regular basis to contribute their conceptual questions on PeerWise, answer some of the questions provided by their peers, comment on these questions, as well as respond to the comments provided to them. Thus, each PeerWise question underwent multiple iterations before its final version was generated and saved in the database.

Table 2. A PeerWise sequence outlining teacher-candidates' expected participation

Cycle	Week	Expected Tasks
1	1	Author at least six questions and upload them to PeerWise
	2	Answer at least 10 questions designed by your peers Provide comments on at least five of the questions you answered
2	3	Review all relevant comments on your own questions and respond when applicable Modify at least three of your questions to address the comments Author at least three additional questions
	4	Answer at least 10 questions designed by your peers Provide comments on at least five questions you answered
Cycle 2 repeats every two weeks for the duration of the course		

PeerWise is especially suitable for modeling question-driven pedagogy in STEM teacher education. It can supplement Peer Instruction, highlight the difference between plug-and-chug and conceptual questions, and show how conceptual questions enhance student understanding. Another teacher-candidate said:

I think for myself...getting away from the number-crunching questions to conceptual questions, and understanding that even though number-crunching can feel harder, it doesn't really show how well you know something compared to conceptually, and for me PeerWise really helped me realise that.

Moreover, PeerWise requires teacher-candidates to provide constructive feedback and react in a positive and productive way to the feedback provided by peers. It increases teacher-candidates' ability to evaluate their peers' questions and explanations, which is crucial for improving their confidence in designing inquiry-based learning environments (Milner-Bolotin et al., 2016). As educators often adopt and adapt already existing activities to suit their pedagogical goals, an ability to become a "critical consumer" of already existing resources is a valuable skill for novice teachers.

PeerWise collaboration also emphasizes the iterative nature of designing pedagogically effective educational materials. The teachers who realize that teaching in general, and inquiry-based teaching in particular, is an iterative process and it takes time and patience to learn how to teach by inquiry and how to ask questions that can facilitate inquiry, are going to be more resilient in the face of failure. These teachers will be less likely to give up if an inquiry-based activity they have designed does not generate desired results. Another way of helping teacher-candidates to acquire inquiry-based teaching skills is to engage them in activities that utilize real-time data collection and analysis tools.

Data-Driven Inquiry in STEM Teacher Education

In order to support teacher-candidates in acquiring inquiry-based teaching skills as well as positive attitudes about inquiry, it is important to provide them with multiple opportunities to experience authentic technology-enhanced learning environments as students first and then reflect on them as future teachers (Milner-Bolotin et al., 2013; Milner-Bolotin & Moll, 2008). Data collection and analysis technology, such as the Logger Pro data acquisition system, is a great asset for supporting authentic inquiry, reducing the cost of risk-taking and allowing learners to experiment with abstract STEM ideas by connecting them to everyday life (Eijck & Roth, 2009; Milner-Bolotin, 2012; Vernier-Technology, 2016).

For example, producing and interpreting graphs and building bridges between algebraic, graphical and physical representations are often difficult for the students. This is a great opportunity for authentic inquiry activities, where students can pose questions and devise authentic experiments that will help them answer these questions (Eshach, 2014; Ruthven, Deaney, & Hennessy, 2009; Wright, 2005). In order to help students connect physical scenarios with their mathematical and physical representations, STEM teachers can use data collection and analysis tools, also referred to as sensors. For instance, Logger Pro motion detector allows students to represent the 1-D motion of a moving object as an $x(t)$ graph (Vernier-Technology, 2016). Figure 5 shows a graph of motion produced by a student moving

Figure 5. An example an inquiry-based activity where students explore 1-D motion: a student was asked to walk in front of the motion detector in order to match a given (blue) graph with the red one (produced by the walking student)

along a straight line in front of a motion detector. Since the data are collected and displayed in real time, the students can collaborate with peers on pursuing their own "What-if" inquiry questions. Moreover, the technology reduces the cost of failure as the experiments can be easily adjusted, improved and repeated. Similar to Peer Instruction, live data collection and analysis tools provide instantaneous formative assessment that was unimaginable before (Schuster, Undreiu, Adams, Brookes, & Milner-Bolotin, 2009). This process encourages authentic inquiry, collaboration and powerful reasoning.

Engaging STEM teacher-candidates in collaborative data-driven inquiry invites them to think differently about STEM education and the relationships between the inquiry that happens in the classroom and the inquiry that happens in science research labs. Conveniently, in addition to Logger Pro and similar data collection tools, many modern smartphone or tablet apps allow rather sophisticated live data collection and analysis (Maciel, 2015). Creative STEM teachers can use these 21st century tools to promote authentic inquiry in their classrooms. However, in order for this to happen, STEM teachers have to acquire the necessary TPACK (Carr, 2012). Without it, these sophisticated tools will remain underused expensive gadgets and the goal of increased student STEM engagement will take much longer to achieve (Cuban, 2001).

Computer Simulations and Modeling: Virtual Inquiry

The data collection and analysis tools are helpful in promoting inquiry-based STEM education (Milner-Bolotin, 2016c). However, not every STEM concept can be easily demonstrated through a hands-on activity. As teacher-candidates work on designing inquiry-based lessons, they often find it challenging to help students grapple with abstract concepts. For example, the consequences of the electromagnetic induction phenomenon can be demonstrated in a lab, but in order to meaningfully understand this abstract physics concept, a more extensive inquiry is required. As a result, many novice teachers struggle with engaging students in authentic explorations of these abstract concepts in order to support student understanding.

Modern technologies such as computer simulations and dynamic modeling software offer ample opportunities to promote meaningful inquiry of abstract phenomena. We have written about this topic elsewhere (Milner-Bolotin, 2015b, 2015c, 2016b), so in this chapter these technologies will be only mentioned briefly. For example, PhET is a suite of STEM simulations built on solid educational research evidence (Wieman, Adams, Loeblein, & Perkins, 2010). They can be used to help students not only to visualize abstract ideas, but also to conduct independent research investigations in virtual environments akin to the ones conducted in real labs (Finkelstein et al., 2005).

PhET simulations open ample opportunities for developing inquiry-driven thinking, yet they are closed tools. Their designers have deliberately decided what parameters the learners will be able to manipulate. In other words, the decision about the model embedded in the simulation has been made and the learner has no input into it. Computer simulations where those decisions have been made based on educational research (such as PhET) are powerful for beginners. Yet, when the learners have acquired advanced subject knowledge and want to ask more sophisticated questions, the simulation might prove to limit their investigations.

The next level of inquiry is attained when the students create and test their own STEM models. An example of modeling technology that has the potential to support an even higher level of student creativity and empower them to experience STEM in a very personal way is dynamic modeling software, such as GeoGebra (Hohenwarter, 2014). Unlike traditional paper and pencil geometrical or algebraic constructions, where a construction or a graphical representation cannot be changed or manipulated easily, or unlike computer simulations where the underlying scientific models have been chosen by the designers, GeoGebra allows students to develop a language, propose and test their own models, dynamically test their understanding, visualize abstract relationships and eventually foster their STEM intuition (Martinovic, Karadag, & McDougall, 2014).

However, the use of these tools in STEM education does not guarantee inquiry-based learning. As mentioned above, teachers need to acquire inquiry-based teaching skills in order to be able to implement these tools effectively in their classrooms. In order to provide teacher-candidates with multiple opportunities to reflect on their inquiry-based teaching skills and improve them, we implemented the Collaborative Learning Annotation System (CLAS).

Promoting Inquiry Teaching With CLAS

CLAS is an online media player developed at the University of British Columbia (Dang, 2016). It allows uploading, sharing, and commenting on videos stored in the system, with both time-specific and general comments (Figure 6). The participants can respond to specific comments and create commenting threads. The instructor has full control of who has access to the videos. The comments made by the instructor or by the students can be private or public. Most importantly, CLAS is compatible with the videos recorded using smartphones and other common devices, such as iPads or tablets available to teacher-candidates. Thus, no additional special video-recording equipment was needed to use the system.

CLAS was incorporated in the methods courses as a tool to support teacher-candidates' reflection on inquiry-based mini-lessons they taught. Each teacher-candidate had to teach at least four different 10-15 minute long mini-lessons to

Figure 6. A screen shot of a CLAS interface. By clicking on different icons in the Time-specific comments area, one can read and respond to all the comments pertinent to different parts of the video. To find out more about the system visit: http://ets.educ.ubc.ca/clas/

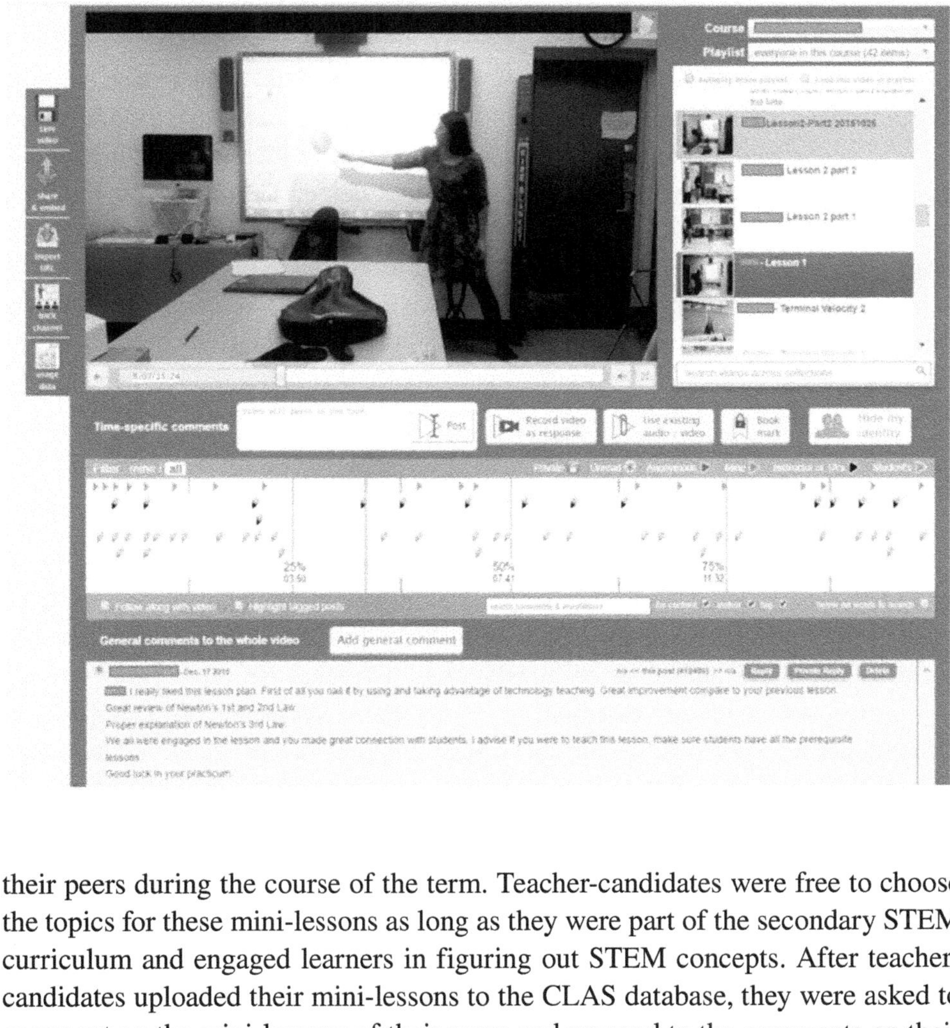

their peers during the course of the term. Teacher-candidates were free to choose the topics for these mini-lessons as long as they were part of the secondary STEM curriculum and engaged learners in figuring out STEM concepts. After teacher-candidates uploaded their mini-lessons to the CLAS database, they were asked to comment on the mini-lessons of their peers and respond to the comments on their own mini-lessons. Moreover, on multiple occasions, teacher-candidates decided to incorporate the comments and to re-teach their lessons. Since all the mini-lessons were recorded, teacher-candidates were able to reflect on their own progress and think of areas for improvement.

In this process they learned to inquire about their own teaching, to provide and receive constructive feedback, and to think about how to support students in figuring out STEM concepts for themselves. The process of watching their own teaching videos for the first time was very revealing for most of the teacher-candidates. This allowed them to reflect on the gap between their own perception of learner engagement in their lessons and the real engagement as manifested in the video. This is especially important for promoting inquiry-based STEM education and student engagement, as it is very common to see how teachers' perceptions of their own inquiry-based teaching differ drastically from the classroom realities experienced by their students.

Lastly, the positive feedback on their teaching provided by their peers and the course instructors allowed teacher-candidates to identify their strengths and gauge their own progress. As many of the teacher-candidates decided to repeat their mini-lessons, reflecting on their teaching using CLAS helped them see their teaching as a work in progress. It also helped them identify their strengths and see their improvement. Passionate and less experienced teachers are often their own worst critiques, so receiving positive and constructive support from colleagues was a much-needed boost for improving teacher-candidates' self-confidence.

USING COLLABORATIVE TECHNOLOGY IN TEACHER EDUCATION

The goal of the study was to explore how using collaborative educational technologies in STEM teacher education can support teacher-candidates in moving towards inquiry-based teaching in their own classrooms. The study was motivated by our desire to provide practical recommendations for teacher educators who want to support teacher-candidates in acquiring inquiry-based skills and promote teachers' positive attitudes towards teaching STEM by inquiry. In this study, we used five different technology-enhanced pedagogies that supported teacher-candidates' collaboration and mutual support, while providing them with multiple opportunities for improvement (Table 1).

The main outcome of our exploration was not identifying the "perfect" technology to support teacher-candidates, but making sure that each one of the technologies used in STEM teacher education has a clear pedagogical purpose – to engage teacher-candidates in inquiry-based learning as students, encourage them to reflect on their experiences as future teachers, and eventually apply these pedagogies during their practicum and consequent teaching. We call this approach a deliberate use of technology (Milner-Bolotin, 2016b). This study has highlighted the value of purposeful implementation of technology in order to promote productive collaboration and peer support among teacher-candidates. Technology helped us to promote inquiry-based teaching during the teacher education courses and supported teacher-candidates in inquiring about their own teaching practices. This helped us to bridge the research-informed technology-enhanced STEM pedagogies to the classroom practices enacted by teacher-candidates during their practicum and post-graduation. We recommend that teacher-educators who want to promote technology-enhanced inquiry-based teaching should model the use of collaborative technologies in their own courses. Modeling research-based pedagogies in collaborative methods courses, where teacher-candidates can learn from and with each other, is important for supporting teacher-candidates' shifts towards these pedagogical practices (Milner-Bolotin et al., 2016).

FUTURE RESEARCH DIRECTIONS

This chapter explored how modern collaborative educational technologies can be used in STEM teacher education in order to help teacher-candidates develop the necessary TPACK, positive attitudes about inquiry-based learning, and confidence in their ability to use inquiry-based pedagogies in their own classrooms. While this chapter explored the deliberate use of technology in STEM methods courses in a teacher education program, we have yet to study how teacher-candidates' experiences during their school practicum and after graduation might influence their attitudes about inquiry-based STEM education and their ability to continue using inquiry in their own classrooms. Supporting teacher-candidates use of technology-enhanced inquiry in teacher education courses is crucial, as teacher-candidates might have limited opportunities during their school practicum to implement these research-informed pedagogies. Teacher-candidates also remain under a lot of pressure during their school practicum, as School and Faculty Advisors are the ones who decide if a teacher-candidate has passed the practicum – a key element of the teacher education program. As one of the teacher-candidates expressed it:

I wasn't aware of the amount of pressure that would come from both the students and my School Advisor...basically in favour of plug and chug physics. I didn't realise how much you're going against when you try and change the way kids do physics. By the end of my practicum I didn't do conceptual physics anymore, because the kids didn't want it, my School Advisor didn't want it, my Faculty Advisor didn't want it. They wanted the kids to get through and enjoy the class more than they wanted them to struggle with the concepts...It depends on the culture of the school... I was very prepared to do conceptual physics, but when my School Advisors and Faculty Advisors told me that wasn't what they wanted, I had no idea what physics education looked like...

Teacher-candidates can become conduits for change, but at the same time, their desire for inquiry-based teaching in a school environment where teachers want to use "plug-and-chug" pedagogy might be a dangerous strategy for them. Thus, more research is required to understand how teacher-candidates can be supported during their practicum in implementing inquiry-based pedagogies.

Lastly, even after graduation novice teachers are more vulnerable than their more experienced colleagues. Novice teachers come to a school culture that might have different attitudes about inquiry than their own. They also have to earn their reputation and address the learning objectives posed by the provincially (or state) mandated curriculum. Therefore, it is important to investigate how novice teachers can be supported in enacting inquiry-based STEM education practices their first years after graduation.

CONCLUSION

This chapter has demonstrated how five different kinds of modern collaborative educational technologies can be implemented in STEM teacher education in order to support teacher-candidates in (1) acquiring relevant inquiry-based teaching skills, (2) forming positive attitudes about inquiry-based STEM education, and (3) building resiliency in the face of initial failure of inquiry-based teaching practices. The technology-enhanced pedagogies explored here include: Peer Instruction, PeerWise, data-driven inquiry-based pedagogy, computer-supported inquiry and collaborative reflection on mini-lessons using CLAS. The use of these technologies in a supportive and collaborative learning environment has highlighted the value of collaboration with peers, reflection on their TPACK and on their teaching, and openness to changing their practice as a result of constructive feedback. Nevertheless, we have to remember that the instructors in STEM methods courses often have little influence on teacher-candidates' experiences during their school practicum. Therefore, in order to support teacher-candidates in acquiring inquiry-based teaching skills, we have to support them during their practicum.

Lastly, this study applied a T-ZPD theoretical framework to teacher education, as much as we consider ZPD in designing learning environments for students. Based on our studies, we have found that teacher-candidates who are provided with multiple opportunities to collaborate with peers will be more likely to acquire an extensive and meaningful TPACK than their counterparts working alone (Milner-Bolotin et al., 2016). Modern technologies can provide multiple constructive opportunities for these collaborations. Thus, we challenge teacher educators to consider how technology can be implemented in STEM teacher education to inspire and support teacher-candidates in implementing inquiry-based technology-enhanced teaching practices in their own classrooms.

ACKNOWLEDGMENT

The author would like to acknowledge Davor Egersdorfer for his thoughtful, detailed, and constructive feedback on the manuscript.

REFERENCES

Abbitt, J. T. (2011). Measuring Technological Pedagogical Content Knowledge in preservice teacher education: A review of current methods and instruments. *Journal of Research on Technology in Education*, *43*(4), 281–300. doi:10.1080/15391523.2011.10782573

Abd-El-Khalick, F., Boujaoude, S., Duschl, R., Lederman, N. G., Mamlok-Naaman, R., Hofstein, A., & Tuan, H.-L. (2004). Inquiry in Science Education: International Perspectives. In E. Krugly-Smolska & P. C. Taylor (Eds.), *Culture and Comparative Studies* (pp. 397–419).

Amy, N. (2016). *Plickers*. Retrieved from https://www.plickers.com/

An, Y.-Y. (2015). Games to promote inquiry learning. In M. J. Spector (Ed.), *The SAGE Encyclopedia of Educational Technology* (Vol. 1, pp. 318–321). Los Angeles, CA: SAGE Publications.

Ball, D. L., Thames, M. H., & Phelps, G. (2008). Content knowledge for teaching: What makes it special? *Journal of Teacher Education*, *59*(5), 389–407. doi:10.1177/0022487108324554

Beatty, I. D., Gerace, W. J., Leonard, W. J., & Dufresne, R. J. (2006). Designing effective questions for classroom response systems teaching. *American Journal of Physics*, *74*(1), 31–39. doi:10.1119/1.2121753

Blömeke, S., & Delaney, S. (2012). Assessment of teacher knowledge across countries: A review of the state of research. *ZDM Mathematics Education*, *44*(3), 223–247. doi:10.1007/s11858-012-0429-7

Bloom, B. S. (1956). *Taxonomy of educational objectives: Cognitive domain* (Vol. 1). New York: Longman.

Bransford, J. D., Brown, A. L., & Cocking, R. R. (2002). *How people learn: Brain, mind, experience, and school*. Washington, DC: The National Academies Press.

British Columbia Ministry of Education. (2015). *Building students success: BC's new curriculum*. Retrieved from https://curriculum.gov.bc.ca/

BSCS. (2008). *Scientists in Science Education: 1958-2008 Innovation Reform Vision*. Retrieved from www.bscs.org

Burridge, P., & Carpenter, C. (2013). Expanding pedagogical horizons: A case study of teacher professional development. *Australian Journal of Teacher Education, 38*(9), 10–24. doi:10.14221/ajte.2013v38n9.3

Carr, J. M. (2012). Does math achievement hAPPen when iPads and game-based learning are incorporated into fifth-grade mathematics instruction? *Journal of Information Technology Education: Research, 11*, 269–286. doi:10.1007/s10639-006-9024-2

Chachashvili-Bolotin, S., Milner-Bolotin, M., & Lissitsa, S. (2016). Examination of factors predicting secondary students' interest in tertiary STEM education. *International Journal of Science Education, 38*(2), 25.

Cole, A. L., & Knowles, J. G. (2000). *Researching teaching: exploring teacher development through reflexive inquiry*. Allyn and Bacon.

Corrigan, D., Gunstone, R., & Dillon, J. (2011). *Approaches to Considering the Professional Knowledge Base of Science Teachers: The Professional Knowledge Base of Science Teaching*. Springer Netherlands. doi:10.1007/978-90-481-3927-9_1

Crawford, B. A. (2007). Learning to teach science as inquiry in the rough and tumble of practice. *Journal of Research in Science Teaching, 44*(4), 613–642. doi:10.1002/tea.20157

Cuban, L. (1990). Reforming again, again, and again. *Educational Researcher, 19*(1), 3–13. doi:10.3102/0013189X019001003

Cuban, L. (2001). *Oversold and underused: Computers in the classroom*. Harvard University Press Review.

Dang, T. (Producer). (2016). *Collaborative Learning Annotation System* [Media player]. Retrieved from http://ets.educ.ubc.ca/clas/

Daniels, H. (2001). Applications of sociocultural activity theory to education. In Vygotsky and Pedagogy (Vol. 1, pp. 96-130). New York: Routledge Falmer.

DeBoer, G. E. (1991). *A History of Ideas in Science Education: Implications for practice* (Vol. 1). New York: Teachers College Press.

Denny, P. (2016). *PeerWise*. Retrieved from http://peerwise.cs.auckland.ac.nz/

Dewey, J. (1938). *Logic: The Theory of Inquiry*. New York: Henry Hold and Company.

Eijck, M., & Roth, W. M. (2009). Authentic science experiences as a vehicle to change students orientations towards science and scientific career choices: Learning from the path followed by Brad. *The Culture of Science Education*, *4*(3), 611–638. doi:10.1007/s11422-009-9183-8

Eshach, H. (2014). The use of intuitive rules in interpreting students difficulties in reading and creating kinematic graphs. *Canadian Journal of Physics*, *92*(1), 1–8. doi:10.1139/cjp-2013-0369

Etkina, E., Karelina, A., Ruibal-Villasenor, M., Rosengrant, D., Jordan, R., & Hmelo-Silver, C. E. (2010). Design and reflection help students develop scientific abilities: Learning in introductory physics laboratories. *Journal of the Learning Sciences*, *19*(1), 54–98. doi:10.1080/10508400903452876

Etkina, E., Planinši, G., & Vollmer, M. (2013). A simple optics experiment to engage students in scientific inquiry. *American Journal of Physics*, *81*(11), 815–222. doi:10.1119/1.4822176

Etkina, E., Van Heuvelen, A., Brookes, D. T., & Mills, D. (2002). Role of Experiments in Physics Instruction - A Process Approach. *The Physics Teacher*, *40*(6), 351–355. doi:10.1119/1.1511592

Feder, T. (2010). Europe reflects on a decade of higher education reforms. *Physics Today*, *63*(May), 24–27. doi:10.1063/1.3431323

Finkelstein, N. D., Adams, W. K., Keller, C. J., Kohl, P. B., Perkins, K. K., Podolefsky, N. S., & LeMaster, R. et al. (2005). When learning about the real world is better done virtually: A study of substituting computer simulations for laboratory equipment. *Physical Review Special Topics - Physics Education Research*, *1*(1), 010103.

Fraser, J. M., Timan, A. L., Miller, K., Dowd, J. E., Tucker, L., & Mazur, E. (2014). Teaching and physics education research: Bridging the gap. *Reports on Progress in Physics*, *77*(032401), 17. PMID:24595011

Hake, R. R. (1998). Interactive-engagement versus traditional methods: A six-thousand-student survey of mechanics test data for introductory physics courses. *American Journal of Physics*, *66*(1), 64–74. doi:10.1119/1.18809

Hake, R. R. (2007). Six lessons from the physics education reform effort. *Latin-American Journal of Physics Education*, *1*(1), 24–31.

Harris, J. B., & Hofer, M. J. (2011). Technological Pedagogical Content Knowledge (TPACK) in action: A descriptive study of secondary teachers curriculum-based, technology-related instructional planning. *Journal of Research on Technology in Education*, *43*(3), 211–229. doi:10.1080/15391523.2011.10782570

Hmelo-Silver, C. E., Duncan, R. G., & Chinn, C. A. (2012). Scaffolding and Achievement in Problem-Based and Inquiry Learning: A Response to Kirschner, Sweller, and Clark (2006). *Educational Psychologist*, *42*(2), 99–107. doi:10.1080/00461520701263368

Hohenwarter, M. (2014). *GeoGebra*. Retrieved from www.geogebra.org

Hourigan, M., & Donaghue, J. (2013). The challenges facing initial teacher education: Irish prospective elementary teachers mathematics subject matter knowledge. *International Journal of Mathematical Education in Science and Technology*, *44*(1), 36–58. doi:10.1080/0020739X.2012.690897

Jonassen, D., & Land, S. (Eds.). (2012). *Theoretical foundations of learning environments* (2nd ed.). New York, NY: Routledge.

Jones, M. G., & Leagon, M. (2014). Science teacher attitudes and beleifs: Reforming practice. In N. G. Lederman & S. K. Abel (Eds.), *Handbook of research on science education* (Vol. 2, pp. 830–847). New York: Routledge.

Koehler, M. J., & Mishra, P. (2015). Technological Pedagogical Content Knowledge. In M. J. Spector (Ed.), *The SAGE encyclopedia of educational technology* (Vol. 2, pp. 782–785). Los Angeles: SAGE Publications.

Koh, J. H. L., & Divaharan, S. (2011). Developing pre-service teachers technology integration expertise through the TPACK: Developing instructional model. *Journal of Educational Computing Research*, *44*(1), 35–58. doi:10.2190/EC.44.1.c

Krajcik, J. S., & Mun, K. (2014). Promises and challenges of using learning technologies to promote student learning of science. In L. Norman G & S. K. Abell (Eds.), Handbook of Research on Science Education (Vol. 2, pp. 337-360). New York: Routledge.

Lasry, N. (2008). Clickers or flashcards: Is there really a difference? *The Physics Teacher*, *46*(5), 242–244. doi:10.1119/1.2895678

Lasry, N., Guillemette, J., Dugdale, M., Bhatnagar, S., Whittaker, C., Charles, E., & Mazur, E. (2014). *You gain some, you lose some: A big data look at how students change conceptions*. Physics Review Special Topics - Physics Education Research.

Lasry, N., Guillemette, J., & Mazur, E. (2014). Two steps forward, one step back. *Nature Physics*, *10*(June), 402–403. doi:10.1038/nphys2988

Lasry, N., Mazur, E., & Watkins, J. (2008). Peer Instruction: From Harvard to the two-year college. *American Journal of Physics*, *76*(11), 1066–1069. doi:10.1119/1.2978182

Lasry, N., Watkins, J., Mazur, E., & Ibrahim, A. (2013). Response times to conceptual questions. *American Journal of Physics*, *81*(9), 703–706. doi:10.1119/1.4812583

Lee, M.-H., & Tsai, C.-C. (2010). Exploring teachers perceived self efficacy and Technological Pedagogical Content Knowledge with respect to educational use of the World Wide Web. *Instructional Science*, *38*(1), 1–21. doi:10.1007/s11251-008-9075-4

Let's Talk Science. (2012). *Spotlight on science learning: A benchmark of Canadian talent*. Retrieved from http://www.letstalkscience.ca/our-research/spotlight.html

Let's Talk Science. (2013). *Spotlight on science learning: The high cost of dropping science and math*. Retrieved from http://www.letstalkscience.ca/our-research/spotlight2013.html

Let's Talk Science. (2015). *Exploring parental influence: Shaping teen decisions regarding science education*. Retrieved from http://www.letstalkscience.ca/research-publications/spotlight-on-science-learning/2015-spotlight.html

Luft, J. A., & Hewson, P. W. (2014). Research on teacher professional development knowledge in science. In N. G. Lederman & S. K. Abel (Eds.), *Handbook of research on science education* (Vol. 2, pp. 889–909). New York: Routledge.

Maciel, T. (2015). Smartphones in the classroom help students see inside the black box. *APS News*, *24*(3), 5–6.

D. Martinovic, Z. Karadag, & D. McDougall (Eds.). (2014). *Proceedings of the Fifth North American GeoGebra Conference: Explorative learning with technology: GeoGebra-NA 2014*. Toronto, ON, Canada: University of Toronto.

Mazur, E. (1997a). Moving the mountain: Impediments to change. *The Physics Teacher*, *35*, 1–4.

Mazur, E. (1997b). *Peer Instruction: User's manual*. Upper Saddle River, NJ: Prentice Hall.

Mazur, E. (1997c). *Understanding or Memorization: Are we teaching the right thing?* Paper presented at the Conference on the Introductory Physics Course on the Occasion of the Retirement of Robert Resnick, New York, NY.

McDermott, L. C., Heron, P. R. L., Shaffer, P. S., & Stetzer, M. R. (2006). Improving the preparation of K-12 teachers through physics education research. *American Journal of Physics, 74*(9), 763–767. doi:10.1119/1.2209244

Milner-Bolotin, M. (2012). Increasing interactivity and authenticity of chemistry instruction through data acquisition systems and other technologies. *Journal of Chemical Education, 89*(4), 477–481. doi:10.1021/ed1008443

Milner-Bolotin, M. (2014a). *Closing the research-practice gap through innovative technology use in STEM teacher education.* Paper presented at the Fifth North American GeoGebra Conference (GeoGebra-NA 2014): Explorative Learning with Technology, Toronto, Canada.

Milner-Bolotin, M. (2014b). Using PeerWise to promote student collaboration on design of conceptual multiple-choice questions. *Physics in Canada, 70*(3), 149–150.

Milner-Bolotin, M. (2015a). Learning physics teaching through collaborative design of conceptual multiple-choice questions. *The Canadian Journal of Action Research, 16*(2), 22–41.

Milner-Bolotin, M. (2015b). Technology-enhanced teacher education for 21st century: Challenges and possibilities. In X. Ge, D. Ifenthaler, & J. M. Spector (Eds.), *Emerging technologies for STEAM education* (pp. 137–156). Springer International Publishing. doi:10.1007/978-3-319-02573-5_8

Milner-Bolotin, M. (2015c). Visual literacy skills in science, technology, engineering, and mathematics education. In M. J. Spector (Ed.), *The SAGE Encyclopedia of Educational Technology* (Vol. 2, pp. 826–830). Thousand Oaks, CA: SAGE Publications.

Milner-Bolotin, M. (2016a). *Mathematics and Science Teaching and Learning through Technologies.* Retrieved from http://scienceres-edcp-educ.sites.olt.ubc.ca/

Milner-Bolotin, M. (2016b). Promoting Deliberate Pedagogical Thinking with Technology in physics teacher education: A teacher-educator's journey. In T. G. Ryan & K. A. McLeod (Eds.), *The Physics Educator: Tacit Praxes and Untold Stories* (pp. 112–141). Champaign, IL: Common Ground and The Learner.

Milner-Bolotin, M. (2016c). Rethinking technology-enhanced physics teacher education: From theory to practice. *Canadian Journal of Science, Mathematics and Technology Education, 16*(3), 284–295. doi:10.1080/14926156.2015.1119334

Milner-Bolotin, M., Egersdorfer, D., & Vinayagam, M. (2016). Investigating the effect of question-driven pedagogy on the development of physics teacher-candidates' Pedagogical Content Knowledge. *Physics Review Special Topics: Physics Education Research, 12*(020128), 1–16.

Milner-Bolotin, M., Fisher, H., & MacDonald, A. (2013). Modeling active engagement pedagogy through classroom response systems in a physics teacher education course. *LUMAT: Research and Practice in Math, Science and Technology Education, 1*(5), 523–542.

Milner-Bolotin, M., Kotlicki, A., & Rieger, G. (2007). Can students learn from lecture demonstrations: The role and place of interactive lecture experiments in large introductory science courses. *Journal of College Science Teaching, 36*(4), 45–49.

Milner-Bolotin, M., & Moll, R. F. (2008). Physics exam problems reconsidered: Using Logger Pro technology to evaluate student understanding of physics. *The Physics Teacher, 46*(8), 494–500. doi:10.1119/1.2999067

Minner, D. D., Levy, A. J., & Century, J. (2010). Inquiry-based science instruction–What is it and does it matter? Results from a research synthesis years 1984 to 2002. *Journal of Research in Science Teaching, 47*(4), 474–496. doi:10.1002/tea.20347

Mishra, P., & Koehler, M. J. (2006). Technological Pedagogical Content Knowledge: A Framework for teacher knowledge. *Teachers College Record, 108*(6), 1017–1054. doi:10.1111/j.1467-9620.2006.00684.x

National Research Council. (2012). *A Framework for K-12 Science Education: Practices, Crosscutting Concepts, and Core Ideas*. Author.

National Research Council. (2013). *Next Generation Science Standards: For States, by States*. Washington, DC: The National Academies Press.

Niess, M. L. (2005). Preparing teachers to teach science and mathematics with technology: Developing a technology pedagogical content knowledge. *Teaching and Teacher Education, 21*(5), 509–523. doi:10.1016/j.tate.2005.03.006

OECD. (2014). *PISA 2012 Results: What students know and can do – student performance in mathematics reading and science* Retrieved from http://www.oecd.org/pisa/keyfindings/pisa-2012-results.htm

Olson, S., & Loucks-Horsley, S. (Eds.). (2000). *Inquiry and the National Science Education Standards: A Guide for Teaching and Learning*. National Academies Press.

Pollock, S. J. (2004). *No Single Cause: Learning Gains, Student Attitudes, and the Impacts of Multiple Effective Reforms.* Paper presented at the Physics Education Research Conference, Sacramento, CA. doi:10.1063/1.2084720

Ruthven, K., Deaney, R., & Hennessy, S. (2009). Using graphing software to teach about algebraic forms: A study of technology-supported practice in secondary-school mathematics. *Educational Studies in Mathematics, 71*(3), 279–297. doi:10.1007/s10649-008-9176-7

Schmidt, W. H., Blömeke, S., Tatto, M. T., Hsieh, F.-J., Cogan, L. S., Houang, R. T., & Schwille, J. (2011). *Teacher education matters: A study of middle school mathematics teacher preparation in six countries.* New York: Teachers College Press.

Schuster, D., Undreiu, A., Adams, B., Brookes, D., & Milner-Bolotin, M. (2009). Motion-Matching: A Challenge Game to Generate Motion Concepts. *The Physics Teacher, 47*(7), 381–385. doi:10.1119/1.3225502

Shelton, B. E. (2015). TPCK (Technological Pedagogical Content Knowledge): Implications for 21st century teacher education. In M. J. Spector (Ed.), *The SAGE Encyclopedia of Educational Technology* (Vol. 2, pp. 785–790). Los Angeles, CA: SAGE Publications.

Shulman, L. S. (1986a). Paradigms and research programs in the study of teaching: A contemporary perspective. In M. C. Wirrtock (Ed.), *Handbook of Research on Teaching* (pp. 3–36). New York: Collier Macmillan.

Shulman, L. S. (1986b). Those who understand: Knowledge growth in teaching. *Educational Researcher, 15*(2), 4–14. doi:10.3102/0013189X015002004

Tan, Y. S. M., & Caleon, I. S. (2016). Dealing with the ambiguities of science inquiry. *The Physics Teacher, 54*(March), 168–172. doi:10.1119/1.4942141

Vernier-Technology. (2016). *Logger Pro* (Version 3.6.1). Portland, OR: Vernier Technology. Retrieved from www.vernier.com

Vygotsky, L. (1978). *Mind in Society: The Development of Higher Psychological Processes*. Cambridge, MA: Harvard University Press.

Wieman, C. E. (2012). Applying new research to improve science education. *Issues in Science and Technology*, (Fall): 1–9.

Wieman, C. E., Adams, W. K., Loeblein, P., & Perkins, K. K. (2010). Teaching physics using PhET simulations. *The Physics Teacher, 48*(April), 225–227. doi:10.1119/1.3361987

Wieman, C. E., & Perkins, K. (2005). Transforming physics education. *Physics Today*, *58*(11), 36–42. doi:10.1063/1.2155756

Wright, D. (2005). Graphical Calculators: Tools for Mathematical Thinking. In S. Johnston-Wilder & D. Pimm (Eds.), *Teaching Secondary Mathematics with ICT* (pp. 145–158). New York: Open University Press.

KEY TERMS AND DEFINITIONS

Collaborative Learning Annotation System (CLAS): An online system that enables learners to collaborate on uploaded videos.

Collaborative Teacher Inquiry: Teacher collaboration focused on inquiring about their own practices and student learning.

Computer Simulations and Real Data Collection and Analysis: Enable students to model natural phenomena either virtually or in real-life through data collection and analysis.

Inquiry-Based STEM Learning: Learning of STEM disciplines that encourages learners to ask questions, test hypotheses and ideas, and conduct experiments. It is often conducted in a collaborative environment.

Pedagogical Content Knowledge (PCK): Teacher's knowledge expressed as an overlap of subject knowledge and pedagogical knowledge. PCK concept was introduced by Lee Shulman in 1986.

Peer Instruction: An interactive data-driven pedagogy popularized by Harvard Professor Eric Mazur in the early 1990s.

PeerWise: An online collaborative platform used for learner collaboration on multiple-choice questions.

Teacher Zone of Proximal Development (T-ZPD): The difference between the teachers' individual TPACK and their TPACK resulting from their collaboration with peers.

Technological Pedagogical and Content Knowledge (TPACK): An expansion of the PCK that includes the relevant knowledge of subject-specific technologies. TPACK was introduced by Koehler and Mishra in the 1990s.

Chapter 11
Project-Based MOOC:
Enhancing Knowledge Construction and Motivation to Learn

Miri Barak
Technion – Israel Institute of Technology, Israel

Abeer Watted
Technion – Israel Institute of Technology, Israel

ABSTRACT

This chapter describes a project-based massive open online course (MOOC) in nanotechnology and nanosensors that was offered in two languages: English and Arabic. A mixed methods research was conducted to examine the role of project-based learning in the process of knowledge construction and motivation to learn a MOOC. The study compared between three groups of science and engineering students: English MOOCers, Arabic MOOCers, and university students. Findings indicated positive attitudes about learning in a project-based MOOC, especially with relation to gaining work experience. Findings also indicated that in a project-based setting, MOOC participants were mostly driven by a desire to establish connections with peers, whereas university students were mostly motivated by their interest in the subject matter. Arabic MOOCers, who were less successful in solving ill-defined problems, narrowed the gap, and at the end of the course received similar grades in the final project.

DOI: 10.4018/978-1-5225-2525-7.ch011

INTRODUCTION

In recent years, there is a growing drive among higher education institutions to develop distance education programs, in general (Barak, 2012) and massive open online courses (MOOCs) in specific (Barak, Watted & Hasick, 2016; Kizilcec, & Schneider, 2015). MOOCs are web-based courses, designed to provide free and accessible high-quality education to the masses (Barak et al., 2016; Cormier & Siemens, 2010). The term was first coined in 2008 to describe a course, titled: "Connectivism and Connective Knowledge", offered online for free by the University of Manitoba (Cormier & Siemens, 2010). MOOCs are free courses, open to all participants from all ages and countries. They do not require preliminary qualifications or prerequisite courses. Most MOOCs include short segments of video lectures arranged according to the course topics. They also include an e-book or related articles, discussion forums, and learning tasks that participants are required to submit. Because masses of participants (thousands or even tenths of thousands) take these courses, the learning tasks are either automatically graded (e.g. closed-ended quizzes) or peer-graded (e.g. open-ended essays, written assignments).

In 2012, new MOOC platforms were developed, utilizing advanced information and communication technologies and automated computing to support the large number of participants who sought interest in taking the open courses. While in May 2012, there were 26 MOOCs, offered on various platforms, in January 2016, their number reached more than 4,500. The largest MOOC platforms are Coursera, Udacity, edX, and FutureLearn, all founded in 2012. Coursera and Udacity emerged from Stanford University, edX was founded by Harvard University and MIT, and FutureLearn is a private company owned by the UK's Open University. Today, Coursera platform offers more than 1500 courses in a variety of subject matter areas, from humanities to sciences and engineering (see https://www.coursera.org).

Following this trend, the Technion – Israel institute of Technology, launched in 2014 a course on nanotechnology and nanosensors in a MOOC format, which is running until this day. The course was developed by Prof. Hossam Haick, from the Faculty of Chemical Engineering, a renowned researcher in the field of nanotechnology. The nanotechnology and nanosensors course was the world's first MOOC in this unique field, and the first that was simultaneously delivered in two languages: English and Arabic. In fact, it was the first MOOC delivered in the Arabic language. Our goal was twofold. First, to reach out to a wide range of people worldwide, even those who live in countries with no diplomatic relations with Israel. Second, to provide an example - a model for promoting sociocultural learning in the context of engineering education, by integrating project-based learning, multicultural teamwork, and peer assessment into the curriculum.

Following previous work on project-based learning in higher education (Barak & Dori, 2005), the nanotechnology and nanosensors MOOC's learning assignments included features related to the use of an authentic driving question, a community of inquiry, and the use of technology-based cognitive tools. Project-based learning involves learners in hands-on authentic investigation (Barak & Dori, 2005; Krajcik, McNeil & Reiser, 2007; Thomas, 2000). In order for an activity to be considered as project-based learning, it has to involve the construction of knowledge, that is, the development of new ideas, understandings, and/or skills, on the part of learners. This raises a question regarding the role of project-based MOOCs in the process of knowledge construction and motivation to learn among science and engineering students.

In light of the aforesaid, this chapter includes four parts. First, it presents the study's background and a literature review; second, it describes the nanotechnology and nanosensors MOOC and its innovative instructional design; third, it presents a study conducted to examine MOOC participants' knowledge construction and; fourth, it presents our suggestions for future research and conclusions.

BACKGROUND

The Sociocultural Theory

Our study is rooted in the sociocultural theory that emphasizes the importance of culture and social context to the cognitive development of learners. Socioculturalism is one of the prominent learning theories positioned on the interface between educational psychology and sociology. It articulates the important contribution that society and culture has on individual cognitive development. Guided by the work of Vygotsky (1987), the sociocultural theory maintains that cognitive development occurs while learners are engaged in social activities (Lemke, 2001; Ragupathi, 2014). It poses an opposition to cognitive theories that perceive education as merely a systematic knowledge construction process, disconnected from the context and surroundings in which it occurs (Lemke, 2001; Wertsch, 1991). It realizes learning as an interactive, contextual, and shared process, therefore encourages collaborative teamwork among learners (Ragupathi, 2014; Vygotsky, 1978).

The sociocultural theory maintains three major propositions for promoting meaningful learning. The first proposition is related to the notion of scaffolding in the process of learning. Based on the zone of proximal development (ZPD) model (Vygotsky, 1978), scaffolding is perceived as an imperative process for reducing the distance between a learner's actual and potential developmental level (Lemke, 2001; Wertsch, 1991). The second proposition asserts that learning occurs while using mediator aids (Vygotsky, 1978; Wertsch, 1991). Mediator aids include physical tools such as graphics, linguistic tools, or symbolic tools. Wertsch, (1991) identified writing, diagrams schemes, mnemonic techniques, and computers, as mediators that facilitate cognitive development of learners. The third proposition calls attention to the profound role that cultural background plays in the learning process (Lemke, 2001). The sociocultural theory maintains that learners are not 'blank slates'; they bring with them a set of ideas and belief systems, adopted from the social and cultural group to which they belong (Ragupathi, 2014). In our study, MOOC students were encouraged to work on projects in small multicultural groups, supporting each other through peer-scaffolding and online discourse.

While cognitive constructivists believe that learning takes place through personal incentives and alone, social constructivists argue that learning is conducted through cultural context and interactions with others (Ragupathi, 2014). Learning in collaboration, sharing information, and working in teams are easily enabled by web 2.0 platforms and cloud applications (Barak, 2017; Barak et al., 2009). The rapid development of such online technologies offers new opportunities for education. One of the most significant innovations is the development of massive open online courses (MOOC) (Barak et al., 2016; Cormier & Siemens, 2010; Kizilcec, & Schneider, 2015).

Massive Open Online Courses

Since they were first introduced in 2008, massive open online courses (MOOCs) are basing their position as a significant development in the area of online distance education (Barak et al., 2016; Onah, Sinclair & Boyatt, 2014). Massive open online courses (MOOCs) are web-based learning environments in which participants, worldwide, can acquire knowledge for free without any commitment or prior requirements. They provide high-quality learning materials offered by well-known lecturers from hundreds of universities. Contrary to traditional online courses which charge tuition, carry credit, and limit enrollment to a few dozen to ensure interaction with instructors, MOOCs are usually free, credit-less, and massive. Since they were first launched, MOOCs' popularity is growing rapidly among adult learners. For example, in March 2013, there were three million enrollers in Coursera (a MOOC platform); eight months later, there were about seven million enrollers, and at the end of 2015, there were more than seventeen million enrollers from all over the world. This educational phenomenon is perceived as a 'disruptive innovation' to traditional higher education among many educationist, stakeholders, and policy makers. However, there is yet little evidence regarding the way and manner in which MOOCs promote learning and cognitive development among adult learners.

MOOCs have various advantages over traditional in-class courses, such as accessibility (i.e., everyone have free access to course contents and assignments), flexibility in time and place (i.e., learners can study a MOOC whenever and wherever convenient), and multi-cultural interactions (i.e., learners can communicate with peers without geographical or political borders). Realizing the significant potential of MOOCs, studies have explored different aspects, such as attitudes toward MOOCs, motivation to learn, and effectiveness of different types of MOOC (Barak et al., 2016; Breslow, et al., 2013; Kizilcec, & Schneider, 2015).

Literature on MOOCs indicates an ongoing debate about their educational value. Some studies indicated high positive satisfaction among MOOC participants; while others point out difficulties and challenges (Jordan, 2014; Kizilcec, & Schneider, 2015; McNutt, 2013). The more autonomous, diverse, and open the course is, the greater the potential for students' learning to be limited by the lack of structure and support. Accordingly, the more open and self-directed the course is, the more students seek to engage in collaborative learning (Barak et al., 2016). The increasing number of MOOC-related research can be described in two "waves". The 'first wave' focused mainly on the problem of attrition and dropout rates in MOOCs (Jordan, 2014; Onah, Sinclair & Boyatt, 2014). The 'second wave' examined MOOC enrollers' motivation, engagement patterns, and collaborative learning in large groups (Ferguson & Clow, 2015; Kizilcec & Schneider, 2015). However, little attention has been devoted to the study of project-based MOOCs and their impact on students' learning outcomes.

Project-Based Learning

Project-based learning (PBL) is an instructional approach that actively engages students at solving real-life, open-ended problems in an authentic context (Barak & Dori, 2005; Krajcik & Blumenfeld, 2008; Thomas, 2000). It is conceptualized as a pedagogical model that organizes learning around challenging questions that involve learners in critical thinking, problem-solving, decision making, and investigative activities. It provides learners with the opportunity to work autonomously over a certain length of time, concluding in a presentation, written essay, and/or an artifact (Barak & Dori, 2005; Krajcik, McNeil & Reiser, 2007). PBL applies hand-on, student-centered activities, involving authentic learning and assessment. It underlines explicit educational goals, and in most cases, it involves small group learning and personal reflection (Krajcik & Blumenfeld, 2008; Thomas, 2000).

Many studies examined the PBL approach, endorsing its merits for producing better learning outcomes than the traditional approach (Jurow, 2005; Thomas, 2000). In the science education realm, PBL as found to foster meaningful learning and better understanding of scientific concepts and phenomena (Krajcik & Blumenfeld, 2008; Krajcik, McNeil & Reiser, 2007). Educators endorse PBL as a prominent instructional approach for connecting science and technology to students' daily lives as well as promoting scientific literacy and engineering knowledge (Jurow, 2005; Krajcik & Blumenfeld, 2008). Science educators contend that PBL should become an integral component in the curriculum of science and engineering courses (Morales, Bang, & Andre, 2013). To make this approach applicable, students must be autonomous to some degree and take responsibility in carrying out the projects (Morales et al., 2013; Thomas, 2000). By planning the work needed to achieve the projects' objectives, reading relevant literature, discuss possibly solutions, choosing the best solution, and implementing it, learners procure competencies, within the context of their sociocultural background and abilities (Thomas, 2000).

While most studies indicate that hands-on assignments facilitate students' interest and performance, there are some cases in which the learning gains are limited (Krajcik & Blumenfeld, 2008). This can be explained by ineffective project management, interpersonal tensions during group work, and a deficiency in systematic work processes, leading to inadequate time organization, and consequently a shortage of time for discussions and reflections (Jurow, 2005; Krajcik & Blumenfeld, 2008). With the lack of management and communication skills, PBL activities may become time consuming and inefficient (Krajcik & Blumenfeld, 2008). Therefore, the use of advanced communication and information technologies is essential for successful BPL (Barak & Dori, 2005).

In the past decade, studies indicated the important role of web-based online technologies in facilitating PBL management and implementation (Dominguez & Jaimo, 2010; Rooij, 2009). Online technologies are universally accepted in the educational community as learning tools that scaffold knowledge construction (Barak, 2017; Barak & Ziv, 2013; Morales et al., 2013; Rooij, 2009). They were found to promote meaningful learning by enhancing students' involvement and conceptual understanding of scientific concepts (Barak & Ziv, 2013; Bell, Maeng & Binns, 2013). These are carried out by applying digital simulations of experiments which might not be experienced otherwise, collaborative learning, collecting and organizing data, and preparing reports while fostering peer assessment (Barak & Dori, 2005; Barak, 2017; Barak & Ziv, 2013; Bell et al., 2013). Experts in project management recommend the use of online applications for designing and organizing projects (Dominguez & Jaimo, 2010; Rooij, 2009).

DESIGNING A PROJECT-BASED MOOC IN TWO LANGAUGES

The Technion's Nanotechnology and Nanosensors course, taught by Prof. Hossam Haick, was delivered in parallel as two MOOCs -- in English and Arabic, by the same teaching team. All learning materials were prepared in both languages, including the course guidelines, presentation slides, learning assignments, and video lectures. The MOOCs were purposefully designed to meet the same high quality and academic level of the face-to-face course. They introduce fundamental principles of nanotechnology and nanomaterials and their vital role in new sensing properties and applications. They discuss broad, interdisciplinary areas that encompass chemistry, physics, biology, materials science, and electrical engineering. The Arabic MOOC was developed in addition to the English version to provide access to high-quality education to a wider range of people. Thus, using the web as a 'bridge' across geographical and political borders. Today, both the English and Arabic Nanotechnology and Nanosensors MOOCs are hosted on Coursera platform: https://www.coursera.org/course/nanosensors.

The MOOCs are designed as a ten weeks course, divided into two sessions (part 1 and 2), of five weeks each. They include lecture videos, e-books, weekly quizzes and discussion forums, where students can introduce themselves, ask content-related questions, seek for technical support, provide answers, and share ideas. The MOOCs are designed according to the project-based learning approach. Accordingly, each course is divided into two parts. In the first part, students work individually on open assignments, applying their knowledge of nanotechnology to topics of personal interest. Then, in the second part, they are required to work on a final project in small multicultural groups. The open assignments are set to induce innovative ideas that can then be elaborated in final project; for example:

Give an example for an innovative nano-device that you would like to have on a daily basis. Explain why it is important to you. Notice that it can be, for example: a sensor in your body (detecting or protecting from a disease, improving abilities, enhancing you senses), a sensor in digital/technical devices (smartphone, pad, TV, car), a sensor in your home or office (simplifying your workload) etc.

The Final project encouraged students to work in groups of three or four, preferably from different nationalities. The students were given the following instructions:

You will be asked to write a final project that accounts for 60% of the course total grade. It focuses on the utilization of nanotechnology and nanosensors to imitate a specific sense that relates to human senses: vision, hearing, smell, or touch. Other non-specific senses such as balance and pain can be included as well. The final project should include, but not confined to: (a) multidisciplinary presentation and discussion of the overall design approach, (b) characterization, (c) fabrication process, and (d) application of the targeted artificial sense. The project is conducted in collaboration, in groups of 3-to-4 students; preferable, multicultural and interdisciplinary groups in order to gain new and interesting ideas. Your answers will be graded by peers, according to a grading rubric.

Due to the large numbers of MOOC participants, the grading of the open assignments and the final project are based on peer-assessment. MOOC participants were asked to grade the work of their peers according to a ridged scoring rubric that was designed by the course teaching team. Each student was required to grade at least three assignments and three final projects in order to receive their own grades.

A STUDY ON KNOWLEDGE CONSTRUCTION AND MOTIVATION TO LEARN IN MOOCS

Research Goal and Participants

Following the development and delivery of a massive open online courses (MOOCs) in nanotechnology and nanosensors, we conducted a research that was set to examine the role of project-based learning in the process of knowledge construction and motivation to learn a MOOC. The study compared between three groups of science and engineering students: English MOOCers, Arabic MOOCers, and university students. This raised the following research questions:

1. Are there differences between MOOC participants (English and Arabic) and university students in:
    a. The topics of their projects?
    b. Familiarity with the subject matter and work experience?
    c. Content knowledge and understanding of the learning materials?
2. What motivate students to learn a project-based MOOC?

The study compared between three groups of students, as follows:

- **English MOOC Participants:** Represent a heterogeneous group of learners - a combination of participants with diverse academic, professional, and cultural backgrounds.
- **Arabic MOOC Participants:** Represent a semi-heterogeneous group of learners - a combination of participants from different countries and academic background (informal education, undergraduates, or graduates), but with a similar culture.

- **University Students:** Represent a homogenous group of learners - students that were accepted to the Technion according to their high achievements in the matriculation and psychometric examinations.

To recruit research participants, we sent a message to all the participants who enrolled to the *Nanotechnology and Nanosensors* MOOCs in English and Arabic, encouraging them to take part in our study. Participation was voluntary and the participants could withdraw at any given point. Although many answered the pre-questionnaire, less than 10% answered the post-questionnaires. Therefore, our sample includes 140 participants, distributed as follows: 58 English MOOCers, 36 Arabic MOOCers, and 46 university students. The participants' demographics and academic background, by courses, are presented in Table 1.

Chi-Square test indicated no statistically significant differences between the research groups in respect to gender distribution. However, significant differences were indicated for Age - the university students were younger than the English and Arabic MOOC participants (χ^2 (4) = 11.77, p = .019); Level of education - the percentage of graduates among the university students was lower (χ^2 (4) = 110.11, p = .000); Academic field - most of the university students studied engineering, while

Table 1. Participant's demographics and academic background, by courses

Demographics and academic background		% English MOOC (N= 58)	% Arabic MOOC (N= 36)	% University students (N= 46)
Gender	Male	79	78	79
	Female	21	22	21
Age	25 and younger	31	36	43
	26-35	36	39	57
	36 or older	33	25	-
Level of education	Post-secondary	17	27	-
	Undergraduate	44	35	89
	Graduates	39	38	11
Academic field	Engineering/Business	40	28	82
	Science/Medicine	60	72	18
Occupation status	Students	42	47	89
	Industry worker	21	15	11
	Researchers	17	9	-
	Academic lecturers	10	21	-
	Other	10	8	-

the distribution between engineering and sciences among the MOOC participants was somewhat equal (χ^2 (2) = 20.48, p = .000); Occupation status - only few university students worked in the industry (χ^2 (85) = 17.49, p = .000); and Geographical location - all the university students were located in the same country, while the English MOOC participants were from all continents, and the Arabic MOOC participants were from four continents (χ^2 (10) = 76.83, p = .000).

Method

The study was based on the 'mixed method' model, applying the parallel convergence design in which quantitative and qualitative data is collected and analyzed simultaneously throughout the study (Creswell, 2014). *The quantitative study* was conducted according to the quasi-experimental design (Campbell & Stanley, 1963). Data were collected via a 'professional content knowledge questionnaire' that was administered before and after the MOOCs. The questionnaire included two parts: a) two closed-ended questions, indicating participants' familiarity with the subject matter and work experience, on a three-point scale; b) eight multiple-choice questions, indicating participants' knowledge and understanding of the course contents. The questionnaire was validated through an inter-rater process, resulting in a CVR (content validity ratio) of 1.00, which indicates full agreement among the five assessors. The assessors panel included three experts in nanotechnology (one professor and two PhD students) and two experts in science education (one professor and one PhD student). The quantitative data were statistically analyzed by the SPSS system using the descriptive statistics and general linear model tests (Field, 2009). In addition to the questionnaire, we analyzed and compared students' grades in the quizzes, open assignments, and final project.

The qualitative study was conducted according to the general inductive content analysis approach (Hsieh & Shannon, 2005; Thomas, 2006). Data were collected from the final projects, forum posts, and e-mail messages to examine students' learning outcomes and motivation to learn the MOOC. The analysis and coding process included four main steps. First, we compiled all the messages (email and forums) into one coherent file according to a chronological and hierarchical order: a single message, a thread, and a forum that is composed of several threads. Second, three independent researchers read it rigorously and marked relevant text segments that indicated, explicitly or implicitly, participants' motivation to learn. Third, the marked text segments were rearranged in new paragraphs according to thematic relations. We identified emerging categories and made inferences. Fourth, we used the constant comparative method (Strauss & Corbin, 1990) to compare each text assigned to a category with each of those already assigned to that category, through inter-coder assessment agreement of three researchers, with Kalpha = .87, which is relatively high (Hayes & Krippendorff, 2007).

Findings

The 'Findings' section includes four parts, each provides data for one of the research questions or sub-questions. First, we present the topics that the participants chose for their final project. Second, we discuss participants' familiarity with the subject matter and their work experience. Third, we present findings related to participants' content knowledge gain. Finally, we present evidence related to what motivates them to learn a MOOC.

The Topics of the Projects

This section presents the analysis of participants' projects, as were submitted at the end of the courses. Data indicated that although the participants were encouraged to work on the projects in small groups of 3 or 4, some preferred to work in pairs or even individually. Among the English MOOC participants, 7% worked individually, 14% in pairs, with an average of 2.8 members in a group. Among the Arabic MOOC participants, 11% worked individually, 17% in pairs, with an average of 2.4 members in a group. Among the university students, 13% worked individually, 22% in pairs, with an average of 2.3 members in a group. The number of participants, by number of group members in each course is presented in Table 2.

The findings presented in Table 2 show that most English MOOC participants preferred working in groups of three or four members, and the Arabic MOOC participants preferred working in groups of three. Regarding the university students, many worked in groups of three, but many others preferred to work in pairs or individually.

Table 2. The number of participants, by number of members in a project group in each course

		No. of members in a project group				No. of groups/ projects ($N_{Proj.} = 56$)	No. of participants (N = 140)
		1*	2	3	4		
Research group	English MOOC	4	4	6	7	21	58
	Arabic MOOC	4	3	6	2	15	36
	Univ. Students	6	5	6	3	20	46

* Individual learners

Overall, we analyzed 56 projects that described the fabrication and utilization of nanotechnology and nanosensors for imitating senses, such as: sight, hearing, taste, smell, and touch. The projects included many scientific and engineering fields, such as food processing, environment, medicine, military, industry, and daily-life devices. Some projects presented innovative ideas that included the integration of several senses; for example: sensing food-borne pathogens through taste and smell-imitating nanosensors. The projects' distribution across topics and research groups is presented in Figure 1.

Figure 1 shows that most of the participants' projects, from the three groups, focused on the design of nanosensors for touch (e.g. sensing capabilities of the human gut), smell (e.g. electrical sensor for cancer detection), and sight (e.g. artificial retina). Hearing (e.g. nano-based human prosthetic auditory sense) and taste (nanosensor for detecting flavors) were less popular topics. The English MOOCers and the university students presented similar trends in choosing their project's topic. In addition, 6% or 5% of their projects were highly innovative, including the integration of several devices to fabricate a nanosensor that imitates two or more senses. A different picture was identified in the Arabic MOOC projects. None focused on hearing devices and none included the integration of several senses.

As described above, the projects targeted many scientific and engineering fields. In the field of food engineering, there were projects that focused on: taste and smell-imitating nanosensor for sensing food-borne pathogens; artificial nose and tongue based on colorimetric sensor arrays for the evaluation of green tea; and nanosensor for detecting toxicity in food. In the field of environmental engineering, there were projects that focused on nanosensor array device for detection and control of harmful

Figure 1. The percentage of projects for each topic, by research group

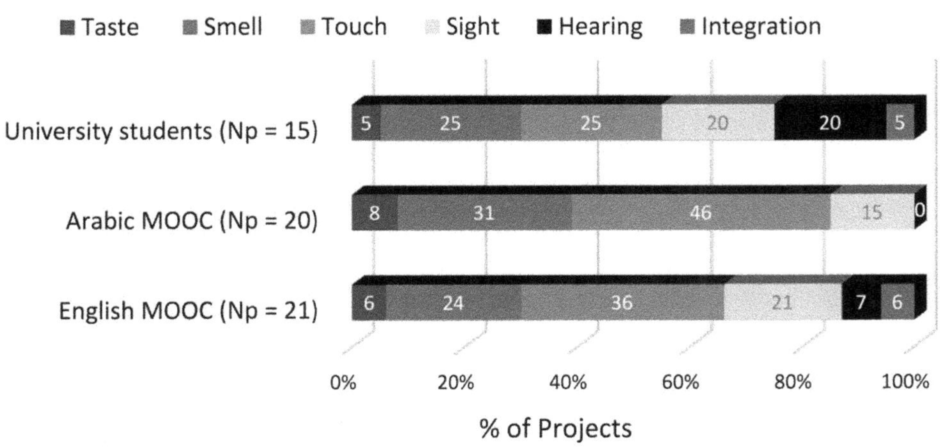

microbes; an electronic nose that is able to identify gaseous chemicals compounds; a biosensor for oxygen detection for the prevention of "dead zones", hypoxic (low-oxygen) areas in oceans and in large lakes. In the medical field, the projects focused on nanosensors for lung disease prevention; Alzheimer biosensor; shoes for the blind; and nanotechnology-based tactile sensor to provide sense of touch to the disabled. In the field of military and law enforcement, the projects focused on nasal implant that can identify explosive or toxin particulates in the air. In the industry field, the projects focused on nanosensors for improving oil and gas detection. For daily-life, the projects focused on sensors for locating lost devices and remote controls.

Familiarity With the Subject Matter and Work Experience

This section includes two parts. First, it discusses participants' familiarity with the subject matter (i.e. concepts and principles of nanosensors); then, it presents the students' assertions about their work experience. In both parts, the assertions of the English MOOCers, the Arabic MOOCers, and the university students, before and after the project-based course, are described. With relation to participants' familiarity with the subject matter, before and after the project-based course, findings indicated a decrease among the English MOOCers in the percentage of those who asserted that they are mostly new to nanotechnology. They asserted an increase in the percentage of 'somewhat familiar' and 'very familiar', but the difference was not statistically significant. The Arabic MOOCers asserted only a slight decrease in being mostly new to nanotechnology, with no significant familiarity gain. The third group - the university students, indicated a similar trend, with no statistical significant difference between the pre- and post-assertions (Figure 2).

Figure 2. Participants' familiarity with the subject matter, before and after the course

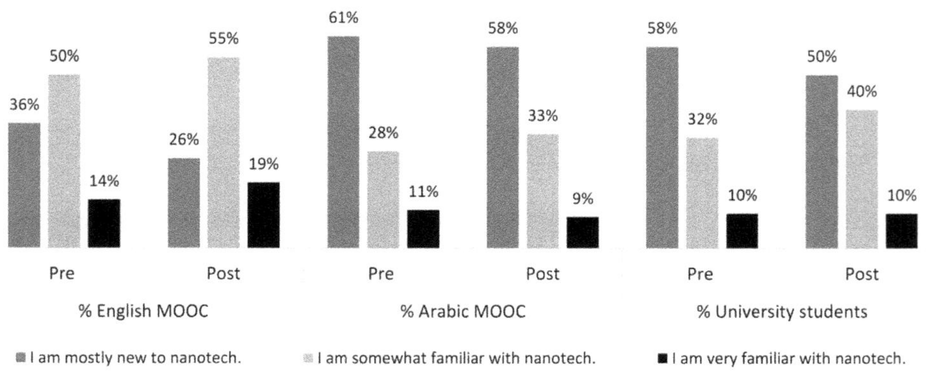

With relation to participants' work experience with nanosensors, the English MOOCers indicated a statistically significant decrease in the percentage of students who indicated that they have no work experience in the nanotechnology field ($Z = -2.00$, $p = .046$), and a statistically significant increase in the percentage of participants who asserted to have some work experience ($Z = -2.18$, $p = .029$). Similar trends were indicated among the Arabic MOOCers and university students, with no statistical significant difference. Since the university students were undergraduates, junior or freshman, none indicated having significant work experience (Figure 3).

Overall, the findings presented in Figures 2 and 3 indicate that the English MOOC participants experienced higher familiarity gain of nanotechnology concepts and principles, compare to their counterparts. In addition, they benefited from the project-based MOOC by gaining work experience in nanosensors. Regarding the Arabic MOOCers, while they did not assert high familiarity gains, they benefited from the project-based MOOC by gaining some work experience. The university students asserted somewhat familiarity gain, with no work experience.

Content Knowledge and Understanding of the Learning Materials

Our next step was to examine participants' percentage of correct answers to the multiple-choice questions, indicating their knowledge and understanding of the course contents, before and after the course. Data indicated that prior to the course, the percentage of correct answers ranged from 0% to 67%. Data also indicated that topics such as: dimensionality of nanowires, the four range of structures, and quantum dots, were familiar to some of the students prior to taking the course; whereas the topic of synthesizing highly oriented nanostructures was less familiar. At the end of the course, the percentage of correct answers ranged from 25% to 89% (Table 3).

Figure 3. Participants' work experience with nanosensors, before and after the course

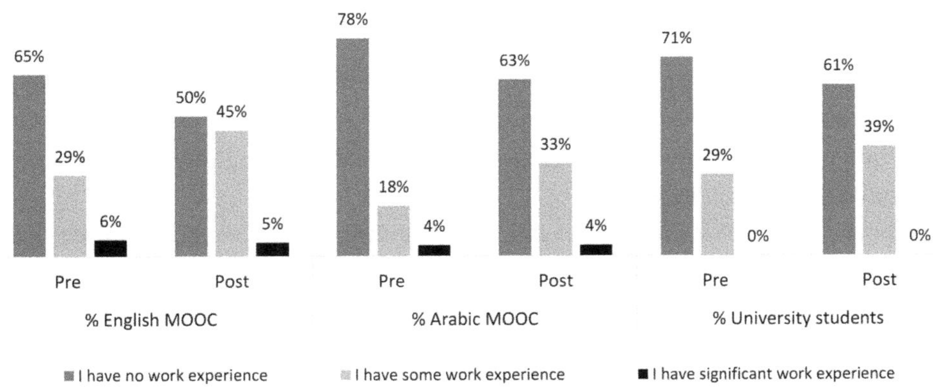

Table 3. Participants' percentages of correct answers in the pre- and post-questionnaires

Knowledge in Nanotechnology and Nanosensors		% Correct answers		
		English MOOC	Arabic MOOC	Univ. students
1. What is the dimensionality of nanowires?	Pre	45	31	54
	Post	78	39	89
2. When discussing nanoscience, what is the four range of structures?	Pre	53	33	50
	Post	84	69	86
3. What is the best description of "bottom up" fabrication?	Pre	59	27	39
	Post	85	42	89
4. Which would be a better characteristic of a sensor?	Pre	29	11	43
	Post	59	28	79
5. What are quantum dots?	Pre	67	28	54
	Post	81	47	75
6. What distinguishes between the different types of metal oxide sensors?	Pre	29	17	46
	Post	85	36	78
7. What is the best way for synthesizing highly oriented nanostructures?	Pre	5	0	4
	Post	38	25	46
8. What is the main difference between flexible skin-like sensors and the human skin?	Pre	29	25	32
	Post	78	33	71

Analysis of Covariance (ANCOVA) test was conducted to compare between the three groups for the post-scores' total means, while correcting for the pre-scores. Data indicated that overall, the English MOOC participants and the university students received higher post-scores compared to their counterparts in the Arabic MOOC (M_{Eng} = 73.49, SD_{Eng} = 18.74; M_{Univ} = 76.78, SD_{Univ} = 21.71; M_{Arab} = 44.79, SD_{Arab} = 25.94). These differences were found to be statistically significant ($F(2, 118) = 15.87$, $p = .000$, $\eta_p^2 = .21$).

In addition to analyzing participants' responses to the questionnaires, we examined and compared their grades in the quizzes, open assignments, final project, and final grades. Table 4 presents mean grades and standard deviations of the learning assignments for each group.

Table 4 shows that the participants from the three groups received similar high grades in the weekly quizzes (around 94 points) and in the final project (around 90 points). Analysis of Variance (ANOVA) test indicated no statistically significant differences between the three groups in the quizzes, final project, and final course

Table 4. Mean grades and standard deviations of the learning assignments, by research group

	Mean Grades* (SD)		
Learning Assignments	**English MOOC (N= 58)**	**Arabic MOOC (N= 36)**	**Univ. students (N= 46)**
Quizzes (mean of 10)	94.56 (4.92)	94.89 (4.09)	94.17 (4.36)
Open assignments (mean of 3)	75.75 (22.14)	50.88 (14.19)	82.32 (10.21)
Final project	88.90 (11.73)	90.95 (12.75)	90.75 (7.14)
Final course grades	87.40 (8.05)	83.72 (11.08)	89.75 (10.44)

* On a scale of 0 to 100

grades. The only statistically significant difference was indicated for the open assignments; the Arabic MOOC participants received statistically significant lower grades, compared to their counterparts ($F(2, 102) = 16.63$, $p = .000$, $\eta_p^2 = .25$). The relatively low grades in the open assignments may suggest that science and engineering students are less familiar with open and ill-defined assignments.

Participants Motivation to Learn a Project-Based MOOC

In order to examine participants' motivation to learn a project-based MOOC, we conducted a general inductive content analysis of participants' online forums and email correspondence. Data indicated that the MOOCs (English and Arabic) participants' were more motivated to take the course in order to meet people with similar expertise, share ideas, and collaborate, compared to university students. The later were more motivated by their interest in the contents and desire to receive credit points. The following examples present English MOOCers' assertions about the importance of networking and being part of a community of people with similar interests.

K.K., an environmental engineer from the US. He uses gas sensing devices for pollutant measurements. At the end of the English MOOC, after submitting the final project, he wrote an email message to the teaching assistant:

The gas sensor nanotechnology is amazing and holds great promise for pollution monitoring. The project was special as I was able to tie it to my work as an environmental engineer and "met" 3 friendly and knowledgeable colleagues from all over the world to collaborate with.

S.P., is an engineer from India who specializes in microelectronics. He participated in the English MOOC, and wrote the following post:

Over the last 2 years I have done about six online courses. The most interesting part of MOOC has always been the discussion forums where like-minded people exchange ideas. MOOC has also brought interested people from different geographical locations together on one platform of learning, as far as working for the final project was concerned. We had real good discussions; this reinforced the concepts dealt with. More importantly I made good friends due to this.

E.B., is an engineer from the US who used the discussion forums to present his startup company and to invite other participants to contribute ideas for advancing his business:

I am currently running my own business based on Gamification mobile apps. I am very interested in developing EDA VLSI tools for programmable nano-chips, Embedded Nanosensor Systems & CGH via photonic-multi-walled nanotubes.

The Arabic MOOCers were characterized by their motivation to communicate with people with similar interests and apply knowledge for the benefit of others. For example, M.K. an Engineer from Egypt who participated in the Arabic MOOC sent a message to the course lecturer, reading:

It wasn't that easy, but your course taught me many things about nanomaterials and nanosensors which I wish to apply in my career as an engineer...I'd like to learn more for the benefit of other people.

R.T., an engineer from the Syria who participated in the Arabic MOOC, used the online forums to invite other participants to share knowledge:

I am R.T. from Syria, I studied applied chemistry and I have a master's degree in inorganic chemistry. I'm looking forward to collaborating with you in this course, and exchanging information.

The university students were mostly driven by their interest in the subject matter and their desire to advance their personal knowledge in the field of nanotechnology and nanosensors. For example, S.O. is an undergraduate student at the Technion, he asserted that:

I was not very active in the online forums, but I looked at the messages people posted every now and then and saw there were lots of interesting discussions. Because we are from the Technion, our group found it easier to meet face-to-face and discuss issues related to the course. However, I believe that people from other countries found the online forums very useful for asking questions and sharing ideas.

K.L., a material engineering university student, asserted:

I need to add one more course to my academic curriculum in order to close all my academic assignments in undergraduate studies. This course is approved by the faculty administration and it sounds interesting and engaging.

O.H., a chemical engineering university student, asserted:

This course is part of my degree in chemical engineering... I took this course because I find the subject interesting and because this course was recommended by my fellow students.

D.H., an electrical engineer who is studying for an MSc degree, asserted:

I'm a master's degree student and I'm taking this course to accomplish my degree. I also work in the industry in a full time job as an electro-optical engineer and this course will give me a brief introduction to the world of Nano-Sensors.

The examples above indicate that the university students were mostly motivated by their interest in the subject matter.

FUTURE RESEARCH DIRECTIONS

The *Nanotechnology and Nanosensors* MOOC can be thought of as a showcase for similar courses in a variety of science and engineering disciplines. Lecturers in higher education can create interactive lessons while introducing scientific concepts to people all over the world. The *Nanotechnology and Nanosensors* MOOC offers learning materials that can be freely accessed not only by scientists and engineers, but also by high school teachers and students who are inquisitive about how the next generation of technological devices will look like. This study provides some insights about the advantages and challenges of becoming the next generation science educators in higher education, in an era of global changes. It provides an interesting view on cultural and epistemological aspects of learning in a massive online open course.

By concentrating university outreach efforts, the study can develop a strong positive interaction with the concepts of nanoscience, sensing capabilities, and education, thus make significant impact in the science education realm. The knowledge brought to and developed in the proposed study, could heavily impact the curricular elements of MOOCs in science and engineering, ranging from freshman science for non-majors to advanced courses on nanotechnology, materials, and electronic devices/sensors. These efforts could encourage tailoring the overall of educational programs to meet current specific needs and to respond to the evolution of higher education (in the form of MOOCs). The strong education foundation, which characterize the activities and outcomes of the proposed study, might have a direct and measurable positive impact on students' learning in the 21st century and the way they interact and acquire knowledge.

The current study is based on the belief that science should be practiced not only to satisfy one's inquisitive drive and curiosity, but also as an idealistic endeavor in the service of society, since many of the projects can be further developed and implemented. In addition, the bilingual MOOC follows the idea that education is a powerful agent for enhancing changes in both mind and heart, emphasizing the notion that people from all over the world share similar ideas and interest. The practice and the appreciation of the scientific process and progress should not be confined to a group of people isolated in an "ivory tower," of the academia, but should be accessible to all layers of society. Essentially, it is important to realize that science, like other disciplines governing the fabric of society, is an integral part of modern civilization. Furthermore, learning in a massive, online, open course that encourages international and multicultural group work, can result in cooperation between people and companies from different countries. New markets can be identified and new needs can be discovered leading to global cooperation. Transnational collaboration while designing innovative nano-devices can result in its actual production in the future. These aspects of project-based MOOC should be further examined to indicate their effect on science and engineering education.

CONCLUSION

This chapter describes the development of a massive open online course (MOOC) in nanotechnology and nanosensors, in two languages - English and Arabic, at the Technion- Israel Institute of Technology. The nanotechnology and nanosensors course was the world's first MOOC in this unique field, and the first that was simultaneously delivered in two languages: English and Arabic. In fact, it was the first MOOC delivered in the Arabic language. Our goal was to appeal to a wide range of people, from all over the world. The development of the MOOCs served as a trigger for promoting project-based learning among science and engineering students. This initiated a study that examined the role of project-based MOOCs in the process of knowledge construction and motivation to learn, comparing between the English MOOC participants, Arabic MOOC participants, and university students.

The nanotechnology and nanosensors is unique in that it encourages students to learn in small international and multicultural groups on project-based assignments that require innovative thinking. It is also unique in its innovative topics, open-ended assignments, and peer-grading activities. Students' projects discuss scientific and engineering issues that are in the frontier of science and technology research, such as: tailoring nanomaterials and related nanosensors for important near future applications (e.g., medicine, food, security), and far future applications (e.g., nano-robots or nano-labs that travel inside the human body, flexible nanosensors that can serve as a skin for robots, etc.). Since the Arabic MOOC was delivered by an Israeli university, it served as a powerful educational catalyst, bringing together people from different cultures and nations to work on collaborative learning assignments (Barak et al., 2016).

The findings of our study indicated that participants from the three research groups (English MOOCers, Arabic MOOCers, and university students) asserted positive attitudes towards learning a project-based MOOC, especially with relation to gaining work experience. Differences between the groups were indicated in their assertions about familiarity with the subject matter. Our findings indicated that while the English MOOC participants indicated a strong increase in familiarity with the learning materials, the Arabic MOOC participants and the university students indicated a mild increase. This result was surprising since the participants from the three research groups received similar high grades in the weekly quizzes and in the final projects. We can only speculate that the Arabic group's responses were influenced by the fact that most of them were studying a MOOC for the first time. They were struggling with technical problems that probably led to negative attitudes and low feeling of success. Regarding the university students' low attitudes, we can speculate that this has resulted from the fact that they are high achievers with a strong individual drive. Since project-based learning requires teamwork and being dependent on others, their high expectations about their personal success might not have been fully met.

Although the participants from the three research groups received similar grades in the quizzes and the final project, the Arabic MOOC participants received lower scores on the open assignments, compared to their counterparts. This may suggest that at the beginning of the course, the Arabic science and engineering students were less familiar with solving ill-structured problems. Indeed, many science and engineering courses are still oriented toward lecturing, emphasizing science and math fundamentals rather than engineering applications (Felder & Brent, 2005). According to Felder and Brent (2005), traditional engineering education emphasize objective analysis rather than exploration of ideas and creative problem solving. However, once the Arabic MOOC participants have gained experience in solving open assignments, they were able to apply learning strategies and thinking skills, which were evident in the high-level projects they developed at the end of the course.

It is important to note that the open assignments and the final project were graded through a peer assessment (grading) system. Peer assessment is a pedagogical method, generally referring to a process wherein students take part in evaluating the quality of their classmates' learning outcomes (Barak & Rafaeli, 2004; Falchikov & Goldfinch, 2000). A continuously growing body of research underlines the value of peer assessment for the learning process (Falchikov & Goldfinch, 2000; Hanrahan & Isaacs, 2001; Li & Gao, 2016). It was documented as an effective educational strategy for increasing motivation (Hanrahan & Isaacs, 2001), promoting student learning (Li & Gao, 2016), gaining positive effects on students achievements (Barak & Rafaeli, 2004), and increasing students' engagement (Falchikov & Goldfinch, 2000). In the peer assessment process, students are exposed to the work of their peers and can learn new ideas and also learn from mistakes. They take the role of the 'assessor', which involves higher order cognitive skills such as critical and argumentative thinking. This may explain the positive learning outcomes that were achieved in our study.

Research on students' motivation to learn in MOOCs is gaining much interest (Barak et al., 2016). Studies indicated that online learners are more intrinsically motivated than their on-campus counterparts (Shroff, Vogel & Coombes, 2008) and that their intrinsic motivation is positively related to their learning performance (Cho & Heron, 2015). In the context of MOOCs, because it is an open and free learning environment, participants tend to choose only segments of the learning environment, following their goals and interests (Kizilcec & Schneider, 2015; Wang & Baker, 2015). The findings presented in this study indicated that in a project-based learning environment, MOOC participants are more likely to be motivated by a desire to establish international connections, compared to university students. A closer look at the data indicated that the English MOOCers were motivated by a desire to be informed about innovations in nanotechnology. They were seeking to learn and share new ways for solving scientific and engineering problems they encounter in a laboratory or the industry. Arabic MOOCers were motivated by a desire to communicate with people with similar interests and apply knowledge for the benefit of others in their society; while the university students were mostly driven by their interest in the subject matter and their desire to advance their personal knowledge.

The findings presented above may explain the results that indicated that many university students preferred to work individually or in pairs. This is interesting since compared to MOOC participants, university students are more familiar with each other (from other courses) and have better opportunities to work in close proximity (on-campus). However, preferences toward individual work is not new in science and engineering education. Studies have indicated that in many science and engineering universities, instruction is oriented toward lecturing and individual assignments rather than active class involvement and cooperative learning (Barak, 2017; Felder & Brent, 2005).

Alternatively, MOOC participants (English and Arabic) were more inclined to work in groups. To do so, they had to overcome time differences, geographical distances, and cultural differences. They were required to invest much effort in planning and executing their learning and working process. The cultural differences and the lack of early acquaintance required MOOC participants to be flexible, build trust, and construct special relationships to carry out the project on time and in the best way. Thus, it is apparent that geographical distance and lack of familiarity among learners do not impede on their preference and ability to work in a group, if they are highly motivated. The differences between MOOC participants and university students in their preference about group work can be explained by the fact that many MOOC participants are scientist and engineers who are already working in the industry or laboratories (Barak et al., 2016; Breslow et al., 2013). Thus, their understanding of and experience in group work is greater than that of the university students.

Overall, this study contributes to the teaching and learning of science and engineering in higher education by examining aspects of effective learning in an international context. It reveals cultural and motivational incentives and barriers in the settings of a massive open online course. The *Nanotechnology and Nanosensors* course appeals to scientists and engineers from different backgrounds and cultures, who are interested in promoting their knowledge in nanomaterial. Contrary to most MOOCs in science and engineering that base their grading on examinations, this course encouraged people to work collaboratively on innovative projects, focusing on real-world problems.

ACKNOWLEDGMENT

The authors gratefully acknowledge the Israeli Ministry of Science, Technology, and Space (grant # 3-10841) for the generous support. The Authors gratefully acknowledge the generous support of the Technion's President Foundation. The authors would like to thank Prof. Hossam Haick, Dr. Meital Bar-Segev, and Dr. Nisreen Shehada for their valuable contribution to this study.

REFERENCES

Barak, M. (2012). Distance education: Towards an organizational and cultural change in higher education. *The Journal of Enterprising Communities: People and Places in the Global Economy*, *6*(2), 124–137. doi:10.1108/17506201211228930

Barak, M. (2017). Science teacher education in the twenty-first century: A pedagogical framework for technology-integrated social constructivism. *Research in Science Education*, *47*(2). doi: 10.1007/s11165-015-9501-y

Barak, M., & Dori, Y. (2005). Enhancing undergraduate students chemistry understanding through project-based learning in an IT environment. *Science Education*, *89*(1), 117–139. doi:10.1002/sce.20027

Barak, M., Herscovitz, O., Kaberman, Z., & Dori, Y. J. (2009). MOSAICA: A Web-2.0 based system for the preservation and presentation of cultural heritage. *Computers & Education*, *53*(3), 841–852. doi:10.1016/j.compedu.2009.05.004

Barak, M., & Rafaeli, S. (2004). Online question-posing and peer-assessment as means for Web-based knowledge sharing. *International Journal of Human-Computer Studies*, *61*(1), 84–103. doi:10.1016/j.ijhcs.2003.12.005

Barak, M., Watted, A., & Haick, H. (2016). Motivation to learn in massive open online courses: Examining aspects of language and social engagement. *Computers & Education*, *94*, 49–60. doi:10.1016/j.compedu.2015.11.010

Barak, M., & Ziv, S. (2013). Wandering: A Web-based platform for the creation of location-based interactive learning objects. *Computers & Education*, *62*, 159–170. doi:10.1016/j.compedu.2012.10.015

Bell, R. L., Maeng, J. L., & Binns, I. C. (2013). Learning in context: Technology integration in a teacher preparation program informed by situated learning theory. *Journal of Research in Science Teaching*, *50*(3), 348–379. doi:10.1002/tea.21075

Breslow, L., Pritchard, D. E., DeBoer, J., Stump, G. S., Ho, A. D., & Seaton, D. T. (2013). Studying learning in the worldwide classroom: Research into edX's first MOOC. *Research & Practice in Assessment*, *8*, 13–25.

Bruner, J. S. (1990). *Acts of Meaning*. Cambridge, MA: Harvard University Press.

Campbell, D. T., & Stanley, J. C. (1963). Experimental and Quasi-Experimental Designs for Research on Teaching. In N. L. Gage (Ed.), *Handbook of Research on Teaching*. Chicago: Rand McNally.

Cho, M.-H., & Heron, M. L. (2015). Self-regulated learning: The role of motivation, emotion, and use of learning strategies in students learning experiences in a self-paced online mathematics course. *Distance Education, 36*(1), 80–99. doi:10.1080/01587919.2015.1019963

Cooper, S., & Sahami, M. (2013). Reflections on Stanfords MOOCs. *Communications of the ACM, 56*(2), 28–30. doi:10.1145/2408776.2408787

Cormier, D., & Siemens, G. (2010). Through the open door: Open courses as research, learning & engagement. *EDUCAUSE Review, 45*(4), 30–39.

Creswell, J. W. (2014). *Research Design: Qualitative, Quantitative, and Mixed Methods Approaches*. SAGE Publications.

Dominguez, C., & Jaimo, A. (2010). Data based learning: A project based approach organized through a course management system. *Computers & Education, 55*(3), 1312–1320. doi:10.1016/j.compedu.2010.06.001

Falchikov, N., & Goldfinch, J. (2000). Student peer assessment in higher education: A meta-analysis comparing peer and teacher marks. *Review of Educational Research, 70*(3), 287–322. doi:10.3102/00346543070003287

Felder, R. M., & Brent, R. (2005). Understanding Student Differences. *The Journal of Engineering Education, 94*(1), 57–72. doi:10.1002/j.2168-9830.2005.tb00829.x

Ferguson, R., & Clow, D. (2015). Examining engagement: analysing learner subpopulations in massive open online courses (MOOCs). In *The 5th International Learning Analytics and Knowledge Conference (LAK15)*. Poughkeepsie, NY: ACM.

Field, A. P. (2009). *Discovering statistics using SPSS: and sex and drugs and rock 'n' roll* (3rd ed.). London: Sage Publications.

Hanrahan, J., & Isaacs, G. (2001). Assessing self- and peer assessment: The students views. *Higher Education Research & Development, 20*(1), 53–70. doi:10.1080/07294360123776

Hayes, A. F., & Krippendorff, K. (2007). Answering the call for a standard reliability measure for coding data. *Communication Methods and Measures, 1*(1), 77–89. doi:10.1080/19312450709336664

Hsieh, H. F., & Shannon, S. E. (2005). Three approaches to qualitative content analysis. *Qualitative Health Research, 15*(9), 1277–1288. doi:10.1177/1049732305276687 PMID:16204405

Jordan, K. (2014). Initial Trends in Enrolment and Completion of Massive Open Online Courses. *International Review of Research in Open and Distance Learning*, *15*(1), 133–160. doi:10.19173/irrodl.v15i1.1651

Jurow, A. S. (2005). Shifting engagements in figured worlds: Middle school mathematics students participation in an architectural design project. *Journal of the Learning Sciences*, *14*(1), 35–67. doi:10.1207/s15327809jls1401_3

Kizilcec, R. F., & Schneider, E. (2015). Motivation as a lens to understand online learners: Toward data-driven design with the OLEI scale. *ACM Transactions on Computer-Human Interactions*, *22*(2). http://dx.doi.org./10.1145/2699735

Krajcik, J., & Blumenfeld, P. C. (2008). Project-based learning. In L. M. Given (Ed.), *The Sage Encyclopedia of Qualitative Research* (pp. 422–423). Sage.

Krajcik, J., McNeil, K., & Reiser, B. J. (2007). Learning-goals-driven design model: Developing curriculum materials that align with national standards and incorporate project-based pedagogy. *Science Education*, *92*(1), 1–32. doi:10.1002/sce.20240

Lemke, J. L. (2001). Articulating communities: Sociocultural perspectives on science education. *Journal of Research in Science Teaching*, *38*(3), 296–316. doi:10.1002/1098-2736(200103)38:3<296::AID-TEA1007>3.0.CO;2-R

Li, L., & Gao, F. (2016). The effect of peer assessment on project performance of students at different learning levels. *Assessment & Evaluation in Higher Education*, *41*(6), 885–900. doi:10.1080/02602938.2015.1048185

McNutt, M. (2013). Bricks and MOOCs. *Science*, *342*(6157), 402. doi:10.1126/science.1246943 PMID:24159014

Morales, T. M., Bang, E. J., & Andre, T. (2013). A one-year case study: Understanding the rich potential of project-based learning in a virtual reality class for high school students. *Journal of Science Education and Technology*, *22*(5), 791–806. doi:10.1007/s10956-012-9431-7

Onah, D. F. O., Sinclair, J., & Boyatt, R. (2014). Dropout rates of massive open online courses: behavioral patterns. Proceedings of EDULEARN14, 5825-5834.

Ragupathi, K. (2014). Virtually Vygotsky: Using technology to scaffold student learning. *Technology in Pedagogy*, *20*, 1–9.

Rooij, S. W. (2009). Scaffolding project-based learning with the project management body of knowledge (PMBOK). *Computers & Education*, *52*(1), 210–219. doi:10.1016/j.compedu.2008.07.012

Shroff, R. H., Vogel, D. R., & Coombes, J. (2008). Assessing individual-level factors supporting student intrinsic motivation in online discussions: A qualitative study. *Journal of Information Systems Education, 19*(1), 111–125.

Strauss, A., & Corbin, J. (1990). *Basics of Qualitative Research: Grounded Theory Procedures and Techniques*. Newbury Park, CA: Sage Publications.

Thomas, D. R. (2006). A General Inductive Approach for Analyzing Qualitative Evaluation Data. *The American Journal of Evaluation, 27*(2), 237–246. doi:10.1177/1098214005283748

Thomas, J. W. (2000). *A review of research on project-based learning executive summary*. San Rafael, CA: The Autodesk Foundation. Retrieved April 13, 2014 from http://www.bobpearlman.org/BestPractices/PBL_Research.pdf

Vygotsky, L. S. (1978). *Mind in Society: Interaction between learning and development*. Cambridge, MA: Harvard University Press.

Wang, Y., & Baker, R. (2015). Content or platform: Why do students complete MOOCs? *MERLOT Journal of Online Learning and Teaching, 11*(1), 17–30.

Wertsch, J. (1991). *Voices of the mind: A sociocultural approach to mediated action*. Cambridge, MA: Harvard University Press.

KEY TERMS AND DEFINITIONS

Knowledge Construction: A cognitive process that involves exploration and results in the generation of new understanding.

Massive Open Online Course (MOOC): A web-based course, designed to provide free and accessible high-quality education to the masses.

Motivation to Learn: An internal or external drive that stimulates learning and gives it meaning.

Nanosensors: Chemical, biological, or mechanical sensory points used to convey information about or detect the presence of nanoparticles.

Nanotechnology: A branch of science and engineering that deals with particles of less than 100 nanometers, focusing on the manipulation of atoms and molecules.

Project-Based Learning (PBL): A pedagogical model that organizes learning around challenging questions that involve real-world problems and investigative activities.

Compilation of References

Abbitt, J. T. (2011). Measuring Technological Pedagogical Content Knowledge in preservice teacher education: A review of current methods and instruments. *Journal of Research on Technology in Education, 43*(4), 281–300. doi:10.1080/15391523.2011.10782573

Abd-El-Khalick, F., Boujaoude, S., Duschl, R., Lederman, N. G., Mamlok-Naaman, R., Hofstein, A., & Tuan, H. L. (2004). Inquiry in science education: International perspectives. *Science Education, 88*(3), 397–419. doi:10.1002/sce.10118

Abd-El-Khalick, F., Boujaoude, S., Duschl, R., Lederman, N. G., Mamlok-Naaman, R., Hofstein, A., & Tuan, H.-L. (2004). Inquiry in Science Education: International Perspectives. In E. Krugly-Smolska & P. C. Taylor (Eds.), *Culture and Comparative Studies* (pp. 397–419).

Abramovich, S. (2005). Early algebra with graphics software as a Type II application of technology. In C. D. Maddux & D. L. Johnson (Eds.), Classroom integration of Type II uses of technology in education (pp. 21-33). New York, NY: The Haworth Press.

Abramovich, S. (2011). Technology as a context for secondary pre-teachers' use of algebraic inequalities in a capstone course. In *Proceedings of Society for Information Technology & Teacher Education International Conference 2011*. Chesapeake, VA: Association for the Advancement of Computing in Education.

Abramovich, S. (2014). On the duality of transition from visual to symbolic in the teaching of school mathematics. *Open Mathematical Education Notes, 4*, 9–18.

Abramovich, S. (2016). Do computers enable problem solving or just make it "easy"? *Computer Tools in Education, 2*, 45–54.

Abramovich, S., & Cho, E. K. (2009). Mathematics, computers, and young children as a research-oriented learning environment for a teacher candidate. *Asia Pacific Education Review, 10*(2), 247–259. doi:10.1007/s12564-009-9015-2

Abramovich, S., & Connell, M. L. (2015). Digital fabrication and hidden inequalities: Connecting procedural, factual and conceptual knowledge. *International Journal of Technology in Teaching and Learning, 11*(2), 76–89.

Ainsworth, S. (2006). DefT: A conceptual framework for considering learning with multiple representations. *Learning and Instruction*, *16*(3), 183–198. doi:10.1016/j.learninstruc.2006.03.001

Al-Alwani, A. (2005). *Barriers to Integrating Information Technology in Saudi Arabia Science Education* (Doctoral dissertation). The University of Kansas.

Alexander, R. (2010). *Children, their world, their education: Final report and recommendations of the Cambridge Primary Review*. Abingdon, UK: Routledge.

Almerich, G., Orellana, N., Suarez-Rodriguez, J., & Diaz-Garcia, I. (2016). Teacher 's information and communication technology competences: A structural approach. *Computers & Education*, *100*, 110–125. doi:10.1016/j.compedu.2016.05.002

Alozie, N. M., Moje, E. B., & Krajcik, J. S. (2010). An analysis of the supports and constraints for scientific discussion in high school project-based science. *Science Education*, *94*, 395–427.

American Academy of Pediatrics (AAP), Council on Communications and Media. (2016). Media and Young Minds. *Pediatrics*, *138*(5), 2591. PMID:27940793

American Association for the Advancement of Science (AAAS). (1993). *Benchmarks for Science Literacy*. Oxford University Press.

Amy, N. (2016). *Plickers*. Retrieved from https://www.plickers.com/

Andrews-Hanna, J. R., Smallwood, J., & Spreng, R. N. (2014). The default network and self-generated thought: Component processes, dynamic control, and clinical relevance. *Annals of the New York Academy of Sciences*, *1316*, 29–52.

Andrews, T. M., Leonard, M. J., Colgrove, C. A., & Kalinowski, S. T. (2011). Active learning not associated with student learning in a random sample of college biology courses. *CBE Life Sciences Education*, *10*(4), 394–405. doi:10.1187/cbe.11-07-0061 PMID:22135373

Angeli, C., & Valanides, N. (2009). Epistemological and methodological issues for the conceptualization, development, and assessment of ICT–TPCK: Advances in technological pedagogical content knowledge (TPCK). *Computers & Education*, *52*(1), 154–168. doi:10.1016/j.compedu.2008.07.006

Annetta, L. A., Frazier, W. M., Folta, E., Holmes, S. Y., Lamb, R., & Cheng, M. T. (2013). Science teacher efficacy and extrinsic factors toward professional development using video games in a design-based research model: The next generation of STEM learning. *Journal of Science Education and Technology*, *22*(1), 47–61. doi:10.1007/s10956-012-9375-y

Anthony, G. (1996). When mathematics students fail to use appropriate learning strategies. *Mathematics Education Research Journal*, *8*(1), 23–37. doi:10.1007/BF03217287

An, Y.-Y. (2015). Games to promote inquiry learning. In M. J. Spector (Ed.), *The SAGE Encyclopedia of Educational Technology* (Vol. 1, pp. 318–321). Los Angeles, CA: SAGE Publications.

Archambault, L. M., & Barnett, J. H. (2010). Revisiting technological pedagogical content knowledge: Exploring the TPACK framework. *Computers & Education*, *55*(4), 1656–1662. doi:10.1016/j.compedu.2010.07.009

Ardac, D., & Sezen, A. H. (2002). Effectiveness of computer-based chemistry instruction in enhancing the learning of content and variable control under guided versus unguided conditions. *Journal of Science Education and Technology*, *11*(1), 39–48. doi:10.1023/A:1013995314094

Aronin, S., & Floyd, K. K. (2013). Using an iPad in inclusive preschool classrooms to introduce STEM concepts. *Teaching Exceptional Children*, *45*(4), 34–39. doi:10.1177/004005991304500404

Ault, M., Craig-Hare, J., Ellis, J. D., Bulgren, J., Kretschmer, I., & Frey, B. B. (2016). Using Reason Racer to support argumentation in middle school science instruction. In M. J. Urban & D. A. Falvo (Eds.), *Improving K-12 STEM Education Outcomes through Technological Integration* (pp. 399–431). doi:10.4018/978-1-4666-9616-7.ch018

Ault, M., Craig-Hare, J., Frey, B., Ellis, J. D., & Bulgren, J. (2015). The effectiveness of Reason Racer, a game designed to engage middle school students in scientific argumentation. *Journal of Research on Technology in Education*, *47*(1), 21–40. doi:10.1080/15391523.2015.967542

Aydin, E., & Cagiltay, N. (2012). A new RF and microwave engineering course enriched with advanced technologies. *Computer Applications in Engineering Education*, *20*(4), 634–645. doi:10.1002/cae.20432

Badia, A., Meneses, J., & Sigales, C. (2013). Teachers Perceptions of Factors Affecting the Educational Use of ICT in Technology-Rich Classrooms. *Electronic Journal of Research in Educational Psychology*, *11*(3), 787–808. doi:10.14204/ejrep.31.13053

Baheti, R., & Gill, H. (2011). Cyber-physical systems. *The Impact of Control Technology*, *12*, 161–166.

Bair, R., & Stafford, T. (2016). Digital literacy. *Training & Development*, *43*(3), 10–11.

Ball, D. L., & Bass, H. (2000). Interweaving content and pedagogy in teaching and learning to teach: Knowing and using mathematics. In J. Boaler (Ed.), *Multiple perspectives on the teaching and learning of mathematics* (pp. 83–104). Westport, CT: Ablex.

Ball, D. L., & Bass, H. (2003). Making mathematics reasonable in school. In J. Kilpatrick, W. G. Martin, & D. Schifter (Eds.), *A Research companion to principles and standards for school mathematics* (pp. 27–44). Reston, VA: National Council of Teachers of Mathematics.

Ball, D. L., & Cohen, D. K. (1999). Developing practice, developing practitioners: Toward a practice-based theory of professional education. In G. Sykes & L. Darling-Hammond (Eds.), *Teaching as the learning profession: Handbook of policy and practice* (pp. 3–32). San Francisco: Jossey Bass.

Ball, D. L., Thames, M. H., & Phelps, G. (2008). Content knowledge for teaching: What makes it special? *Journal of Teacher Education*, *59*(5), 389–407. doi:10.1177/0022487108324554

Banchi, H., & Bell, R. (2008). The Many Levels of Inquiry. *Science and Children*, *46*(2), 26–29.

Bandura, A. (2007). *L'auto-efficacité*. Bruxelles, Belgium: de Boeck.

Bandura, A. (1993). Perceived self-efficacy in cognitive development and functioning. *Educational Psychologist*, *28*(2), 117–148. doi:10.1207/s15326985ep2802_3

Banilower, E. R., Smith, P. S., Weiss, I. R., Malzahn, K. A., Campbell, K. M., & Weis, A. M. (2013). *Report of the 2012 national survey of science and mathematics education*. Chapel Hill, NC: Horizon Research, Inc.

Barabási, A.-L. (2016). Network Science. Cambridge University Press. Retrieved from http://barabasi.com/networksciencebook/

Barak, M. (2017). Science teacher education in the twenty-first century: A pedagogical framework for technology-integrated social constructivism. *Research in Science Education*, *47*(2). doi:10.1007/s11165-015-9501-y

Barak, M. (2012). Distance education: Towards an organizational and cultural change in higher education. *The Journal of Enterprising Communities: People and Places in the Global Economy*, *6*(2), 124–137. doi:10.1108/17506201211228930

Barak, M., & Dori, Y. (2005). Enhancing undergraduate students chemistry understanding through project-based learning in an IT environment. *Science Education*, *89*(1), 117–139. doi:10.1002/sce.20027

Barak, M., Herscovitz, O., Kaberman, Z., & Dori, Y. J. (2009). MOSAICA: A Web-2.0 based system for the preservation and presentation of cultural heritage. *Computers & Education*, *53*(3), 841–852. doi:10.1016/j.compedu.2009.05.004

Barak, M., & Rafaeli, S. (2004). Online question-posing and peer-assessment as means for Web-based knowledge sharing. *International Journal of Human-Computer Studies*, *61*(1), 84–103. doi:10.1016/j.ijhcs.2003.12.005

Barak, M., Watted, A., & Haick, H. (2016). Motivation to learn in massive open online courses: Examining aspects of language and social engagement. *Computers & Education*, *94*, 49–60. doi:10.1016/j.compedu.2015.11.010

Barak, M., & Ziv, S. (2013). Wandering: A Web-based platform for the creation of location-based interactive learning objects. *Computers & Education*, *62*, 159–170. doi:10.1016/j.compedu.2012.10.015

Barron, B., & Darling-Hammond, L. (2008). *Teaching for meaningful learning: A review of research on inquiry-based and cooperative learning*. George Lucas Educational Foundation.

Barrow, J. D. (2007). *New theories of everything: the quest for ultimate explanation*. Oxford, UK: Oxford University Press.

Barrow, L. H. (2006). A brief history of inquiry: From Dewey to standards. *Journal of Science Teacher Education, 17*(3), 265–278. doi:10.1007/s10972-006-9008-5

Barsalou, L. W. (2015). Situated conceptualization: Theory and applications. In Y. Coello & M. H. Fischer (Eds.), Foundations of embodied cognition: Vol. 1. Perceptual and emotional embodiment (pp. 11-37). East Sussex, UK: Psychology Press.

Barsalou, L. W. (2008). Grounded cognition. *Annual Review of Psychology, 59*(1), 617–645. doi:10.1146/annurev.psych.59.103006.093639 PMID:17705682

Barsalou, L. W., Santos, A., Simmons, W. K., & Wilson, C. D. (2008). Language and simulation in conceptual processing. In M. De Vega, A. M. Glenberg, & A. C. A. Graesser (Eds.), *Symbols, Embodiment, and Meaning* (pp. 245–283). Oxford, UK: Oxford University Press. doi:10.1093/acprof:oso/9780199217274.003.0013

Bastien, C. (1997). *Les connaissances de l'enfant à l'adulte*. Paris, France: Armand Colin.

Bastien, C., & Bastien-Toniazzo, M. (2004). *Apprendre à l'école*. Paris, France: Armand Colin.

Baviskar, S. N., Hartle, R. T., & Whitney, T. (2009). Essential criteria to characterize constructivist teaching: Derived from a review of the literature and applied to five constructivist-teaching method articles. *International Journal of Science Education, 31*(4), 541–550. doi:10.1080/09500690701731121

Beatty, I. D., Gerace, W. J., Leonard, W. J., & Dufresne, R. J. (2006). Designing effective questions for classroom response systems teaching. *American Journal of Physics, 74*(1), 31–39. doi:10.1119/1.2121753

Belland, B. R., Walker, A. E., Kim, N. J., & Lefler, M. O. (2016). Synthetizing results from empirical research on computer-based scaffolding in STEM education: A meta-analysis. *Review of Educational Research, 20*(10), 1–36.

Bell, R. L., Maeng, J. L., & Binns, I. C. (2013). Learning in context: Technology integration in a teacher preparation program informed by situated learning theory. *Journal of Research in Science Teaching, 50*(3), 348–379. doi:10.1002/tea.21075

Bell, R., Smetana, L., & Binns, I. (2005). Simplifying inquiry instruction. *Science Teacher (Normal, Ill.), 72*(7), 30–34.

Belshaw, D. (2011). *The Essential Elements of Digital Literacies*. Retrieved from http://dougbelshaw.com/ebooks/digilit/

Bennett, F. (2002). The future of computer technology in K-12 education. *Phi Delta Kappan, 83*(8), 621–626. doi:10.1177/003172170208300811

Bensaude-Vincent, B., & Newman, W. R. (2007). *The artificial and the natural: an evolving polarity*. MIT Press.

Berenfeld, B. (2010). Leveraging technology for a new science learning paradigm. In University Leadership: Bringing Technology-Enabled Education to Learners of All Ages. Cambridge, MA: Learning International Networks Consortium (LINC).

Berenfeld, B. (1999). The Internet in Our Classrooms: Teaching Tomorrow's Skills for Tomorrow's World. In *Science and Environment Education. Views from Developing Countries* (pp. 215–235). Washington, DC: The World Bank.

Bieda, K. N. (2010). Enacting proof-related tasks in middle school mathematics: Challenges and opportunities. *Journal for Research in Mathematics Education*, *41*(4), 351–382.

Billington, C. (2016). *How digital technology can support early language and literacy outcomes in early years settings: A review of the literature*. London: National Literacy Trust.

Binder, J. R., & Desai, R. H. (2011). The neurobiology of semantic memory. *Trends in Cognitive Sciences*, *15*(11), 527–536. doi:10.1016/j.tics.2011.10.001 PMID:22001867

BioInformatics. (2007). *Scientists using social media to inform lab purchasing decisions*. Retrieved from: https://www.gene2drug.com/product/digital-marketing-to-life-scientists-social-media-and-email-best-practices

BioInformatics. (2016). *Social media and life scientists in 2016*. Retrieved from: https://www.gene2drug.com/social-media-life-scientists-2016

Black, J. B. (2010). An embodied/grounded cognition perspective on educational technology. In M. S. Khine & I. M. Saleh (Eds.), *New Science of Learning. Cognition, Computers and Collaboration in Education* (pp. 45–52). New York, NY: Springer. doi:10.1007/978-1-4419-5716-0_3

Blackwell, C. K., Lauricella, A. R., & Wartella, E. (2014). Factors influencing digital technology use in early childhood education. *Computers & Education*, *77*, 82–90. doi:10.1016/j.compedu.2014.04.013

Blackwell, C. K., Lauricella, A. R., Wartella, E., Robb, M., & Schomburg, R. (2013). Adoption and use of technology in early education: The interplay of extrinsic barriers and teacher attitudes. *Computers & Education*, *69*, 310–319. doi:10.1016/j.compedu.2013.07.024

Blömeke, S., & Delaney, S. (2012). Assessment of teacher knowledge across countries: A review of the state of research. *ZDM Mathematics Education*, *44*(3), 223–247. doi:10.1007/s11858-012-0429-7

Bloom, B. S. (1956). *Taxonomy of educational objectives: Cognitive domain* (Vol. 1). New York: Longman.

Bodzin, A. M., Fu, Q., Bressler, D., & Vallera, F. L. (2015). Examining the enactment of Web GIS on students geospatial thinking and reasoning and tectonics understanding. *Computers in the Schools*, *32*(1), 63–81. doi:10.1080/07380569.2015.1004278

Bodzin, A., & Park, J. C. (2002). Using a non-restrictive web-based forum to promote reflective discourse with preservice science teachers. *Contemporary Issues in Technology & Teacher Education, 2*(3). Retrieved from http://www.citejournal.org/vol2/iss3/science/article1.cfm

Boeing, G. (2016). Visual Analysis of Nonlinear Dynamical Systems: Chaos, Fractals, Self-Similarity and the Limits of Prediction. *Systems, 4*(4), 37. doi:10.3390/systems4040037

Bøe, M. V., & Henriksen, E. K. (2013). Love it or leave it. Norwegian students motivations and expectations for post-compulsory physics. *Science Education, 97*(4), 550–573. doi:10.1002/sce.21068

Bolstad, T., Gilbert, J., McDowall, S., Bull, A., Boyd, S., & Hipkins, R. (2012). *Supporting future-oriented learning and teaching. A New Zealand perspective*. New Zealand Ministry of Education.

Borgatti, S. P. (2006). Identifying Sets of Key Players in a Social Network. *Computational & Mathematical Organization Theory, 12*(1), 21–34. doi:10.1007/s10588-006-7084-x

Börner, K., Palmer, F., Davis, J. M., Hardy, E., Uzzo, S. M., & Hook, B. J. (2009). Teaching children the structure of science. In *IS&T/SPIE Electronic Imaging* (pp. 724307–724307). International Society for Optics and Photonics. Retrieved from http://proceedings.spiedigitallibrary.org/proceeding.aspx?articleid=1334930

Börner, K., Sanyal, S., & Vespignani, A. (2007). Network science. *Annual Review of Information Science & Technology, 41*(1), 537–607. doi:10.1002/aris.2007.1440410119

Bower, M., & Sturman, D. (2015). What are the educational affordances of wearable technologies? *Computers & Education, 88*, 343–353. doi:10.1016/j.compedu.2015.07.013

Boyd, D. (2014). *It's complicated: The social lives of networked teens*. New Haven, CT: Yale University Press.

Brandes, U., Robins, G., McCranie, A., & Wasserman, S. (2013). What is network science? *Network Science, 1*(1), 1–15. https://doi.org/10.1017/nws.2013.2

Bransford, J. D., Brown, A. L., & Cocking, R. R. (2002). *How people learn: Brain, mind, experience, and school*. Washington, DC: The National Academies Press.

Bransford, J. D., & Donovan, M. S. (2005). Scientific Inquiry and How People Learn. In *How Students Learn: History, Mathematics, and Science in the Classroom* (pp. 397–419). Washington, DC: The National Academies.

Brant, G., Hooper, E., & Sugrue, B. (1991). Which comes first the simulation or the lecture? *Journal of Educational Computing Research, 7*(4), 469–481. doi:10.2190/PWDP-45L8-LHL5-2VX7

Breiner, J. M., Harkness, S. S., Johnson, C. C., & Koehler, C. M. (2012). What is STEM? A discussion about conceptions of STEM in education and partnerships. *School Science and Mathematics, 112*(1), 3–11. doi:10.1111/j.1949-8594.2011.00109.x

Brennan, K., Hernández, A. M., & Resnick, M. (2009). Scratch: creating and sharing interactive media. In *Proceedings of the 9th international conference on Computer supported collaborative learning* (vol. 2, pp. 217–217). International Society of the Learning Sciences. Retrieved from http://dl.acm.org/citation.cfm?id=1599503.1599576

Breslow, L., Pritchard, D. E., DeBoer, J., Stump, G. S., Ho, A. D., & Seaton, D. T. (2013). Studying learning in the worldwide classroom: Research into edX's first MOOC. *Research & Practice in Assessment*, *8*, 13–25.

Brey, P. (2014). From reflective to constructive philosophy of technology. *Journal of Engineering Studies*, *6*(2), 129–136.

British Columbia Ministry of Education. (2015). *Building students success: BC's new curriculum*. Retrieved from https://curriculum.gov.bc.ca/

Brown, J. S., Collins, A., & Duguid, P. (1989). Situated Cognition and the Culture of Learning. *Educational Researcher*, *18*(1), 32–42. doi:10.3102/0013189X018001032

Bruner, J. S. (1990). *Acts of Meaning*. Cambridge, MA: Harvard University Press.

Bruner, J. S. (2003). *Making stories: law, literature, life*. Cambridge, MA: Harvard University Press.

Brush, T., & Bannon, S. (1998). Characteristics of technology leaders: A survey of school administration in the United States. *International Studies in Educational Administration*, *26*(2), 47–56.

BSCS. (2008). *Scientists in Science Education: 1958-2008 Innovation Reform Vision*. Retrieved from www.bscs.org

Buchanan, R. (1992). Wicked problems in design thinking. *Design Issues*, *8*(2), 5–21. doi:10.2307/1511637

Buckenmeyer, J. A. (2010). Beyond computers in the classroom: Factors related to technology adoption to enhance teaching and learning. *Contemporary Issues in Education Research*, *3*(4), 27–35. doi:10.19030/cier.v3i4.194

Buckner, R. L., Andrews-Hanna, J. R., & Schacter, D. L. (2008). The Brains Default Network Anatomy, Function, and Relevance to Disease. *Annals of the New York Academy of Sciences*, *1124*(1), 1–38. doi:10.1196/annals.1440.011 PMID:18400922

Bulgren, J. A., Ellis, J. D., & Marquis, J. G. (2014). The use and effectiveness of an argumentation and evaluation intervention in science classes. *Journal of Science Education and Technology*, *23*(1), 82–97. doi:10.1007/s10956-013-9452-x

Bull, G., Knezek, G., & Gibson, D. (2009). Editorial: A rationale for incorporating engineering education into the teacher education curriculum. *Contemporary Issues in Technology & Teacher Education*, *9*(3), 222–225. Retrieved from http://www.editlib.org/p/32411

Burnett, C., & Myers, J. (2002). "Beyond the frame": Exploring children's literacy practices. *Reading Literacy and Language, 36*(2), 56-62.

Burridge, P., & Carpenter, C. (2013). Expanding pedagogical horizons: A case study of teacher professional development. *Australian Journal of Teacher Education, 38*(9), 10–24. doi:10.14221/ajte.2013v38n9.3

Bybee, R. (2011). Scientific and engineering practices in K-12 classrooms, Understanding A framework for K-12 science education. *Science Scope, 35*(4), 6–11, 13.

Bybee, R. W., Taylor, J. A., Gardner, A., Van Scotter, P., Powell, J. C., Westbrook, A., & Landes, N. (2006). *The BSCS 5E Instructional Model: Origins, Effectiveness, and Applications*. Colorado Springs, CO: BSCS.

Cai, , Lu, B., Fan, Z., Indhumathi, C., Lim, K. T., Chan, C. W., & Li, L. et al. (2006). Bio-edutainment: Learning life science through X gaming. *Computers & Graphics, 30*(1), 3–9. doi:10.1016/j.cag.2005.10.003

Calculator Soup, L. (n.d.). *Calculator Index*. Retrieved January 10, 2017, from http://www.calculatorsoup.com/calculators/

Calvino, I. (1988). *Six Memos for the Next Millennium*. Cambridge, MA: Harvard University Press.

Campbell, D. T., & Stanley, J. C. (1963). Experimental and Quasi-Experimental Designs for Research on Teaching. In N. L. Gage (Ed.), *Handbook of Research on Teaching*. Chicago: Rand McNally.

Carlone, H. (2004). The cultural production of science in reform-based physics: Girls access, participation, and resistance. *Journal of Research in Science Teaching, 41*(4), 392–414. doi:10.1002/tea.20006

Carr, D., & Bossomaier, T. (2011). Relativity in a rock field: A study of physics learning with a computer game. *Australasian Journal of Educational Technology, 27*(6), 1042–1067. doi:10.14742/ajet.928

Carr, J. M. (2012). Does math achievement hAPPen when iPads and game-based learning are incorporated into fifth-grade mathematics instruction? *Journal of Information Technology Education: Research, 11*, 269–286. doi:10.1007/s10639-006-9024-2

Carr, M., & Biddlecomb, B. (1998). Metacognition in mathematics from a constructivist perspective. In D. Hacker, J. Dunlosky, & A. Graesser (Eds.), *Metacognition in educational theory and practice* (pp. 69–91). Mahwah, NJ: Erlbaum.

Carter, S. M., & West, M. A. (1998). Reflexivity, Effectiveness, and Mental Health in BBC-TV Production Teams. *Small Group Research, 29*(5), 583–601. doi:10.1177/1046496498295003

Cartwright, N. (1999). *The Dappled World: A Study of the Boundaries of Science*. Cambridge, MA: Cambridge University Press. doi:10.1017/CBO9781139167093

Casasanto, D., & Lupyan, G. (2015). All Concepts are Ad Hoc Concepts. In E. Margolis & S. Laurence (Eds.), *The Conceptual Mind: New Directions in the Study of Concepts* (pp. 543–566). Cambridge, MA: The MIT Press.

Ceppi, G., & Zini, M. (Eds.). (1998). *Children, spaces, relations: Meta-project for an environment for young children.* Reggio Emilia, Italy: Reggio Children.

Chachashvili-Bolotin, S., Milner-Bolotin, M., & Lissitsa, S. (2016). Examination of factors predicting secondary students' interest in tertiary STEM education. *International Journal of Science Education, 38*(2), 25.

Chao, J., Chiu, J. L., DeJaegher, C. J., & Pan, E. A. (2016). Sensor-Augmented Virtual Labs: Using Physical Interactions with Science Simulations to Promote Understanding of Gas Behavior. *Journal of Science Education and Technology, 25*(1), 16–33. doi:10.1007/s10956-015-9574-4

Chao, J., Parker, K., & Fontana, A. (2011). Developing an interactive social media based learning environment. *Issues in Informing Science and Information Technology, 8*, 323–334.

Chen, C.-H., & Wang, K.-C., & Lin, Y.-H. (2015). The Comparison of Solitary and Collaborative Modes of Game-based Learning on Students' Science Learning and Motivation. *Journal of Educational Technology & Society, 18*(2), 237–248.

Chiang, T. H. C., Yang, S. J. H., & Hwang, G. J. (2014). Students online interactive patterns in augmented reality-based inquiry activities. *Computers & Education, 78*, 97–108. doi:10.1016/j.compedu.2014.05.006

Cho, M.-H., & Heron, M. L. (2015). Self-regulated learning: The role of motivation, emotion, and use of learning strategies in students learning experiences in a self-paced online mathematics course. *Distance Education, 36*(1), 80–99. doi:10.1080/01587919.2015.1019963

Christakis, D. A. (2014). Interactive media use at younger than the age of 2 years: Time to rethink the American Academy of Pediatrics guideline? *JAMA Pediatrics, 168*(5), 399–400. doi:10.1001/jamapediatrics.2013.5081 PMID:24615347

Clark, A., & Moss, P. (2001). *Listening to young children: The Mosaic approach.* London: National Children's Bureau for the Joseph Rowntree Foundation.

Clark, D., Tanner-Smith, E., Killingsworth, S., & Bellamy, S. (2013). *Digital Games for Learning: A Systematic Review and Meta-Analysis. Executive Summary.* Menlo Park, CA: SRI International.

Clark, R. C. (2004). Multimedia learning in e-courses. In R. E. Mayer (Ed.), *The Cambridge Handbook of Multimedia Learning* (pp. 589–616). New-York, NY: Cambridge University Press.

Clark, R. C., & Mayer, R. E. (2016). *E-learning and the science of instruction: Proven guidelines for consumers and designers of multimedia learning.* Hoboken, NJ: John Wiley & Sons. doi:10.1002/9781119239086

Cleland, C. (2002). Methodological and epistemic differences between historical science and experimental science. *Philosophy of Science, 69*(3), 474–496. doi:10.1086/342455

Clement, J. J. (2000). Model based learning as a key research area for science education. *International Journal of Science Education, 22*(9), 1041–1053. doi:10.1080/095006900416901

Clement, J. J. (2008). *Creative Model Construction in Scientists and Students. The Role of Imagery, Analogy, and Mental Simulation*. New York, NY: Springer. doi:10.1007/978-1-4020-6712-9

Clement, J. J. (2009). The Role of Imagistic Simulation in Scientific Thought Experiments. *Topics in Cognitive Science, 1*(4), 686–710. doi:10.1111/j.1756-8765.2009.01031.x PMID:25163453

Clements, D. H. (2002). Computers in early childhood mathematics. *Contemporary Issues in Early Childhood, 3*(2), 160–181. doi:10.2304/ciec.2002.3.2.2

Clements, D., & Sarama, J. (2003). Strip mining for gold: Research and policy in educational technology – A response to 'Fool's Gold'. *Educational Technology Review, 11*(1), 7–69.

Cobb, S., & Fraser, D. S. (2004). Multimedia learning in virtual reality. In R. E. Mayer (Ed.), *The Cambridge Handbook of Multimedia Learning* (pp. 525–548). New-York, NY: Cambridge University Press.

Coffman, T., & Klinger, M. (2015). Google glass: using wearable technologies to enhance teaching and learning. *Proceedings of Society for Information Technology & Teacher Education International Conference*, 1777-1780.

Cohen, E. (2002a). *Challenges of information technology education in the 21st century*. IGI Global. Retrieved from http://books.google.com/books?hl=en&lr=&id=Tbt3OE0UT0IC&oi=fnd&pg=PR7&dq=+%22challenges+of+information+technology+education%22&ots=g__Lx9NkCA&sig=SpuoZkU6t-sWbi5q67iYEAIJ1fQ

Cohen, E. (2002b). *Challenges of information technology education in the 21st century*. IGI Global. Retrieved from http://books.google.com/books?hl=en&lr=&id=Tbt3OE0UT0IC&oi=fnd&pg=PR7&dq=+%22challenges+of+information+technology+education%22&ots=g__Lx9NkCA&sig=SpuoZkU6t-sWbi5q67iYEAIJ1fQ

Cole, A. L., & Knowles, J. G. (2000). *Researching teaching: exploring teacher development through reflexive inquiry*. Allyn and Bacon.

Collins, C. J., & Han, J. (2004). Exploring applicant pool quantity and quality: The effects of early recruitment practice strategies, corporate advertising, and firm reputation. *Personnel Psychology, 57*(3), 685–717. doi:10.1111/j.1744-6570.2004.00004.x

Committee on Network Science for Future Army Applications, & National Research Council. (2005). *Network Science*. The National Academies Press. Retrieved from https://www.nap.edu/catalog/11516/network-science

Concord Consortium. (2012). Framing Mixed-Reality Labs. *Newsletter*, 2014.

Connell, L., & Lynott, D. (2014). Principles of representation: Why you can't represent the same concept twice. *Topics in Cognitive Science*, 1–17. PMID:24895329

Connell, M. L. (2001). Actions upon objects: A metaphor for technology enhanced mathematics instruction. In D. Tooke & N. Henderson (Eds.), *Using information technology in mathematics* (pp. 143–171). Binghamton, NY: Haworth Press.

Connor, A. M., Karmokar, S., & Whittington, C. (2015). From STEM to STEAM: Strategies for enhancing engineering and technology education. *International Journal of Engineering Pedagogies*, *5*(2), 37–47. doi:10.3991/ijep.v5i2.4458

Cooper, G., Tindall-Ford, S., Chanderler, P., & Sweller, J. (2001). Learning by imaging. *Journal of Experimental Psychology: Human Perception and Performance: Applied*, *7*(1), 68–82. PMID:11577621

Cooper, S., & Sahami, M. (2013). Reflections on Stanfords MOOCs. *Communications of the ACM*, *56*(2), 28–30. doi:10.1145/2408776.2408787

Cope, C., Staehr, L., & Horan, P. (2002). Towards establishing the best ways to teach and learn about IT. *Challenges of Information Technology Education in the 21st Century*, 57.

Copple, C., & Bredekamp, S. (Eds.). (2009). *Developmentally appropriate practice in early childhood programs serving children from birth through age 8* (3rd ed.). Washington, DC: National Association for Education of Young Children.

Cordes, C., & Miller, E. (Eds.). (2000). *Fool's gold: A critical look at computers in childhood.* Retrieved July 19, 2007, from http://www.allianceforchildhood.net/projects/computers/computers_reports.htm

Cormier, D., & Siemens, G. (2010). Through the open door: Open courses as research, learning & engagement. *EDUCAUSE Review*, *45*(4), 30–39.

Cornell University. (2009). *Digital Literacy*. Retrieved from https://digitalliteracy.cornell.edu/

Corrigan, D., Gunstone, R., & Dillon, J. (2011). *Approaches to Considering the Professional Knowledge Base of Science Teachers: The Professional Knowledge Base of Science Teaching*. Springer Netherlands. doi:10.1007/978-90-481-3927-9_1

Covert, M. W. (2013). Simulating a Living Cell. *Scientific American*, *310*(1), 44–51. doi:10.1038/scientificamerican0114-44 PMID:24616970

Cox, R., & Brna, P. (2016). Twenty years on: Reflections on Supporting the use of external representations in problem solving.... *International Journal of Artificial Intelligence in Education*, *26*(1), 193–204. doi:10.1007/s40593-015-0054-z

Cox, T. (2008). Learning Styles and Students' Attitudes toward the Use of Technology in Higher and Adult Education Classes. *Institute for Learning Styles Journal*, *1*, 1–13.

Craig-Hare, J., Fowler, T., Landever, A., Hobin, L., Parscale, G., & Irminger, T. (2013). *Learning engagement through higher-order thinking skills and technology*. Research paper presented at the annual meeting of the International Society for Technology in Education (ISTE), San Antonio, TX.

Cramer, C., Sheetz, L., Sayama, H., Trunfio, P., Stanley, H. E., & Uzzo, S. (2015). In G. Mangioni, F. Simini, S. M. Uzzo, & D. Wang (Eds.), NetSci High: Bringing Network Science Research to High Schools (pp. 209–218). Complex Networks, VA: Springer International Publishing. Retrieved from http://link.springer.com/chapter/10.1007/978-3-319-16112-9_21

Crawford, B. A. (2007). Learning to teach science as inquiry in the rough and tumble of practice. *Journal of Research in Science Teaching*, *44*(4), 613–642. doi:10.1002/tea.20157

Creswell, J. W. (2014). *Research Design: Qualitative, Quantitative, and Mixed Methods Approaches*. SAGE Publications.

Crompton, H., Burke, D., Gregory, K. H., & Gräbe, C. (2016). The Use of Mobile Learning in Science: A Systematic Review. *Journal of Science Education and Technology*, *25*(2), 149–160. doi:10.1007/s10956-015-9597-x

Cuban, L. (1990). Reforming again, again, and again. *Educational Researcher*, *19*(1), 3–13. doi:10.3102/0013189X019001003

Cuban, L. (2001). *Oversold and underused: Computers in the classroom*. Harvard University Press Review.

Cviko, A., McKenney, S., & Voogt, J. (2014). Teacher roles in designing technology-rich learning activities for early literacy: A cross-case analysis. *Computers & Education*, *72*, 68–79. doi:10.1016/j.compedu.2013.10.014

Czerniak, C. M. (1990, April). *A study of self-efficacy, anxiety, and science knowledge in pre-service elementary teachers*. Paper presented at the annual meeting of the National Association of Research in Science Teaching, Atlanta, GA.

Czerniak, C. M., & Haney, J. J. (1998). The effect of collaborative concept mapping on elementary preservice teachers anxiety, efficacy, and achievement in physical science. *Journal of Science Teacher Education*, *9*(4), 303–320. doi:10.1023/A:1009431400397

D'Angelo, C., Rutstein, D., Harris, C., Bernard, R., Borokhovski, E., & Haertel, G. (2014). *Simulations for STEM Learning: Systematic Review and Meta-Analysis*. Menlo Park, CA: SRI International.

Dabbagh, N., & Kitsantas, A. (2012). Personal learning environments, social media, and self-regulated learning: A natural formula for connecting formal and informal learning. *The Internet and Higher Education*, *15*(1), 3–8. doi:10.1016/j.iheduc.2011.06.002

Dagiéne, V. (2011). The development of ICT competency in pre-service teacher education. *International Journal of Digital Literacy and Digital Competence*, *2*(2), 1–10. doi:10.4018/jdldc.2011040101

Dang, T. (Producer). (2016). *Collaborative Learning Annotation System* [Media player]. Retrieved from http://ets.educ.ubc.ca/clas/

Daniels, H. (2001). Applications of sociocultural activity theory to education. In Vygotsky and Pedagogy (Vol. 1, pp. 96-130). New York: Routledge Falmer.

De Decker, E., De Craemer, M., De Bourdeaudhuij, I., Wijndaele, K., Duvinage, K., Koletzko, B., & Cardon, G. et al. (2012). Influencing factors of screen time in preschool children: An exploration of parents perceptions through focus groups in six European countries. *Obesity Reviews*, *13*(1), 75–84. doi:10.1111/j.1467-789X.2011.00961.x PMID:22309066

de Jong, T. (2006). Technological advances in inquiry learning. *Science*, *312*(5773), 532–533. doi:10.1126/science.1127750 PMID:16645080

De Jong, T., Linn, M. C., & Zacharia, Z. C. (2013). Physical and virtual laboratories in science and engineering education. *Science*, *340*(6130), 305–308. doi:10.1126/science.1230579 PMID:23599479

De Jong, T., Sotiriou, S., & Gillet, D. (2014). Innovations in STEM education: The Go-Lab federation of online labs. *Smart Learning Environments*, *1*(1), 3. doi:10.1186/s40561-014-0003-6

De la Torre, L., Sanchez, J., Dormido, S., Sanchez, J. P., Yuste, M., & Carreras, C. (2011). Two web-based laboratories of the FisL@bs network: Hookes and Snells laws. *European Journal of Physics*, *32*(2), 571–584. doi:10.1088/0143-0807/32/2/027

De Vries, E. (2001). Les logiciels dapprentissage: Panoplie ou éventail? *Revue Française de Pédagogie*, *137*(1), 105–116. doi:10.3406/rfp.2001.2851

De Vries, M. J. (2008). Gilbert Simondon and the dual nature of technical artifacts. *Techné: Research in Philosophy and Technology*, *12*(1), 23–35.

Deák, G. O. (2014). Development of adaptive tool-use in early childhood: Sensorimotor, social, and conceptual factors. *Advances in Child Development and Behavior*, *46*, 149–181. doi:10.1016/B978-0-12-800285-8.00006-6 PMID:24851349

DeBoer, G. E. (1991). *A History of Ideas in Science Education: Implications for practice* (Vol. 1). New York: Teachers College Press.

DeCoito, I. (2014a, May). *Engaging Learning Communities through Digital Literacies in Teacher Education.* Paper presented at the annual meeting of the Canadian Society for the Study of Education, St. Catharines, ON.

DeCoito, I. (in press). Addressing Digital Competencies, Curriculum Development, and Instructional Design in Science Teacher Education. In *Encyclopedia of Information Science and Technology.* Hershey, PA: IGI Global.

DeCoito, I. (2014b). Teaching about the nature of science through digital scientific timelines. In P. D. Morrell & K. Popejoy (Eds.), *A few of our favorite things: Teaching ideas for K-12 science methods instructors*. Sense Publishers.

DeCoito, I., & Richardson, T. (2016). Focusing on Integrated STEM Concepts in a Digital Game. In M. Urban & D. Falvo (Eds.), *Improving K-12 STEM Education* (pp. 1–23). Hershey, PA: IGI Global.

Dennis, A. R., Wixom, B. H., & Vandenberg, R. J. (2001). Understanding Fit and Appropriation Effects in Group Support Systems via Meta-Analysis. *Management Information Systems Quarterly*, *25*(2), 167–192. doi:10.2307/3250928

Denny, P. (2016). *PeerWise*. Retrieved from http://peerwise.cs.auckland.ac.nz/

Desoete, A., Roeyers, H., & Buysse, A. (2001). Metacognition and mathematical problem solving in grade 3. *Journal of Learning Disabilities*, *34*(5), 435–447. doi:10.1177/002221940103400505 PMID:15503592

Dewandre, N. (2011). The Sustainability Concept: Can We Stand Between Catastrophism and Denial? In C. Jaeger, J. D. Tàbara, & J. Jaeger (Eds.), *European Research on Sustainable Development* (pp. 29–34). Springer Berlin Heidelberg. doi:10.1007/978-3-642-19202-9_4

Dewey, J. (1938). *Logic: The Theory of Inquiry*. New York: Henry Hold and Company.

Digital Citizenship. (2016). Retrieved August, 13, 2016, from: www.digitalcitizenship.net

Dillon, A., & Jobst, J. (2004). Multimedia learning with hypermedia. In R. E. Mayer (Ed.), *The Cambridge Handbook of Multimedia Learning* (pp. 1–16). New-York, NY: Cambridge University Press.

DiPetta, T., & Woloshyn, V. E. (2009). Exploring teacher-candidates' experiences, beliefs and to technology as an instructional learning tool following instruction in a "technology rich" classroom. *Journal of Educational Technology*, *6*(1), 54–64.

diSessa, A. A. (1993). Toward an epistemology of physics. *Cognition and Instruction*, *10*(2-3), 105–225. doi:10.1080/07370008.1985.9649008

diSessa, A. A. (2006). A history of conceptual change research: threads and fault lines. In R. K. Sawyer (Ed.), *The Cambridge Handbook of the Learning Sciences* (pp. 265–281). New York: Cambridge University Press.

Dittert, N., & Krannich, D. (2013). Digital fabrication in educational contexts – ideas for a constructionist workshop setting. In J. Walter-Herrmann & C. Büching (Eds.), *FabLab: Of machines, makers and inventors* (pp. 173–180). Bielefeld, Germany: Transcript Verlag. doi:10.14361/transcript.9783839423820.173

Dominguez, A., Saena-de-Navarrete, J., de-Marcos, L., Fernandez-Sanz, L., Pages, C., & Martinez-Herraiz, J.-J. (2013). Gamifying learning experiences: Practical implications and outcomes. *Computers & Education*, *63*, 380–392. doi:10.1016/j.compedu.2012.12.020

Dominguez, C., & Jaimo, A. (2010). Data based learning: A project based approach organized through a course management system. *Computers & Education*, *55*(3), 1312–1320. doi:10.1016/j.compedu.2010.06.001

Donna, J., & Miller, B. G. (2013). Using cloud-computing applications to support collaborative scientific inquiry: Examining pre-service teachers' perceived barriers towards integration. *Canadian Journal of Learning and Technology*, *39*(3), 1–17.

Doppelt, Y., Mehalik, M. M., Schunn, C. D., Silk, E., & Krysinski, D. (2008). *Engagement and achievements: A case study of design-based learning in a science context*. Retrieved from http://scholar.lib.vt.edu/ejournals/JTE/v19n2/doppelt.html

Drigas, A., & Kokkalia, G. (2014). ICTs in Kindergarten. *International Journal of Emerging Technologies in Learning*, *9*(2), 52–58. doi:10.3991/ijet.v9i2.3278

Duch, B. J., Groh, S. E., & Allen, D. E. (2001). *The power of problem-based learning: a practical" how to" for teaching undergraduate courses in any discipline*. Stylus Publishing, LLC. Retrieved from http://books.google.com/books?hl=en&lr=&id=-78ZnGLRacAC&oi=fnd&pg=PR8&dq=%22Problem-based+Learning%22&ots=xnWlgDw3ka&sig=vBRvU2gMlhr8M4enSgcEfbRi51Y

Duchâteau, C. (2001). Synthèse du 5° colloque Hypermédias et Apprentissage. In *Proceedings of 5° colloque Hypermédias et Apprentissage* (pp. 376-382). Grenoble, France: INRP.

Duffy, T. M., & Jonassen, D. H. (1992). *Constructivism and the technology of instruction: A conversation*. Psychology Press. Retrieved from http://books.google.com/books?hl=en&lr=&id=yS310A35FC0C&oi=fnd&pg=PR1&dq=constructivism&ots=1v4VqyqzSh&sig=dwffGNrNiTZc-3VF5omxI7ANk-Q

Duggan, M., Ellison, N. B., Lampe, C., Lenhart, A., & Madden, M. (2015). *Pew Research Center: Internet, Science and Technology*. Retrieved from http://www.pewinternet.org/2015/01/09/demographics-of-key-social-networking-platforms-2

Duit, R., Niedderer, H., & Schecker, H. (2007). Teaching physics. In S. K. Abell & N. G. Lederman (Eds.), *Handbook of Research in Science Education* (pp. 599–629). Mahwah, NJ: Erlbaum.

Dunleavy, M., Dede, C., & Mitchell, R. (2009). Affordances and limitations of immersive participatory augmented reality simulations for teaching and learning. *Journal of Science Education and Technology*, *18*(1), 7–22. doi:10.1007/s10956-008-9119-1

Dunne, D., & Martin, R. (2006). Design thinking and how it will change management education: An interview and discussion. *Academy of Management Learning & Education*, *5*(4), 512–523. doi:10.5465/AMLE.2006.23473212

Duschl, R. A., & Osborne, J. (2002). Supporting and promoting argumentation discourse in science education. *Studies in Science Education*, *38*(1), 39–72. doi:10.1080/03057260208560187

Eijck, M., & Roth, W. M. (2009). Authentic science experiences as a vehicle to change students orientations towards science and scientific career choices: Learning from the path followed by Brad. *The Culture of Science Education*, *4*(3), 611–638. doi:10.1007/s11422-009-9183-8

Engeström, J. (2005). *Why some social network services work and others don't — Or: the case for object-centered sociality*. Retrieved from http://www.zengestrom.com/blog/2005/04/why-some-social-network-services-work-and-others-dont-or-the-case-for-object-centered-sociality.html

Ertmer, P., Ottenbreit-Leftwich, A., Sadik, O., Sendurur, E., & Sendurur, P. (2012). Teacher beliefs and technology integration practices: A critical relationship. *Computers & Education*, *59*(2), 423–435. doi:10.1016/j.compedu.2012.02.001

Eshach, H. (2006). *Science literacy in primary schools and pre-schools*. Dordrecht: Springer Verlag. doi:10.1007/1-4020-4674-X

Eshach, H. (2014). The use of intuitive rules in interpreting students difficulties in reading and creating kinematic graphs. *Canadian Journal of Physics*, *92*(1), 1–8. doi:10.1139/cjp-2013-0369

Ess, C. (2015). The Onlife Manifesto: Philosophical Backgrounds, Media Usages, and the Futures of Democracy and Equality. In L. Floridi (Ed.), *The onlife manifesto* (pp. 89–111). Springer-Verlag GmbH. doi:10.1007/978-3-319-04093-6_14

Etkina, E., Karelina, A., Ruibal-Villasenor, M., Rosengrant, D., Jordan, R., & Hmelo-Silver, C. E. (2010). Design and reflection help students develop scientific abilities: Learning in introductory physics laboratories. *Journal of the Learning Sciences*, *19*(1), 54–98. doi:10.1080/10508400903452876

Etkina, E., Planinši, G., & Vollmer, M. (2013). A simple optics experiment to engage students in scientific inquiry. *American Journal of Physics*, *81*(11), 815–222. doi:10.1119/1.4822176

Etkina, E., Van Heuvelen, A., Brookes, D. T., & Mills, D. (2002). Role of Experiments in Physics Instruction - A Process Approach. *The Physics Teacher*, *40*(6), 351–355. doi:10.1119/1.1511592

Eurydice. (2011). *Chiffres clés de l'utilisation des TIC pour l'apprentissage et l'innovation à l'école en Europe*. Bruxelles, Belgium: Commission Européenne. Retrieved from: http://eacea.ec.europa.eu/education/eurydice/documents/key_data_series/129FR_HI.pdf

Evans, V. (2015). What's in a concept? Analog versus parametric concepts in LCCM Theory. In E. Margolis & S. Laurence (Eds.), *The Conceptual Mind: New Directions in the Study of Concepts* (pp. 251–290). Cambridge, MA: The MIT Press.

Facebook. (2016). *Stats*. Retrieved October 14, 2016, from http://newsroom.fb.com/company-info/

Falchikov, N., & Goldfinch, J. (2000). Student peer assessment in higher education: A meta-analysis comparing peer and teacher marks. *Review of Educational Research*, *70*(3), 287–322. doi:10.3102/00346543070003287

Falk, O., Mang, C., & Woessmann, L. (2015). *Virtually no effect? Different uses of classroom computers and their effect on student achievement*. Bonn, Germany: IZA.

Feder, T. (2010). Europe reflects on a decade of higher education reforms. *Physics Today*, *63*(May), 24–27. doi:10.1063/1.3431323

Felder, R. M., & Brent, R. (2005). Understanding Student Differences. *The Journal of Engineering Education, 94*(1), 57–72. doi:10.1002/j.2168-9830.2005.tb00829.x

Ferguson, R., & Clow, D. (2015). Examining engagement: analysing learner subpopulations in massive open online courses (MOOCs). In *The 5th International Learning Analytics and Knowledge Conference (LAK15)*. Poughkeepsie, NY: ACM.

Field, A. P. (2009). *Discovering statistics using SPSS: and sex and drugs and rock 'n' roll* (3rd ed.). London: Sage Publications.

Finkelstein, N. D., Perkins, K. K., Adams, W., Kohl, P., & Podolefsky, N. (2005). Can Computer Simulations Replace Real Equipment in Undergraduate Laboratories? *PERC Proceedings 2004*. doi:10.1063/1.2084711

Finkelstein, N. D., Adams, W. K., Keller, C. J., Kohl, P. B., Perkins, K. K., Podolefsky, N. S., & LeMaster, R. et al. (2005). When learning about the real world is better done virtually: A study of substituting computer simulations for laboratory equipment. *Physical Review Special Topics - Physics Education Research, 1*(1), 010103.

Floridi, L. (2013). Technologys in-betweeness. *Philosophy & Technology, 26*(2), 111–115. doi:10.1007/s13347-013-0106-y

Floridi, L. (2014). *The fourth revolution: How the infosphere is reshaping human reality*. OUP Oxford.

Floridi, L. (2015). *The onlife manifesto*. Springer-Verlag GmbH. doi:10.1007/978-3-319-04093-6

Fontana, M., & Terna, P. (2015). *From Agent-based models to network analysis (and return): the policy-making perspective* (Department of Economics and Statistics Cognetti de Martiis. Working Paper No. 201507). University of Turin. Retrieved from https://ideas.repec.org/p/uto/dipeco/201507.html

Forbes, C. T. (2009). *Preservice elementary teachers' development of pedagogical design capacity for inquiry–an activity-theoretical perspective* (Unpublished Dissertation). University of Michigan.

Forrester, J. W. (1968). *Principles of Systems*. Cambridge, MA: Productivity Press.

Forte, A., Humphreys, M., & Park, T. H. (2012, June). Grassroots Professional Development: How Teachers Use Twitter. *ICWSM*.

Fosnot, C. T. (1996). *Constructivism. Theory, Perspectives, and Practice*. ERIC. Retrieved from http://www.eric.ed.gov/ERICWebPortal/recordDetail?accno=ED396998

Foucault, M. (1969). *L'Archéologie du savoir*. Gallimard.

Fraser, J. M., Timan, A. L., Miller, K., Dowd, J. E., Tucker, L., & Mazur, E. (2014). Teaching and physics education research: Bridging the gap. *Reports on Progress in Physics, 77*(032401), 17. PMID:24595011

Freeman, L. C. (2000). Visualizing Social Networks. *Journal of Social Structure*, (1). Retrieved from http://www.cmu.edu/joss/content/articles/volume1/Freeman.html

French, L. (2004). Science as the center of a coherent, integrated early childhood curriculum. *Early Childhood Research Quarterly*, *19*(1), 138–149. doi:10.1016/j.ecresq.2004.01.004

Friedler, Y., Nachmias, R., & Linn, M. C. (1990). Learning scientific reasoning skills in microcomputer based laboratories. *Journal of Research in Science Teaching*, *27*(2), 173–192. doi:10.1002/tea.3660270208

Furness, T. A., III, Winn, W., & Yu, R. (1997). The impact of three dimensional immersive virtual environments on modern pedagogy: Global change, VR and learning. In Proceedings of Workshops in Seattle, Washington, and Loughborough, England in May and June 1997.

Gaines, B. R., & Shaw, M. L. G. (2012). Sociocognitive Inquiry. In I.-H. Ting, T.-P. Hong, & L. S.-L. Wang (Eds.), *Social Network Mining, Analysis, and Research Trends: Techniques and Applications* (pp. 35–55). IGI Global. doi:10.4018/978-1-61350-513-7.ch003

Galley, R., Conole, G., Dalziel, J., & Ghiglione, E. (2010). *Cloudworks as a "pedagogical wrapper" for LAMS sequences: supporting the sharing of ideas across professional boundaries and facilitating collaborative design, evaluation and critical reflection.* Presented at the 2010 European LAMS & Learning Design Conference, Oxford, UK. Retrieved from http://oro.open.ac.uk/22328/

Galperin, P., & Talizyna, T. (1979). Contemporary condition of the SSDMA theory. *Vestnik MGU. Psychology (Savannah, Ga.)*, *14*(4), 54–63.

Ganascia, J. G. (2008). "In silico" Experiments: Towards a Computerized Epistemology. *Newsletter on Philosophy and Computers*, *7*(2), 11–15.

Ganascia, J. G. (2015). Views and Examples on Hyper-Connectivity. In L. Floridi (Ed.), *The Onlife Manifesto* (pp. 65–85). Springer International Publishing.

Gauvain, M. (2001). *The social context of cognitive development.* Guilford Press.

Gee, J. P. (2004). *What video games have to teach us about learning and literacy.* Palgrave Macmillan.

Geist, E. A. (2012). A qualitative examination of two year-olds interaction with tablet based interactive technology. *Journal of Instructional Psychology*, *39*(1), 26–35.

Gendler, T. (2013). Imagination. In E. N. Zalta (Ed.), *The Stanford Encyclopedia of Philosophy (Fall 2013 Edition).* Retrieved from http://plato.stanford.edu/archives/fall2013/entries/imagination/

Gentner, D. (2002). Psychology of Mental Models. In N. J. Smelser & P. B. Bates (Eds.), *International Encyclopedia of the Social and Behavioral Sciences* (pp. 9683–9687). Amsterdam: Elsevier Science.

Gerard, L., Varma, K., Corliss, S., & Linn, M. (2011). Professional development for technology- enhanced inquiry science. *Review of Educational Research, 81*(3), 408–448. doi:10.3102/0034654311415121

Gershenfeld, N. (2005). *Fab: The coming revolution on your desktop – from personal computers to personal fabrication*. New York, NY: Basic Books.

Gess-Newsome, J. (1999). Pedagogical content knowledge: An introduction and orientation. In J. Gess-Newsome & N. Lederman (Eds.), *Examining pedagogical content knowledge: The constructs and its implication for science education* (pp. 3–17). Dordrecht, The Netherlands: Kluwer Academic Publishers.

Gibbs, R. W. Jr. (2006). *Embodiment and Cognitive Science*. New York, NY: Cambridge University Press.

Gil-Pérez, D., Vilches, A., Fernández, I., Cachapuz, A., Praia, J., Valdés, P., & Salinas, J. (2005). Technology as Applied Science. *Science & Education, 14*(3-5), 309–320. doi:10.1007/s11191-004-7935-0

Ginestié, J. (2013, 3-6 december). *Giving meaning to the use of tools: some elements of discussion about ICT in the French curriculum of technology Éducation*. Paper presented at the 27th PATT Conference, Christchurch, New-Zealand.

Ginestié, J., & Tricot, A. (2013). Activité d'élèves, activité d'enseignants en éducation scientifique et technologique. *Recherches en Didactique des Sciences et Technologie, 8*, 9–22.

Girvan, M., & Newman, M. E. J. (2002). Community structure in social and biological networks. *Proceedings of the National Academy of Sciences of the United States of America, 99*(12), 7821–7826. doi:10.1073/pnas.122653799 PMID:12060727

Gleick, J. (1987). *Chaos: Making a new science*. Elmsford, NY: Pergamon Press.

Gobert, J. D., & Buckley, B. C. (2000). Introduction to model-based teaching and learning in science education. *International Journal of Science Education, 22*(9), 891–894. doi:10.1080/095006900416839

Gobert, J. D., & Clement, J. J. (1999). Effects of student-generated diagrams versus student-generated summaries on conceptual understanding of causal and dynamic knowledge in plate tectonics. *Journal of Research in Science Teaching, 36*(1), 39–53. doi:10.1002/(SICI)1098-2736(199901)36:1<39::AID-TEA4>3.0.CO;2-I

Gómez Puente, S. M., Van Eijck, M., & Jochems, W. (2011). Towards characterising design-based learning in engineering education: A review of the literature. *European Journal of Engineering Education, 36*(2), 137–149. doi:10.1080/03043797.2011.565116

Gómez-Tejedor, J. A., Castro-Palacio, J. C., & Monsoriu, J. A. (2014). The acoustic Doppler effect applied to the study of linear motions. *European Journal of Physics, 35*(2), 025006. doi:10.1088/0143-0807/35/2/025006

González, M. Á., González, M. Á., Martín, M. E., Llamas, C., Martínez, Ó., Vegas, J., & Hernández, C. (2015). Teaching and Learning Physics with Smartphones. *Journal of Cases on Information Technology*, *17*(1), 31–50. doi:10.4018/JCIT.2015010103

Gopnik, A. (2012). Scientific thinking in young children: Theoretical advances, empirical research, and policy implications. *Science*, *337*(6102), 1623–1627. doi:10.1126/science.1223416 PMID:23019643

Graesser, A. C., Chipman, P., & King, B. G. (2008). Computer-Mediated Technologies. In M. J. Spector, D. Merrill, & J. van Merrienboer (Eds.), *Handbook of Research on Educational Communications and Technology* (pp. 211–224). New-York, NY: Erlbaum.

Graham, C. R. (2011). Theoretical considerations for understanding technological pedagogical content knowledge (TPACK). *Computers & Education*, *57*(3), 1953–1960. doi:10.1016/j.compedu.2011.04.010

Graham, C. R., Burgoyne, N., Cantrell, P., Smith, L., St. Clair, L., & Harris, R. (2009). Measuring the TPACK confidence of inservice science teachers. *TechTrends*, *53*(5), 70–79. doi:10.1007/s11528-009-0328-0

Gray, L., Thomas, N., & Lewis, L. (2010). *Teachers' Use of Educational Technology in U.S. Public Schools: 2009 (NCES 2010-040)*. Washington, DC: National Center for Education Statistics, Institute of Education Sciences, U.S. Department of Education.

Greenhow, C. (2011). Youth, learning, and social media. *Journal of Educational Computing Research*, *45*(2), 139–146. doi:10.2190/EC.45.2.a

Grober, S., Vetter, M., Eckert, B., & Jodl, H. J. (2007). Experimenting from a distance - remote controlled laboratory (RCL). *European Journal of Physics*, *28*(3), 127–141. doi:10.1088/0143-0807/28/3/S12

Grobstein, P. (2005). Revisiting Science in Culture: Science as Story Telling and Story Revising. *Journal of Research Practice, 1*(1).

Gulek, J. C., & Demirtas, H. (2005). Learning with technology: The impact of laptop use on student achievement. *The Journal of Technology, Learning, and Assessment*, *3*(2), 4–38.

Gurtner, A., Tschan, F., Semmer, N. K., & Nägele, C. (2007). Getting groups to develop good strategies: Effects of reflexivity interventions on team process, team performance, and shared mental models. *Organizational Behavior and Human Decision Processes*, *102*(2), 127–142. doi:10.1016/j.obhdp.2006.05.002

Hake, R. R. (1998). Interactive-engagement versus traditional methods: A six-thousand-student survey of mechanics test data for introductory physics courses. *American Journal of Physics*, *66*(1), 64–74. doi:10.1119/1.18809

Hake, R. R. (2007). Six lessons from the physics education reform effort. *Latin-American Journal of Physics Education*, *1*(1), 24–31.

Hall, J., & Chamblee, G. (2013). Teaching algebra and geometry with GeoGebra: Preparing pre-service teachers for middle grades/secondary mathematics curriculum. *Computers in the Schools*, *30*(1-2), 12–29. doi:10.1080/07380569.2013.764276

Halpern, D. F., Graesser, A. C., & Hakel, M. (2007). 25 learning principles to guide pedagogy and the design of learning environments. Washington, WA: Association for Psychological Science Taskforce and Lifelong Learning Work and Home. Retrieved from http://www.psy.memphis.edu/learning/whatweknow/index.html

Han, I., & Black, J. B. (2011). Incorporating haptic feedback in simulation for learning physics. *Computers & Education*, *57*(4), 2281–2290. doi:10.1016/j.compedu.2011.06.012

Hanrahan, J., & Isaacs, G. (2001). Assessing self- and peer assessment: The students views. *Higher Education Research & Development*, *20*(1), 53–70. doi:10.1080/07294360123776

Hansson, L., Hansson, O., Juter, K., & Redfors, A. (2015). Reality-Theoretical Models-Mathematics: A ternary perspective on physics lessons in upper-secondary school. *Science & Education*, *24*(5-6), 615–644. doi:10.1007/s11191-015-9750-1

Harnisch, D. L., & Connell, M. L. (1991). *An Introduction to educational information technology* (3rd ed.). Kawasaki, Japan: NEC Technical College.

Harrington, H. A., Beguerisse-Díaz, M., Rombach, M. P., Keating, L. M., & Porter, M. A. (2013). Commentary: Teach network science to teenagers. *Network Science*, *1*(2), 226–247. https://doi.org/10.1017/nws.2013.11

Harris, J. B., & Hofer, M. J. (2011). Technological Pedagogical Content Knowledge (TPACK) in action: A descriptive study of secondary teachers curriculum-based, technology-related instructional planning. *Journal of Research on Technology in Education*, *43*(3), 211–229. doi:10.1080/15391523.2011.10782570

Hatherly, P. A., Jordan, S. E., & Cayless, A. (2009). Interactive screen experiments: Innovative virtual laboratories for distance learners. *European Journal of Physics*, *30*(4), 751–762. doi:10.1088/0143-0807/30/4/008

Haugland, S. (1999). The newest software that meets the developmental needs of young children. *Early Childhood Education Journal*, *26*(4), 245–254. doi:10.1023/A:1022915706904

Haugland, S. W. (1992). The effect of computer software on preschool children's developmental gains. *Journal of Computing in Early Childhood Education*, *3*(1), 15–30.

Hayes, A. F., & Krippendorff, K. (2007). Answering the call for a standard reliability measure for coding data. *Communication Methods and Measures*, *1*(1), 77–89. doi:10.1080/19312450709336664

Hechter, R. P., & Vermette, L. A. (2014). Tech-savvy science education? Understanding teacher pedagogical practices for integrating technology in K-12 classrooms. *Journal of Computers in Mathematics and Science Teaching*, *33*(1), 27–47.

Hegarty, M. (2004). Mechanical reasoning as mental simulation. *Trends in Cognitive Sciences, 8*(6), 280–285. doi:10.1016/j.tics.2004.04.001 PMID:15165554

Heidegger, M. (1954). The question concerning technology. *Technology and Values: Essential Readings*, 99-113.

Hérold, J.-F. (2014). *Learning with an IT tool: analysis of the student's activity.* Paper presented at the Interaccion Conference, Tenerife, Spain. doi:10.1145/2662253.2662303

Hérold, J.-F. (2008). Conception et réalisation d'un outil informatisé d'aide à l'apprentissage: une approche cognitive de l'apprenant. In M. Sidir, G.-L. Baron, & E. Bruillard (Eds.), *Communication et apprentissage instrumentés en réseau* (pp. 265–276). Paris, France: Hermès Lavoisier.

Hérold, J.-F. (2012). Analyse cognitive de l'activité de l'élève pour une personnalisation d'un environnement numérique d'apprentissage. *STICEF, 19*, 285–307.

Hérold, J.-F., & Ginestié, J. (2011). Help with solving technological problems in project activities. *International Journal of Technology and Design Education, 21*(1), 55–70. doi:10.1007/s10798-009-9106-8

Hey, T., Tansley, S., & Tolle, K. M. (2009). The fourth paradigm: data-intensive scientific discovery (Vol. 1). Redmond, WA: Microsoft Research.

Higgins, S., Xiao, Z., & Katsipataki, M. (2012). *The impact of digital technology on learning: A summary for the education endowment foundation.* Durham, UK: Education Endowment Foundation and Durham University.

Higgins, S., Xiao, Z., & Katsipataki, M. (2012). *The Impact of Digital Technology on Learning: A Summary for the Education Endowment Foundation.* London: EEF.

Hmelo-Silver, C. E. (2004). Problem-based learning: What and how do students learn? *Educational Psychology Review, 16*(3), 235–266. doi:10.1023/B:EDPR.0000034022.16470.f3

Hmelo-Silver, C. E., Duncan, R. G., & Chinn, C. A. (2012). Scaffolding and Achievement in Problem-Based and Inquiry Learning: A Response to Kirschner, Sweller, and Clark (2006). *Educational Psychologist, 42*(2), 99–107. doi:10.1080/00461520701263368

Hobbs Vinluan, M., & Jennifer Hofman, J. (2014). Transforming Out-of-School Time Environments: The Ys Commitment to Healthy Eating and Physical Activity Standards. *American Journal of Health Promotion, 28*(3), 116–118. doi:10.4278/ajhp.28.3s.S116 PMID:24380456

Hochberg, K., Gröber, S., Kuhn, J., & Müller, A. (2014). The spinning disc: Studying radial acceleration and its damping process with smartphone acceleration sensors. *Physics Education, 49*(2), 137–140. doi:10.1088/0031-9120/49/2/137

Hofer, M., & Grandgenett, N. (2012). TPACK development in teacher education: A longitudinal study of preservice teachers in a secondary M.A.Ed. program. *Journal of Research on Technology in Education, 45*(1), 83–106. doi:10.1080/15391523.2012.10782598

Hogarth, S., Bennett, J., Lubben, F., Campbell, B., & Robinson, A. (2006). *ICT in science teaching*. London, UK: EPPI-Centre.

Hohenwarter, M. (2014). *GeoGebra*. Retrieved from www.geogebra.org

Holmes, K., Preston, G., Shaw, K., & Buchanan, R. (2013). Follow Me: Networked Professional Learning for Teachers. *Australian Journal of Teacher Education, 38*(12), 54–65. doi:10.14221/ajte.2013v38n12.4

Holton, G. (1978). *The Scientific Imagination: Case Studies*. Cambridge, UK: Cambridge University Press.

Hourigan, M., & Donaghue, J. (2013). The challenges facing initial teacher education: Irish prospective elementary teachers mathematics subject matter knowledge. *International Journal of Mathematical Education in Science and Technology, 44*(1), 36–58. doi:10.1080/0020739X.2012.690897

Hsieh, H. F., & Shannon, S. E. (2005). Three approaches to qualitative content analysis. *Qualitative Health Research, 15*(9), 1277–1288. doi:10.1177/1049732305276687 PMID:16204405

Hsin, C. T., Li, M. C., & Tsai, C. C. (2014). The Influence of Young Children's Use of Technology on Their Learning: A. *Journal of Educational Technology & Society, 17*(4), 85–99.

Huang, D. (2011). Mobile cloud computing. *IEEE COMSOC Multimedia Communications Technical Committee (MMTC) E-Letter, 6*(10), 27–31.

Hulleman, C., & Harackiewicz, J. M. (2009). Promoting interest and performance in high school science classes. *Science, 326*(5958), 1410–1412. doi:10.1126/science.1177067 PMID:19965759

Hwang, G. J., Tsai, C. C., Chu, H. C., Kinshuk, K., & Chen, C.-Y. (2012). A context-aware ubiquitous learning approach to conducting scientific inquiry activities in a science park. *Australasian Journal of Educational Technology, 28*(5), 931–947. doi:10.14742/ajet.825

Inden, R. B. (2000). *Imagining India. C.* Hurst Co. Publishers.

International Society for Technology in Education (ISTE). (2016). *ISTE standards for students*. Retrieved August 2, 2016, from http://iste.org/standards

Israeli Ministry of Education. (2009). *Tochnit limudim be'mada ve'technologia be'gan hayeladim hamamlachty ve'hamamlachty-dati* [Curriculum of science and technology for state and state-religious kindergartens]. Department for Curriculum Planning and Development. Retrieved July 19, 2010, from http://cms.education.gov.il/educationcms/units/tochniyot_limudim/kdamyesodi/tochniyotlimudim/madatechnologia.htm (in Hebrew)

Israeli Ministry of Education. (2016). *Digital learning environments in preschool*. Retrieved August 5, 2016 from: http://edu.gov.il/minhalpedagogy/preschool/subjects/science-and-technology/digital-media-lib/Pages/digital-environments.aspx (in Hebrew)

Istance, D., & Kools, M. (2013). OECD Work on Technology and Education: Innovative learning environments as an integrating framework. *European Journal of Education*, *48*(1), 43–57. doi:10.1111/ejed.12017

It's Elemental. (n.d.). Retrieved January 10, 2017, from http://education.jlab.org/itselemental/

Jang, S.-J. (2012). Developing a technology-team teaching model to enhance the TPACK of pre-service science teachers. In S. Abramovich (Ed.), *Computers in education* (Vol. 2, pp. 71–90). New York, NY: Nova Science Publishers.

Jewett, J. W. Jr. (2015). Bringing (century-old) technology into the classroom. Part I: Teaching mechanics and thermodynamics with antiques. *The Physics Teacher*, *53*(7), 428–431. doi:10.1119/1.4931013

Jiménez-Aleixandre, M. P., & Erduran, S. (2007). Argumentation in science education: An overview. In *Argumentation in science education* (pp. 3–27). Springer. doi:10.1007/978-1-4020-6670-2_1

Jirout, J., & Klahr, D. (2012). Childrens scientific curiosity: In search of an operational definition of an elusive concept. *Developmental Review*, *32*(2), 125–160. doi:10.1016/j.dr.2012.04.002

Johansson-Sköldberg, U., Woodilla, J., & Çetinkaya, M. (2013). Design thinking: Past, present and possible futures. *Creativity and Innovation Management*, *22*(2), 121–146. doi:10.1111/caim.12023

Johnson, L., Adams Becker, S., Cummins, M., Estrada, V., Freeman, A., & Ludgate, H. (2013). *Technology outlook for community, technical, and junior colleges 2013-2018: An NMC horizon project sector analysis*. Austin, TX: The New Media Consortium. Houston Community College ISBN.

Johnson, L., Adams Becker, S., Estrada, V., & Freeman, A. (2015). *NMC Horizon Report: 2015 K-12 Edition*. Austin, TX: The New Media Consortium.

Johnson, P. (2009). The 21st Century Skills Movement. *Educational Leadership*, *67*(1), 11.

Johnston, J., & Ahtee, M. (2006). Comparing primary student teachers attitudes, subject knowledge and pedagogical content knowledge needs in a physics activity. *Teaching and Teacher Education*, *22*(4), 503–512. doi:10.1016/j.tate.2005.11.015

Jonassen, D. (2000). *Mindtools for schools: Engaging critical thinking with technology*. Columbus, OH: Merrill.

Jonassen, D., & Land, S. (Eds.). (2012). *Theoretical foundations of learning environments* (2nd ed.). New York, NY: Routledge.

Jones, M. G., & Leagon, M. (2014). Science teacher attitudes and beleifs: Reforming practice. In N. G. Lederman & S. K. Abel (Eds.), *Handbook of research on science education* (Vol. 2, pp. 830–847). New York: Routledge.

Jordan, K. (2014). Initial Trends in Enrolment and Completion of Massive Open Online Courses. *International Review of Research in Open and Distance Learning, 15*(1), 133–160. doi:10.19173/irrodl.v15i1.1651

Judge, S., Floyd, K., & Jeffs, T. (2015). *Using mobile media devices and apps to promote young children's learning. In Young Children and Families in the Information Age* (pp. 117–131). Springer Netherlands.

Junco, R., Helbergert, G., & Loken, E. (2011). The effect of Twitter on college student engagement and grades. *Journal of Computer Assisted Learning, 27*(2), 119–132. doi:10.1111/j.1365-2729.2010.00387.x

Jurow, A. S. (2005). Shifting engagements in figured worlds: Middle school mathematics students participation in an architectural design project. *Journal of the Learning Sciences, 14*(1), 35–67. doi:10.1207/s15327809jls1401_3

Kabilan, M. K., Ahmad, N., & Abidin, M. J. Z. (2010). Facebook: An online environment for learning of English in institutions of higher education? *The Internet and Higher Education, 13*(4), 179–187. doi:10.1016/j.iheduc.2010.07.003

Kahneman, D., & Tversky, A. (1982). The simulation heuristic. In Judgment Under Uncertainty: Heuristics and Biases (pp. 201-208). Cambridge University Press. doi:10.1017/CBO9780511809477.015

Kamii, C., & Dominck, A. (1998). The harmful effects of algorithms in grades 1–4. In L. J. Morrow & M. J. Kenney (Eds.), The teaching and learning of algorithms in school mathematics: 1998 Yearbook (pp. 130–140). Reston, VA: The National Council of Teachers of Mathematics.

Kamil, M. L., Borman, G. D., Dole, J., Kral, C. C., Salinger, T., & Torgesen, J. (2008). Improving adolescent literacy: Effective classroom and intervention practices: A Practice Guide (NCEE #2008-4027). Washington, DC: National Center for Education Evaluation and Regional Assistance, Institute of Education Sciences, U.S. Department of Education. Retrieved from http://ies.ed.gov/ncee/wwc

Kaplan, A. M., & Haenlein, M. (2010). Users of the world, unite! The challenges and opportunities of Social Media. *Business Horizons, 53*(1), 59–68. doi:10.1016/j.bushor.2009.09.003

Kapucu, S. (2014). Salient beliefs of pre-service primary school teachers underlying an attitude "liking or disliking physics". *Science Education International, 25*(4), 437–458.

Kaput, J. J. (1992). Technology and mathematics education. In D. A. Grouws (Ed.), *Handbook of research on mathematics teaching and learning* (pp. 515–556). New York, NY: Macmillan.

Karam, R. (2015). Introduction of the thematic issue on the interplay of physics and mathematics. *Science & Education, 24*(5-6), 487–494. doi:10.1007/s11191-015-9763-9

Kärkkäinen, K., & Vincent-Lancrin, S. (2013). Sparking Innovation in STEM Education with Technology and Collaboration: A Case Study of the HP Catalyst Initiative. *OECD Education Working Papers, 91.* 10.1787/5k480sj9k442-en

Kearney, M., Schuck, S., Burden, K., & Aubusson, P. (2012). Viewing mobile learning from a pedagogical perspective. *Research in Learning Technology, 20*(1), 14406. doi:10.3402/rlt.v20i0.14406

Kelleher, C., & Pausch, R. (2007). Using Storytelling to Motivate Programming. *Communications of the ACM, 50*(7), 58–64. doi:10.1145/1272516.1272540

Kelly, S. (2007). Classroom discourse and the distribution of student engagement. *Social Psychology of Education, 10*(3), 331–352. doi:10.1007/s11218-007-9024-0

Kesner Baruch, Y., Spektor-Levy, O., & Mashal, N. (2014). Pre-schoolers verbal and behavioral responses as indicators of attitudes and scientific curiosity. *International Journal of Science and Mathematics Education, 14*(1), 125–148. doi:10.1007/s10763-014-9573-6

Kilpatrick, J., & Quinn, H. (Eds.). (2009). *Science and Mathematics Education.* Washington, DC: National Academy of Education.

Kind, A. (Ed.). (2016). *The Routledge Handbook of Philosophy of Imagination.* Routledge.

King, K. P. (2011). Professional learning in unlikely spaces: Social media and virtual communities as professional development. *iJET, 6*(4), 40-46.

Kirschner, P. A., Sweller, J., & Clark, R. E. (2006). Why minimal guidance during instruction does not work: An analysis of the failure of constructivist, discovery, problem-based, experiential, and inquiry-based teaching. *Educational Psychologist, 41*(2), 75–86. doi:10.1207/s15326985ep4102_1

Kizilcec, R. F., & Schneider, E. (2015). Motivation as a lens to understand online learners: Toward data-driven design with the OLEI scale. *ACM Transactions on Computer-Human Interactions, 22*(2). http://dx.doi.org./10.1145/2699735

Klopfer, E., Scheintaub, H., Huang, W., Wendel, D., & Roque, R. (2009). The Simulation Cycle: Combining Games, Simulations, Engineering and Science Using StarLogo TNG. *E-Learning and Digital Media, 6*(1), 71–96. https://doi.org/10.2304/elea.2009.6.1.71

Klopfer, E., & Squire, K. (2008). Environmental Detectives—the development of an augmented reality platform for environmental simulations. *Educational Technology Research and Development, 56*(2), 203–228. doi:10.1007/s11423-007-9037-6

Knorr-Cetina, K. (1999). *Epistemic cultures: how the sciences make knowledge.* Harvard University Press.

Koehler, M. J., & Mishra, P. (2009). What is technological pedagogical content knowledge? *Contemporary Issues in Technology & Teacher Education, 9*(1), 60–70.

Koehler, M. J., & Mishra, P. (2015). Technological Pedagogical Content Knowledge. In M. J. Spector (Ed.), *The SAGE encyclopedia of educational technology* (Vol. 2, pp. 782–785). Los Angeles: SAGE Publications.

Koehler, M. J., Mishra, P., & Cain, W. (2013). What is technological pedagogical content knowledge (TPACK)? *Journal of Education, 193*(3), 13–19.

Koehler, M. J., Mishra, P., & Yahya, K. (2007). Tracing the development of teacher knowledge in a design seminar: Integrating content, pedagogy, and technology. *Computers & Education, 49*(3), 740–762. doi:10.1016/j.compedu.2005.11.012

Koh, J. H. L., & Divaharan, S. (2011). Developing pre-service teachers technology integration expertise through the TPACK: Developing instructional model. *Journal of Educational Computing Research, 44*(1), 35–58. doi:10.2190/EC.44.1.c

Korpershoek, H., Kuyper, H., & van der Werf, M. P. C. (2015). Differences in students school motivation: A latent class modelling approach. *Social Psychology of Education, 18*(1), 137–163. doi:10.1007/s11218-014-9274-6

Kozma, R. B., Russell, J., Jones, T., Marx, N., & Davis, J. (1996). The use of multiple, linked representations to facilitate science understanding. In *NATO Symposium on International Perspectives on the Psychological Foundations of Technology-Based Learning Environments, Crete, Greece, Jul 1992, and at the 5th EARLI Conference, Aix-en-Provence, France, Sep 1993.* Mahwah, NJ: Lawrence Erlbaum Associates.

Krajcik, J. S., & Mun, K. (2014). Promises and challenges of using learning technologies to promote student learning of science. In L. Norman G & S. K. Abell (Eds.), Handbook of Research on Science Education (Vol. 2, pp. 337-360). New York: Routledge.

Krajcik, J., & Blumenfeld, P. C. (2008). Project-based learning. In L. M. Given (Ed.), *The Sage Encyclopedia of Qualitative Research* (pp. 422–423). Sage.

Krajcik, J., McNeil, K., & Reiser, B. J. (2007). Learning-goals-driven design model: Developing curriculum materials that align with national standards and incorporate project-based pedagogy. *Science Education, 92*(1), 1–32. doi:10.1002/sce.20240

Kramarski, B., & Mevarech, Z. R. (2003). Enhancing mathematical reasoning in the classroom: The effects of cooperative learning and metacognitive training. *American Educational Research Journal, 40*(1), 281–310. doi:10.3102/00028312040001281

Kuech, R. K., & Lunetta, V. N. (2002). Using digital technologies in the science classroom to promote conceptual understanding. *Journal of Computers in Mathematics and Science Teaching, 21*(2), 103–126.

Kuhn, D., Black, J., Keselman, A., & Kaplan, D. (2000). The development of cognitive skills to support inquiry learning. *Cognition and Instruction, 18*(4), 495–523. doi:10.1207/S1532690XCI1804_3

Kuhn, J., & Vogt, P. (2013). Smartphones as experimental tools: Different methods to determine the gravitational acceleration in classroom physics by using everyday devices. *European Journal of Physics Education, 4*(1), 16–27.

Kwok, R., Cheng, S., Ho-Shing Ip, H., & Kong, J. (2011). Design of Affectively Evocative Smart Ambient Media for Learning. *Computers & Education*, *56*(1), 101–111. doi:10.1016/j.compedu.2010.08.015

Labov, J. B., Reid, A. H., & Yamamoto, K. R. (2010). Integrated biology and undergraduate science education: A new biology education for the twenty-first century? *CBE Life Sciences Education*, *9*(1), 10–16. doi:10.1187/cbe.09-12-0092 PMID:20194802

Landriscina, F. (2015). The role of mental simulation in understanding and in creating scientific concepts. In F. Corni (Ed.), *Proceedings from the 3d Conference on Innovation in Science Education in Primary School and Kindergarten*. University of Modena and Reggio Emilia.

Landriscina, F. (2009). Simulation and Learning: The Role of Mental Models. *Journal of E-Learning and Knowledge Society*, *5*(2), 23–32.

Landriscina, F. (2012). Simulation and Learning: The Role of Mental Models. In N. M. Seel (Ed.), *Encyclopedia of the Sciences of Learning* (pp. 3072–3075). New York, NY: Springer.

Landriscina, F. (2013). *Simulation and Learning. A Model-Centered Approach*. New York, NY: Springer.

Laru, J., Järvelä, S., & Clariana, R. B. (2012). Supporting collaborative inquiry during a biology field trip with mobile peer-to-peer tools for learning: A case study with K-12 learners. *Interactive Learning Environments*, *20*(2), 103–117. doi:10.1080/10494821003771350

Lasry, N. (2008). Clickers or flashcards: Is there really a difference? *The Physics Teacher*, *46*(5), 242–244. doi:10.1119/1.2895678

Lasry, N., Guillemette, J., Dugdale, M., Bhatnagar, S., Whittaker, C., Charles, E., & Mazur, E. (2014). *You gain some, you lose some: A big data look at how students change conceptions*. Physics Review Special Topics - Physics Education Research.

Lasry, N., Guillemette, J., & Mazur, E. (2014). Two steps forward, one step back. *Nature Physics*, *10*(June), 402–403. doi:10.1038/nphys2988

Lasry, N., Mazur, E., & Watkins, J. (2008). Peer Instruction: From Harvard to the two-year college. *American Journal of Physics*, *76*(11), 1066–1069. doi:10.1119/1.2978182

Lasry, N., Watkins, J., Mazur, E., & Ibrahim, A. (2013). Response times to conceptual questions. *American Journal of Physics*, *81*(9), 703–706. doi:10.1119/1.4812583

Laugksch, R. C. (2000). Scientific literacy: A conceptual overview. *Science Education*, *84*(1), 71–94. doi:10.1002/(SICI)1098-237X(200001)84:1<71::AID-SCE6>3.0.CO;2-C

Lauricella, A. R., Wartella, E., & Rideout, V. J. (2015). Young childrens screen time: The complex role of parent and child factors. *Journal of Applied Developmental Psychology*, *36*, 11–17. doi:10.1016/j.appdev.2014.12.001

Lave, J., & Wenger, E. (1991). *Situated learning: legitimate peripheral participation*. Cambridge, UK: Cambridge University Press. doi:10.1017/CBO9780511815355

Lawson, A. (2003). The nature and development of hypothetico-predictive argumentation with implications for science teaching. *International Journal of Science Education, 25*(11), 1387–1408. doi:10.1080/09500690320000521117

Lazonder, A. W., & Harmsen, R. (2016). Meta-Analysis of Inquiry-Based Learning Effects of Guidance. *Review of Educational Research.*

Lee, M.-H., & Tsai, C.-C. (2010). Exploring teachers perceived self efficacy and Technological Pedagogical Content Knowledge with respect to educational use of the World Wide Web. *Instructional Science, 38*(1), 1–21. doi:10.1007/s11251-008-9075-4

Lee, S., Koh, S., Yen, D., & Tang, H.-L. (2002). Perception gaps between IS academics and IS practitioners: An exploratory study. *Information & Management, 40*(1), 51–61. doi:10.1016/S0378-7206(01)00132-X

Lee, Y.-J. (2011). Utilizing formative assessments to guide student learning in an interactive learning environment. *Journal of Educational Technology Systems, 39*(3), 245–260. doi:10.2190/ET.39.3.c

Leinonen, T., & Durall-Gazulla, E. (2014). Design thinking and collaborative learning. *Comunicar, 21*(42). Retrieved from http://eprints.rclis.org/20862/

Lelas, S. (1993). Science as technology. *The British Journal for the Philosophy of Science, 44*(3), 423–442. doi:10.1093/bjps/44.3.423

Lemke, J. (2004). *Why study digital game worlds? Notes toward a basic research agenda for learning technologies.* Retrieved from http://www-personal.umich.edu/~jaylemke/games.htm

Lemke, J. L. (2000). Across the scales of time: Artifacts, activities, and meanings in ecosocial systems. *Mind, Culture, and Activity, 7*(4), 273–290. doi:10.1207/S15327884MCA0704_03

Lemke, J. L. (2001). Articulating communities: Sociocultural perspectives on science education. *Journal of Research in Science Teaching, 38*(3), 296–316. doi:10.1002/1098-2736(200103)38:3<296::AID-TEA1007>3.0.CO;2-R

Lenhart, A. (2015). *Pew Research Center: Teens, Social Media and Technology Overview.* Retrieved from http://www.pewinternet.org/2015/04/09/mobile-access-shifts-social-media-use-and-other-online-activities

Leonov G. A., & Kuznetsov N. V. (2013). Hidden attractors in dynamical systems. From hidden oscillations in Hilbert–Kolmogorov, Aizerman, and Kalman problems to hidden chaotic attractor in Chua circuits. *International Journal of Bifurcation and Chaos, 23.* DOI: 10.1142/S0218127413300024

Leontiev, A. N. (1981). The problem of activity in psychology. In J. V. Wertsch (Ed.), *The concept of activity in Soviet psychology* (pp. 37–72). Armonk, NY: M. E. Sharpe.

Lerner, R. M. (2014). *Agent-Based Modeling as a Social Activity*. Northwestern University. Retrieved from http://gradworks.umi.com/36/69/3669272.html

Lerner, C., & Barr, R. (2014). Screen sense: Setting the record straight – Research-based guidelines for screen use for children under 3 years old. *Zero to Three*.

Let's Talk Science. (2012). *Spotlight on science learning: A benchmark of Canadian talent*. Retrieved from http://www.letstalkscience.ca/our-research/spotlight.html

Let's Talk Science. (2013). *Spotlight on science learning: The high cost of dropping science and math*. Retrieved from http://www.letstalkscience.ca/our-research/spotlight2013.html

Let's Talk Science. (2015). *Exploring parental influence: Shaping teen decisions regarding science education*. Retrieved from http://www.letstalkscience.ca/research-publications/spotlight-on-science-learning/2015-spotlight.html

Levine, T. H., & Marcus, A. S. (2010). How the structure and focus of teachers collaborative activities facilitate and constrain teacher learning. *Teaching and Teacher Education: An International Journal of Research and Studies*, *26*(3), 389–398. doi:10.1016/j.tate.2009.03.001

Levin, I. (2016). Cyber-physical systems as a cultural phenomenon. *International Journal of Design Sciences & Technology*, *22*(1).

Levin, I. (2016). Cyber-physical Systems as a Cultural Phenomenon. *International Journal of Design Sciences and Technology*, *22*(1), 67–80.

Li, L., & Gao, F. (2016). The effect of peer assessment on project performance of students at different learning levels. *Assessment & Evaluation in Higher Education*, *41*(6), 885–900. doi:10.1080/02602938.2015.1048185

Lindgren, R., Tscholl, M., Wang, S., & Johnson, E. (2016). Enhancing learning and engagement through embodied interaction within a mixed reality simulation. *Computers & Education*, *95*, 174–187. doi:10.1016/j.compedu.2016.01.001

Linebarger, D. L., & Vaala, S. E. (2010). Screen media and language development in infants and toddlers: An ecological perspective. *Developmental Review*, *30*(2), 176–202. doi:10.1016/j.dr.2010.03.006

Lin, M. H., & Lee, J. Y. (2015). Data-driven learning: Changing the teaching of grammar in EFL classes. *ELT Journal*, *69*(3), 264–274. doi:10.1093/elt/ccv010

Linnenbrink, E. A. (2006). Emotion research in education: Theoretical and methodological perspectives on the integration of affect, motivation and cognition. *Educational Psychology Review*, *18*(4), 307–314. doi:10.1007/s10648-006-9028-x

Linn, M. C., Clark, D., & Slotta, J. D. (2003). WISE design for knowledge integration. *Science Education*, *87*(4), 517–538. doi:10.1002/sce.10086

Lipponen, L., Rajala, A., Hilppö, J., & Paananen, M. (2015). Exploring the foundations of visual methods used in research with children. *European Early Childhood Education Research Journal*, 1–11.

Littleton, K., & Light, P. (Eds.). (1999). *Learning with computers: Analyzing productive interaction.* Psychology Press. doi:10.4324/9780203282311

Livingstone, S. (2012). Critical reflections on the benefits of ICT in education. *Oxford Review of Education*, *38*(1), 9–24. doi:10.1080/03054985.2011.577938

Louwerse, M. M., & Jeuniaux, P. (2010). The linguistic and embodied nature of conceptual processing. *Cognition*, *114*(1), 96–104. doi:10.1016/j.cognition.2009.09.002 PMID:19818435

Luft, J. A., & Hewson, P. W. (2014). Research on teacher professional development knowledge in science. In N. G. Lederman & S. K. Abel (Eds.), *Handbook of research on science education* (Vol. 2, pp. 889–909). New York: Routledge.

Lukowicz, P., Poxrucker, A., Weppner, J., Bischke, B., Kuhn, J., & Hirth, M. (2015). Glass-physics: Using Google Glass to Support High School Physics Experiments. *Proceedings of the 2015 ACM International Symposium on Wearable Computers*, 151-154. doi:10.1145/2802083.2808407

Luna Scott, C. (2015). *The Futures of Learning 3: what kind of pedagogies for the 21st century?* Paris: UNESCO Education Research and Foresight.

Lynott, D., & Connell, L. (2010). Embodied conceptual combination. *Frontiers in Psychology*, *1*, 1–14. doi:10.3389/fpsyg.2010.00212 PMID:21833267

Lyons, C. D., & Tredwell, C. T. (2015). Steps to implementing technology in inclusive early childhood programs. *Computers in the Schools*, *32*(2), 152–166. doi:10.1080/07380569.2015.1038976

Maass, K., & Doorman, M. (2013). A model for a widespread implementation of inquiry-based learning. *ZDM*, *45*(6), 887–899. doi:10.1007/s11858-013-0505-7

Maciel, T. (2015). Smartphones in the classroom help students see inside the black box. *APS News*, *24*(3), 5–6.

MacKenzie, I. S. (1988). Issues and Methods in the Microcomputer-Based Lab. *Journal of Computers in Mathematics and Science Teaching*, *7*(3), 12–18.

Maddux, C. D. (1984). Educational microcomputing: The need for research. *Computers in the Schools*, *1*(1), 35–41. doi:10.1300/J025v01n01_04

Markic, S., Eilks, I., & Valanides, N. (2008). Developing a tool to evaluate differences in beliefs about science teaching and learning among freshman science student teachers from different science teaching domains: A case study. *Eurasia Journal of Mathematics, Science & Technology Education*, *4*(2), 109–120.

Marshall, J. A., & Young, E. S. (2006). Preservice teachers theory development in physical and simulated environments. *Journal of Research in Science Teaching*, *43*(9), 907–937. doi:10.1002/tea.20124

Martens, B. K., Daly, E. J., Begeny, J. C., & VanDerHeyden, A. (2011). Behavioral approaches to education. In W. W. Fisher, C. C. Piazza, & H. S. Roane (Eds.), *Handbook of applied behavior analysis* (pp. 385–401). New York, NY: The Guilford Press.

Martin-Hansen, L. (2002). Defining Inquiry: Exploring the many types of inquiry in the science classroom. *Science Teacher (Normal, Ill.)*, *69*(2), 34–37.

D. Martinovic, Z. Karadag, & D. McDougall (Eds.). (2014). *Proceedings of the Fifth North American GeoGebra Conference: Explorative learning with technology: GeoGebra-NA 2014*. Toronto, ON, Canada: University of Toronto.

Mason, R. A., & Just, M. A. (2016). Neural Representations of Physics Concepts. *Psychological Science*, *27*(6), 904–913. doi:10.1177/0956797616641941 PMID:27113732

Mayer, R. E. (2005). *Cambridge Handbook of Multimedia Learning*. New-York, NY: Cambridge University Press. doi:10.1017/CBO9780511816819

Mayer, R. E. (2010). Learning with technology. In H. Dumont, D. Istance, & D. Benavides (Eds.), *The nature of learning: using research to inspire practice* (pp. 173–198). New-York, NY: OECD. doi:10.1787/9789264086487-10-en

Mayes, R., & Koballa, T. R. Jr. (2012). Exploring the science framework. *Science and Children*, *50*(4), 8–15.

Maziewski, A., Dobrogowski, W., & Zablotskii, V. (2007). GloLab: Creating a global Internet-accessible laboratory. *Physics Education*, *42*(1), 72–75. doi:10.1088/0031-9120/42/1/009

Mazur, E. (1997c). *Understanding or Memorization: Are we teaching the right thing?* Paper presented at the Conference on the Introductory Physics Course on the Occasion of the Retirement of Robert Resnick, New York, NY.

Mazur, E. (1997a). Moving the mountain: Impediments to change. *The Physics Teacher*, *35*, 1–4.

Mazur, E. (1997b). *Peer Instruction: User's manual*. Upper Saddle River, NJ: Prentice Hall.

McCarrick, K., & Li, X. (2007). Buried treasure: The impact of computer use on young children's social, cognitive, language development and motivation. *AACE Journal*, *15*(1), 73–95.

McComas, W. (1998). The Principal Elements of the Nature of Science: Dispelling the Myths. In W. McComas (Ed.), *The Nature of Science in Science Education: Rationales and Strategies* (pp. 53–72). Dordrecht: Kluwer Academic Publishers.

McConnell, W. J., & Dickerson, D. (2013). Making connections between the digital and physical world. *Science and Children*, *51*(4), 38–43. doi:10.2505/4/sc13_051_04_38

McDermott, L. C., Heron, P. R. L., Shaffer, P. S., & Stetzer, M. R. (2006). Improving the preparation of K-12 teachers through physics education research. *American Journal of Physics, 74*(9), 763–767. doi:10.1119/1.2209244

McNutt, M. (2013). Bricks and MOOCs. *Science, 342*(6157), 402. doi:10.1126/science.1246943 PMID:24159014

Mehra, A., Borgatti, A., Soltis, S., Floyd, T., Halgin, D. S., Brandon, O., & Lopez-Kidwell, V. (2014). Imaginary Worlds: Using Visual Network Scales to Capture Perceptions of Social Networks. In *Contemporary Perspectives on Organizational Social Networks* (Vol. 40, pp. 315–336). Emerald Group Publishing Limited. Retrieved from http://www.emeraldinsight.com/doi/abs/10.1108/S0733-558X%282014%290000040016

Mell, P., & Grance, T. (2010). The NIST definition of cloud computing. *Communications of the ACM, 53*(6), 50.

Mercer, N., Warwick, P., Kershner, R., & Staarman, J. K. (2010). Can the interactive whiteboard help to provide dialogic space for childrens collaborative activity? *Language and Education, 24*(5), 367–384. doi:10.1080/09500781003642460

Merriam-Webster's Learner's Dictionary. (2016). Retrieved from http://www.merriam-webster.com/dictionary/social%20media

Mevarech, Z. R. (1999). Effects of metacognitive training embedded in cooperative settings on mathematical problem solving. *The Journal of Educational Research, 92*(4), 195–205. doi:10.1080/00220679909597597

Mevarech, Z. R., & Kramarski, B. (1997). IMPROVE: A multidimensional method for teaching mathematics in heterogeneous classrooms. *American Educational Research Journal, 34*(2), 365–394. doi:10.3102/00028312034002365

Michko, G. M. (2007). *A Meta-analysis of the Effects of Teaching and Learning with Technology on Student Outcomes in Undergraduate Engineering Education.* Houston, TX: University of Houston.

Middleton, H. (2008). *Researching technology education. Methods and techniques.* Rotterdam, The Netherlands: Sense Publishers.

Miller, A. I. (1986). *Imagery in Scientific Thought.* Cambridge, MA: The MIT Press.

Miller, K. W., & Voas, J. (2010). Ethics and the Cloud. *IT Professional, 12*(5), 4–5. doi:10.1109/MITP.2010.129

Milner-Bolotin, M. (2014a). *Closing the research-practice gap through innovative technology use in STEM teacher education.* Paper presented at the Fifth North American GeoGebra Conference (GeoGebra-NA 2014): Explorative Learning with Technology, Toronto, Canada.

Milner-Bolotin, M. (2016a). *Mathematics and Science Teaching and Learning through Technologies.* Retrieved from http://scienceres-edcp-educ.sites.olt.ubc.ca/

Milner-Bolotin, M. (2012). Increasing interactivity and authenticity of chemistry instruction through data acquisition systems and other technologies. *Journal of Chemical Education, 89*(4), 477–481. doi:10.1021/ed1008443

Milner-Bolotin, M. (2014b). Using PeerWise to promote student collaboration on design of conceptual multiple-choice questions. *Physics in Canada, 70*(3), 149–150.

Milner-Bolotin, M. (2015a). Learning physics teaching through collaborative design of conceptual multiple-choice questions. *The Canadian Journal of Action Research, 16*(2), 22–41.

Milner-Bolotin, M. (2015b). Technology-enhanced teacher education for 21st century: Challenges and possibilities. In X. Ge, D. Ifenthaler, & J. M. Spector (Eds.), *Emerging technologies for STEAM education* (pp. 137–156). Springer International Publishing. doi:10.1007/978-3-319-02573-5_8

Milner-Bolotin, M. (2015c). Visual literacy skills in science, technology, engineering, and mathematics education. In M. J. Spector (Ed.), *The SAGE Encyclopedia of Educational Technology* (Vol. 2, pp. 826–830). Thousand Oaks, CA: SAGE Publications.

Milner-Bolotin, M. (2016b). Promoting Deliberate Pedagogical Thinking with Technology in physics teacher education: A teacher-educator's journey. In T. G. Ryan & K. A. McLeod (Eds.), *The Physics Educator: Tacit Praxes and Untold Stories* (pp. 112–141). Champaign, IL: Common Ground and The Learner.

Milner-Bolotin, M. (2016c). Rethinking technology-enhanced physics teacher education: From theory to practice. *Canadian Journal of Science, Mathematics and Technology Education, 16*(3), 284–295. doi:10.1080/14926156.2015.1119334

Milner-Bolotin, M., Egersdorfer, D., & Vinayagam, M. (2016). Investigating the effect of question-driven pedagogy on the development of physics teacher-candidates' Pedagogical Content Knowledge. *Physics Review Special Topics: Physics Education Research, 12*(020128), 1–16.

Milner-Bolotin, M., Fisher, H., & MacDonald, A. (2013). Modeling active engagement pedagogy through classroom response systems in a physics teacher education course. *LUMAT: Research and Practice in Math, Science and Technology Education, 1*(5), 523–542.

Milner-Bolotin, M., Kotlicki, A., & Rieger, G. (2007). Can students learn from lecture demonstrations: The role and place of interactive lecture experiments in large introductory science courses. *Journal of College Science Teaching, 36*(4), 45–49.

Milner-Bolotin, M., & Moll, R. F. (2008). Physics exam problems reconsidered: Using Logger Pro technology to evaluate student understanding of physics. *The Physics Teacher, 46*(8), 494–500. doi:10.1119/1.2999067

Milton, P. (2015). *Shifting Minds 3.0: Redefining the Learning Landscape in Canada*. C21 Canada.

Minner, D. D., Levy, A. J., & Century, J. (2010). Inquiry-based science instruction–What is it and does it matter? Results from a research synthesis years 1984 to 2002. *Journal of Research in Science Teaching, 47*(4), 474–496. doi:10.1002/tea.20347

Mishra, P., & Koehler, M. J. (2006). Technological pedagogical content knowledge: A framework for teacher knowledge. *Teachers College Record, 108*(6), 1017–1054. doi:10.1111/j.1467-9620.2006.00684.x

Moazzen, I., Hansen, T., Miller, M., Wild, P., Hadwin, A., & Jackson, L. (2013). Literature Review on Engineering Design Assessment Tools. *Proceedings of the Canadian Engineering Education Association.* Retrieved from http://library.queensu.ca/ojs/index.php/PCEEA/article/view/4864

Mokros, J. R., & Tinker, R. F. (1987). The impact of microcomputer-based labs on childrens ability to interpret graphs. *Journal of Research in Science Teaching, 24*(4), 369–383. doi:10.1002/tea.3660240408

Monaghan, J. M., & Clement, J. (1999). Use of a computer simulation to develop mental simulations for understanding relative motion concepts. *International Journal of Science Education, 21*(9), 921–944. doi:10.1080/095006999290237

Moody, J., Mcfarl, D., & Bender-demoll, S. (2005). Dynamic Network Visualization. *American Journal of Sociology, 110*(4), 1206–1241. doi:10.1086/421509

Morales, T. M., Bang, E. J., & Andre, T. (2013). A one-year case study: Understanding the rich potential of project-based learning in a virtual reality class for high school students. *Journal of Science Education and Technology, 22*(5), 791–806. doi:10.1007/s10956-012-9431-7

Moreno, J. L. (1934). *Who Shall Survive? A New Approach to the Problem of Human Interrelations.* Nervous and Mental Disease Publishing Company. doi:10.1037/10648-000

Moreno, R. (2004). Multimedia learning with animated pedagogical agents. In R. E. Mayer (Ed.), *The Cambridge Handbook of Multimedia Learning* (pp. 507–524). New-York, NY: Cambridge University Press.

Mouza, C. (2005). Using technology to enhance early childhood learning: The 100 days of school project. *Educational Research and Evaluation, 11*(6), 513–528. doi:10.1080/13803610500254808

Muehrer, R., Jenson, J., Friedberg, J., & Husain, N. (2012). Challenges and opportunities: Using a science-based video game in secondary school settings. *Cultural Studies of Science Education, 7*(4), 783–805. doi:10.1007/s11422-012-9409-z

Nake, F. (2013). Considering algorithmics and aesthetics. In J. Walter-Herrmann & C. Büching (Eds.), *FabLab: Of machines, makers and inventors* (pp. 79–92). Bielefeld, Germany: Transcript Verlag. doi:10.14361/transcript.9783839423820.79

National Academies of Sciences, Engineering, and Medicine. (2016). *Science literacy: Concepts, contexts, and consequences.* Washington, DC: The National Academies Press. doi:10.17226/23595

National Academy of Engineering (NAE) & National Research Council (NRC). (2014). STEM integration in K-12 education: Status, prospects, and an agenda for research. Washington, DC: The National Academies Press.

National Association for the Education of Young Children (NAEYC). (2012). *Technology and Interactive Media as Tools in Early Childhood Programs Serving Children from Birth through Age 8*. Position Statement. NAEYC and the Fred Rogers Center. Retrieved August 10 2015, from http://www.naeyc.org/files/naeyc/file/positions/PS_technology_WEB2.pdf

National Education Association (NEA). (2010). *Preparing 21st Century students for a global society*. National Education Association.Retrieved November 12, 2016 from http://www.nea.org/assets/docs/A-Guide-to-Four-Cs.pdf

National Governors Association Center for Best Practices Council of Chief State School Officers. (2010). *Common Core State Standards (CCSS)*. National Governors Association Center for Best Practices, Council of Chief State School Officers.

National Research Council (NRC). (2000). *How people learn. Brain, mind, experience, and school*. Washington, DC: The National Academies Press.

National Research Council (NRC). (2000). *Inquiry and the national science education standards*. Washington, DC: National Academy Press.

National Research Council (NRC). (2012). *A Framework for K-12 Science Education: Practices, Crosscutting Concepts, and Core Ideas*. Committee on a Conceptual Framework for New K-12 Science Education Standards. Board on Science Education, Division of Behavioral and Social Sciences and Education. Washington, DC: The National Academies Press.

National Research Council (NRC). (2012). *A framework for K-12 science education: Practices, crosscutting concepts, and core ideas*. Washington, DC: The National Academies Press.

National Research Council. (2003). Improving undergraduate instruction in science, technology, engineering, and mathematics: Report of a workshop. Steering Committee on Criteria and Benchmarks for Increased Learning from Undergraduate STEM Instruction. In R. A. McCray, R. DeHaan, & J. A. Schuck (Eds.), Committee on Undergraduate Science Education, Center for Education, Division of Behavioral and Social Sciences and Education. Washington, DC: The National Academies Press.

National Research Council. (2007). *Taking science to school: Learning and teaching science in grades K-8*. Washington, DC: The National Academies Press.

National Research Council. (2011). Learning Science Through Computer Games and Simulations. Washington, DC: The National Academies Press.

National Research Council. (2011). *Successful STEM Education: A workshop summary. A. Beatty, Rapporteur. Committee on Highly Successful Schools or Programs for K-12 STEM Education, Board on Science Education and Board on Testing and Assessment. Division of Behavioral and Social Sciences and Education*. Washington, DC: The National Academies Press.

National Research Council. (2012). *A Framework for K-12 science education: Practices, crosscutting concepts, and core ideas. Committee on a Conceptual Framework for New K-12 Science Education Standards. Board on Science Education, Division of Behavioral and Social Sciences and Education*. Washington, DC: The National Academies Press.

National Research Council. (2012). *A Framework for K-12 Science Education: Practices, Crosscutting Concepts, and Core Ideas*. Washington, DC: The National Academies Press.

National Research Council. (2013). *Monitoring progress toward successful K-12 STEM education: A nation advancing? Committee on the Evaluation Framework for Successful K-12 STEM Education. Board on Science Education and Board on Testing and Assessment, Division of Behavioral and Social Sciences and Education*. Washington, DC: The National Academies Press.

National Research Council. (2013). *Next Generation Science Standards: For States, by States*. Washington, DC: The National Academies Press.

Natural Science and Engineering Research Council of Canada. (2010). *Women in Science and Engineering in Canada*. Ottawa, Canada: Author.

Nersessian, N. J. (2008). *Creating Scientific Concepts*. Cambridge, MA: The MIT Press.

NGSS Lead States. (2013). *Next generation science standards: For states, by states. The Need for Standards*. Retrieved November 12, 2016 from http://www.nextgenscience.org/need-standards

NGSS Lead States. (2013). *Next Generation Science Standards: For States, By States*. Washington, DC: The National Academies Press.

Ng, W. (2011). Why digital literacy is important for science teaching and learning. *Teaching Science*, *57*(4), 26–32.

Nicolaou, C. T., Nicolaidou, I. A., Zacharia, Z. C., & Constantinou, C. P. (2007). Enhancing fourth graders' ability to interpret graphical representations through the use of microcomputer-based labs implemented within an inquiry-based activity sequence. *Journal of Computers in Mathematics and Science Teaching*, *26*(1), 75.

Nielsen, K. H. (2013). Scientific communication and the nature of science. *Science & Education*, *22*(9), 2067–2086. doi:10.1007/s11191-012-9475-3

Nielsen, W., Miller, A. K., & Hoban, G. (2015). Science teachers response to the digital education revolution. *Journal of Science Education and Technology*, *24*(4), 417–431. doi:10.1007/s10956-014-9527-3

Niess, M. L. (2005). Preparing teachers to teach science and mathematics with technology: Developing a technology pedagogical content knowledge. *Teaching and Teacher Education*, *21*(5), 509–523. doi:10.1016/j.tate.2005.03.006

Nikolopoulou, K., & Gialamas, V. (2009). Investigating pre-service early childhood teachers views and intentions about integrating and using computers in early childhood settings: Compilation of an instrument. *Technology, Pedagogy and Education*, *18*(2), 201–219. doi:10.1080/14759390903003837

Nikolopoulou, K., & Gialamas, V. (2015a). Barriers to the integration of computers in early childhood settings: Teachers perceptions. *Education and Information Technologies*, *20*(2), 285–301. doi:10.1007/s10639-013-9281-9

Nikolopoulou, K., & Gialamas, V. (2015b). ICT and play in preschool: Early childhood teachers beliefs and confidence. *International Journal of Early Years Education*, *23*(4), 409–425. doi:10.1080/09669760.2015.1078727

Nir-Gal, O., & Klein, P. (1999). The Use of the Computer in Early Childhood with or without Adult Mediation–the role of the teacher in the computerized environment. [Hebrew]. *Dapim*, *29*, 76–99.

Nordine, J. (2011). Motivating calculus-based kinematics instruction with Super Mario Bros. *The Physics Teacher*, *49*(6), 380–382. doi:10.1119/1.3628271

Norman, D. (1983). Some observations on mental models. In D. Gentner & A. Stevens (Eds.), *Mental Models* (pp. 7–14). Hillsdale, NJ: Lawrence Erlbaum Associates.

Nuemann, M. M., & Nuemann, D. L. (2014). Touch screen tablets and emergent literacy. *Early Childhood Education Journal*, *42*(4), 231–239. doi:10.1007/s10643-013-0608-3

O'Malley, C., Vavoula, G., Glew, J., Taylor, J., Sharples, M., Lefrere, P., Lonsdale, P., Naismith, L., & Waycott, J. (2005). *Guidelines for learning/teaching/tutoring in a mobile environment.* Public deliverable from the MOBILearn project (D.4.1).

OCDE. (2015). *Students, Computers and Learning: Making the Connection*. New-York, NY: OCDE Publishing. doi:10.1787/9789264239555-en

OECD. (2014). *PISA 2012 Results: What students know and can do – student performance in mathematics reading and science* Retrieved from http://www.oecd.org/pisa/keyfindings/pisa-2012-results.htm

Olson, G. M., Zimmerman, A., & Bos, N. (2008). *Scientific collaboration on the Internet.* The MIT Press. doi:10.7551/mitpress/9780262151207.001.0001

Olson, S., & Loucks-Horsley, S. (Eds.). (2000). *Inquiry and the National Science Education Standards: A Guide for Teaching and Learning*. National Academies Press.

Onah, D. F. O., Sinclair, J., & Boyatt, R. (2014). Dropout rates of massive open online courses: behavioral patterns. Proceedings of EDULEARN14, 5825-5834.

Onuoha-Cajetan. (2007). *Meta-analysis of the effectiveness of computer-based laboratory versus traditional hands-on laboratory in college and pre-college science instructions* (Unpublished doctoral dissertation). Minneapolis Capella University, Minneapolis, MN.

Orr, J., Kamdar, D., Lewis Presser, A., & Vahey, P. (2015). Next Generation Preschool Math: Researchers, public media, teachers, and children team up to identify the best practices for tablet technology in the classroom. In D. Rutledge & D. Slykhuis (Eds.), *Proceedings of Society for Information Technology & Teacher Education International Conference 2015* (pp. 2022-2027). Chesapeake, VA: Association for the Advancement of Computing in Education (AACE). Retrieved November 12, 2016 from https://www.learntechlib.org/p/150279

Osborne, J., Simon, S., Christodoulou, A., Howell-Richardson, C., & Richardson, K. (2013). Learning to argue: A study of four schools and their attempt to develop the use of argumentation as a common instructional practice and its impact on students. *Journal of Research in Science Teaching*, *50*(3), 315–347. doi:10.1002/tea.21073

Papert, S., & Harel, I. (1991). Situating Constructionism. In *Constructionism* (pp. 193–206). Ablex Publishing Corporation. Retrieved from http://www.papert.org/articles/SituatingConstructionism.html

Papert, S. (1980). *Mindstorms: Children, computers, and powerful ideas*. Basic Books, Inc.

Partnership for 21st Century Skills. (2008). *A report and mile guide for 21st Century skills*. Retrieved October 15, 2010, from Http://www.21stCenturyskills.org/downloads/P21_Report.pdf

Partnership for 21st Century Skills. (2009). *Twenty-first Century Skills Map*. National Science Teachers Association.

Pashler, H., Bain, P. M., Bottge, B. A., Graesser, A., Koedinger, K., McDaniel, M., & Metcalfe, J. (2007). *Organizing instruction and study to improve student learning. IES Practice Guide (NCER 2007-2004)*. Jessup, MD: National Center for Education Research.

Pask, C. (2003). Mathematics and the science of analogies. *American Journal of Physics*, *71*(6), 526–534. doi:10.1119/1.1545761

Patarakin, E., & Visser, L. (2012). New Tools for Learning - The Use of Wiki's. In L. Visser, Y. L. Visser, R. Amirault, & M. Simonsom (Eds.), Trends and Issues in Distance Education 2nd Edition International Perspectives (pp. 287–299). Information Age Publishing.

Patrick, H., Mantzicopoulos, P., & Samarapungavan, A. (2009). Motivation for learning science in kindergarten: Is there a gender gap and does integrated inquiry and literacy instruction make a difference. *Journal of Research in Science Teaching*, *46*(2), 166–191. doi:10.1002/tea.20276

Pea, R. D. (2002). Learning science through collaborative visualization over the Internet. In N. Ringertz (Ed.), *Nobel Symposium: Virtual museums and public understanding of science and culture*. Stockholm, Sweden: Nobel Academy Press.

Pea, R. D., Gomez, L. M., Edelson, D. C., Fishman, B. J., Gordin, D. N., & O'Neill, D. K. (1997). Science education as a driver of cyberspace technology development. In K. C. Cohen (Ed.), *Internet links for science education* (pp. 189–220). New York, NY: Plenum. doi:10.1007/978-1-4615-5909-2_12

Pedro, F. (2012). Trusting the unknown: the effect of technology use in education. In S. Dutta & B. Bilbao-Osorio (Eds.), *Global Information Technology Report. Living in a Hyperconnected World* (pp. 135–148). Geneva, Switzerland: SRO-Kunding.

Pellegrino, J. W., & Hilton, M. L. (2013). *Education for life and work: Developing transferable knowledge and skills in the 21st century. National Research Council of the National Academies.* Washington, D.C.: National Academies Press.

Penuel, W. R., & Means, B. (2004). Implementation variation and fidelity in an inquiry science program: Analysis of GLOBE data reporting patterns. *Journal of Research in Science Teaching*, *41*(3), 294–315. doi:10.1002/tea.20002

Periodic Table. (n.d.). Retrieved January 10, 2017, from http://www.printablee.com/postpic/2015/10/periodic-table-with-oxidation-numbers_400959.png

Perkins, D. (2009). *Making learning whole: How seven principles of teaching can transform education.* San Francisco, CA: Jossey-Bass.

Perkins, K., Adams, W., Dubson, M., Finkelstein, N., Reid, S., Wieman, C., & LeMaster, R. (2006). PhET: Interactive simulations for teaching and learning physics. *The Physics Teacher*, *44*(1), 18–23. doi:10.1119/1.2150754

Perrin, A. (2015). *Pew Research Center: Internet, Science and Technology.* Retrieved from http://www.pewinternet.org/2015/10/08/social-networking-usage-2005-2015

Peters, M. A. (2006). Towards philosophy of technology in education: Mapping the field. In *The international handbook of virtual learning environments* (pp. 95–116). Springer Netherlands. doi:10.1007/978-1-4020-3803-7_3

Pierri, E., Karatrantou, A., & Panagiotakopoulos, C. (2008). Exploring the phenomenon of change of phase of pure substances using the Microcomputer-Based-Laboratory (MBL) system. *Chemistry Education Research and Practice*, *9*(3), 234–239. doi:10.1039/B812412B

Plass, J. L., Homer, B. D., & Hayward, E. (2009). Design factors for educationally effective animations and simulations. *Journal of Computing in Higher Education*, *21*(1), 31–61. doi:10.1007/s12528-009-9011-x

Pollock, S. J. (2004). *No Single Cause: Learning Gains, Student Attitudes, and the Impacts of Multiple Effective Reforms.* Paper presented at the Physics Education Research Conference, Sacramento, CA. doi:10.1063/1.2084720

Prestridge, S. (2009). Teachers talk in professional development activity that supports change in their pedagogical beliefs and practices. *Teacher Development*, *13*(1), 43–55. doi:10.1080/13664530902858493

Psycharis, S. (2013). Exploring the effects of the computational experiment approach to the epistemic beliefs, the motivation, the use of modeling indicators and conceptual understanding in three different computational learning environments. *Journal of Education and Training Studies*, *1*(1), 69–87. doi:10.11114/jets.v1i1.32

Puente, S. M. G., van Eijck, M., & Jochems, W. (n.d.). A sampled literature review of design-based learning approaches: a search for key characteristics. *International Journal of Technology and Design Education*, 1–16.

Puentedura, R. R. (2011). The iPad and Research Design (Data / Research Sessions I, II, III). In Leveraging Learning: The iPad in Primary Grades. Auburn, ME: Academic Press.

Punie, Y., Zinnbauer, D., & Cabrera, M. (2006). *A Review of the Impact of ICT on Learning*. Bruxelles, Belgium: Joint Research Centre.

Qablan, A. M., Abuloum, A., & Al-Ruz, J. A. (2009). Effective integration of ICT in Jordanian schools: An analysis of pedagogical and contextual impediments in the science classroom. *Journal of Science Education and Technology*, *18*(3), 291–300. doi:10.1007/s10956-009-9151-9

Quitadamo, I., Kurtz, M., Cornell, C., Griffith, L., Hancock, J., & Egbert, B. (2011). Critical-thinking grudge match: Biology vs. chemistry - examining factors that affect thinking skill in non-majors science. *Journal of College Science Teaching*, *40*(3), 19–25.

Radder, H. (2009). The philosophy of scientific experimentation: a review. *Automated Experimentation, 1*(2), 2-9.

Radetsky, J. S., Schumacher, J., & Zuckerman, B. (2015). Mobile and interactive media use by young children: The good, the bad, and the unknown. *Pediatrics*, *135*(1), 1–3. doi:10.1542/peds.2014-2251 PMID:25548323

Radicchi, F., Castellano, C., Cecconi, F., Loreto, V., & Parisi, D. (2004). Defining and identifying communities in networks. *Proceedings of the National Academy of Sciences of the United States of America*, *101*(9), 2658–2663. doi:10.1073/pnas.0400054101 PMID:14981240

Raes, A., Schellens, T., & De Wever, B. (2014). Web-based collaborative inquiry to bridge gaps in secondary science education. *Journal of the Learning Sciences*, *23*(3), 316–347. doi:10.1080/10508406.2013.836656

Ragupathi, K. (2014). Virtually Vygotsky: Using technology to scaffold student learning. *Technology in Pedagogy*, *20*, 1–9.

Razzouk, R., & Shute, V. (2012). What is design thinking and why is it important? *Review of Educational Research*, *82*(3), 330–348. doi:10.3102/0034654312457429

Redish, E. F., Saul, J. M., & Steinberg, R. N. (1998). Student expectations in introductory physics. *American Journal of Physics*, *66*(3), 212–224. doi:10.1119/1.18847

Redish, E., & Kuo, E. (2015). Language of physics, language of math: Disciplinary culture and dynamic epistemology. *Science & Education*, *24*(5-6), 561–590. doi:10.1007/s11191-015-9749-7

Reid, N., & Yang, M. J. (2002). Open-ended problem solving in school chemistry: A preliminary investigation. *International Journal of Science Education*, *24*(12), 1313–1332. doi:10.1080/09500690210163189

Remmer, A. M., & Jernstedt, C. (1982). Comparative effectiveness of simulation games in secondary and college level instruction: A meta-analysis. *Psychological Reports, 51*(3), 742. doi:10.2466/pr0.1982.51.3.742

Resnick, M. (1997). *Turtles, termites, and traffic jams: explorations in massively parallel microworlds.* MIT Press.

Resnick, M. (2003). Thinking Like a Tree (and Other Forms of Ecological Thinking. *International Journal of Computers for Mathematical Learning, 8*(1), 43–62. doi:10.1023/A:1025632719774

Reynolds, R., & Caperton, I. H. (2009). Comparison of Middle School, High School and Community College Students' Wiki Activity in Globaloria-West Virginia: (Pilot Year-two). In *Proceedings of the 5th International Symposium on Wikis and Open Collaboration* (p. 29:1–29:2). New York, NY: ACM. https://doi.org/10.1145/1641309.1641350

Reynolds, M. A., Jackson Cotwright, C., Polhamus, B., Gertel-Rosenberg, A., & Chang, D. (2013). Obesity prevention in the early care and education setting: Successful initiatives across a spectrum of opportunities. *The Journal of Law, Medicine & Ethics, 41*(2), 8–18. doi:10.1111/jlme.12104 PMID:24446993

Richardson, T., & DeCoito, I. (2017, January). *Teachers and technology in STEM education – Present practice and future directions.* Paper presented at the annual meeting of the Association for Science Teacher Education, Des Moines, IA.

Rieber, L. P. (2004). Multimedia learning in games, simulations and microworlds. In R. E. Mayer (Ed.), *The Cambridge Handbook of Multimedia Learning* (pp. 549–568). New-York, NY: Cambridge University Press.

Rikala, J., & Kankaanranta, M. (2014). The Nature Tour mobile learning application. Implementing the mobile application in Finnish early childhood education settings. In *CSEDU 2014: Proceedings of the 6th International Conference on Computer Supported Education* (vol. 3). SCITEPRESS-Science and Technology Publications.

Ritter, S., Anderson, J. R., Koedinger, K. R., & Corbett, A. (2007). Cognitive Tutor: Applied research in mathematics education. *Psychonomic Bulletin & Review, 14*(2), 249–255. doi:10.3758/BF03194060 PMID:17694909

Robson, J. (2016). Engagement in structured social space: An investigation of teachers online peer-to-peer interaction. *Learning, Media and Technology, 41*(1), 119–139. doi:10.1080/17439884.2015.1102743

Rocke, A. J. (2010). *Image and Reality. Kekulé, Kopp, and the Scientific Imagination.* Chicago, London: The University of Chicago Press. doi:10.7208/chicago/9780226723358.001.0001

Romanenko, V., Neverov, V., & Nikitina, G. (2012). Computer-assisted laboratory instruction at the tertiary level. In S. Abramovich (Ed.), *Computers in education* (Vol. 2, pp. 91–105). New York, NY: Nova Science Publishers.

Rondon, S., Sassi, F., & deAndrade, C. (2013). Computer game-based and traditional learning method: A comparison regarding students' knowledge retention. *BMC Medical Education*, *13*(30), 1–8. PMID:23442203

Rooij, S. W. (2009). Scaffolding project-based learning with the project management body of knowledge (PMBOK). *Computers & Education*, *52*(1), 210–219. doi:10.1016/j.compedu.2008.07.012

Roskos, K., & Neuman, S. B. (2011). The classroom environment. *The Reading Teacher*, *65*(2), 110–114. doi:10.1002/TRTR.01021

Rowland, A., Ellis, J., Craig-Hare, J., Ault, M., Bradley, B., & Bulgren, J. (2015). *Social media for learning: Platform preferences, digital citizenship and authenticity*. Research paper presented at the annual meeting of the International Society for Technology in Education (ISTE), Philadelphia, PA.

Ruggiero, D., & Mong, C. J. (2015). The teacher technology integration experience: Practice and reflection in the classroom. *Journal of Information Technology Education Research*, *14*, 161–178.

Rumelhart, D. E., Smolensky, P., McClelland, J. L., & Hinton, G. E. (1986). Schemata and Sequential Thought Processes in PDP models. In J.L. McClelland, D.E. Rumelhart, & The PDP Research Group (Eds.), Parallel Distributed Processing. Explorations in the Microstructure of Cognition: Vol. 2. Psychological and Biological Models (pp. 7–57). Cambridge, MA: The MIT Press.

Ruthven, K., Deaney, R., & Hennessy, S. (2009). Using graphing software to teach about algebraic forms: A study of technology-supported practice in secondary-school mathematics. *Educational Studies in Mathematics*, *71*(3), 279–297. doi:10.1007/s10649-008-9176-7

Sadler, P. M., Sonnert, G., Hazari, Z., & Tai, R. (2012). Stability and volatility of STEM career interest in high school: A gender study. *Science Education*, *96*(3), 411–427. doi:10.1002/sce.21007

Salomon, G., Perkins, D. N., & Globerson, T. (1991). Partners in cognition: Extending human intelligence with intelligent technologies. *Educational Researcher*, *20*(3), 2–9. doi:10.3102/0013189X020003002

Samarapungavan, A., Mantzicopoulos, P., & Patrick, H. (2008). Learning science through inquiry in kindergarten. *Science Education*, *92*(5), 868–908. doi:10.1002/sce.20275

Sánchez, A., & Brändle, C. (2014). *More network science for teenagers*. Retrieved from http://arxiv.org/abs/1403.3618

Sasseville, B. (2004). Integrating Information and Communication Technology in classroom: A comparative discourse analysis. *Canadian Journal of Learning and Technology*, *30*(2), 5–27.

Savery, J. R., & Duffy, T. M. (1995). Problem based learning: An instructional model and its constructivist framework. *Educational Technology*, *35*(5), 31–38.

Schank, P., & Kozma, R. (2002). Learning chemistry through the use of a representation-based knowledge building environment. *Journal of Computers in Mathematics and Science Teaching*, *21*(3), 253–279.

Schauer, F., Ožvoldová, M., & Lustig, F. (2009). Integrated e-Learning - New Strategy of Cognition of Real World in Teaching Physics. *World Innovations in Engineering Education and Research*, *9*, 119–135.

Schippers, M. C., Den Hartog, D. N., & Koopman, P. L. (2007). Reflexivity in Teams: A Measure and Correlates. *Applied Psychology*, *56*(2), 189–211. doi:10.1111/j.1464-0597.2006.00250.x

Schmidt, W. H., Blömeke, S., Tatto, M. T., Hsieh, F.-J., Cogan, L. S., Houang, R. T., & Schwille, J. (2011). *Teacher education matters: A study of middle school mathematics teacher preparation in six countries*. New York: Teachers College Press.

Schneider, D., Class, B., Benetos, K., Lange, M., Schneider, D., Class, B., … Lange, M. (2012). *Requirements for learning scenario and learning process analytics*. Presented at the World Conference on Educational Multimedia, Hypermedia and Telecommunications. Retrieved from http://edutechwiki.unige.ch/en/Learning_process_analytics

Schneider, M., & Stern, E. (2010). The cognitive perspective on learning: Ten cornerstone findings. In OCDE (Ed.), The nature of learning: Using research to inspire practice (pp. 69-90). Paris, France: OCDE.

Schoenfeld, A. H. (1985). *Mathematical problem solving*. New York: Academic press.

Schroeder, C. M., Scott, T. P., Tolson, H., Huang, T. Y., & Lee, Y. H. (2007). A meta-analysis of national research: Effects of teaching strategies on student achievement in science in the United States. *Journal of Research in Science Teaching*, *44*(10), 1436–1460. doi:10.1002/tea.20212

Schuster, D., Undreiu, A., Adams, B., Brookes, D., & Milner-Bolotin, M. (2009). Motion-Matching: A Challenge Game to Generate Motion Concepts. *The Physics Teacher*, *47*(7), 381–385. doi:10.1119/1.3225502

Schwab, J. J. (1960). Inquiry, the science teacher, and the educator. *The School Review*, *68*(2), 176–195. doi:10.1086/442536

Schwab, J. J. (1962). The teaching of science as enquiry. In J. J. Schwab & P. Brabdwein (Eds.), *The Teaching of Science*. Cambridge, MA: Harvard University Press.

Schwartz, D. L. (1999). The productive agency that drives collaborative learning. In I. P. Dillenbourg (Ed.), *Collaborative learning: Cognitive and computational approaches* (pp. 197–218). Elsevier Science/Permagon.

Schwartz, D. L., & Black, J. B. (1996). Shuttling between depictive models and abstract rules: Induction and fall-back. *Cognitive Science*, *20*(4), 457–497. doi:10.1207/s15516709cog2004_1

Scott, C. (2012). An investigation of science, technology, engineering and mathematics (STEM) focused high schools in the US. *Journal of STEM Education Innovations and Research, 13*(5), 30–39.

Scott, S. (2012). Go ahead. Be social: Using social media to enhance the twenty-first century classroom. *Distance Learning, 9*(2), 54.

Seel, N. (2003). Model-Centered Learning and Instruction. *Tech., Inst. Cognition and Learning, 1*, 59–85.

Seel, N. (2012). Dynamic Modeling and Analogies. In N. Seel (Ed.), *Encyclopedia of the Sciences of Learning* (pp. 1051–1054). New York, NY: Springer.

Segers, E., & Verhoeven, L. (2002). Multimedia support of early literacy learning. *Computers & Education, 39*(3), 207–221. doi:10.1016/S0360-1315(02)00034-9

Seiter, E. (2008). Practicing at home: Computers, pianos, and cultural capital. In T. McPherson (Ed.), *Digital youth, innovations, and the unexpected* (pp. 27–52). Cambridge, MA: MIT Press.

Séjourné, A. (2002). Concepts of hypermedia in physics: The importance of resources from the point of view of learning. *Educational Media International, 39*(3-4), 317–329. doi:10.1080/09523980210166035

Sengupta, P., & Wilensky, U. (2009). Learning electricity with NIELS: Thinking with electrons and thinking in levels. *International Journal of Computers for Mathematical Learning, 14*(1), 21–50. doi:10.1007/s10758-009-9144-z

Settlage, J. (2007). Demythologizing science teacher education: Conquering the false ideal of open inquiry. *Journal of Science Teacher Education, 18*(4), 461–467. doi:10.1007/s10972-007-9060-9

Sfard, A. (1994). Reification as the birth of metaphor. *For the Learning of Mathematics, 14*(1), 44–55.

Shams, L., & Seitz, A. R. (2008). Benefits of multisensory learning. *Trends in Cognitive Sciences, 12*(11), 411–417. doi:10.1016/j.tics.2008.07.006 PMID:18805039

Sharples, M., Taylor, J., & Vavoula, G. (2005). Towards a theory of mobile learning. *Proceedings of mLearn*, 1(1), 1-9.

Sheetz, L., Dunham, V., & Cooper, J. (2015). Professional development for network science as a multi-disciplinary curriculum tool. In *2015 IEEE Integrated STEM Education Conference (ISEC)* (pp. 178–182). https://doi.org/10.1109/ISECon.2015.7119919

Shelton, B. E. (2015). TPCK (Technological Pedagogical Content Knowledge): Implications for 21st century teacher education. In M. J. Spector (Ed.), *The SAGE Encyclopedia of Educational Technology* (Vol. 2, pp. 785–790). Los Angeles, CA: SAGE Publications.

Sheppard, S. D., Macatangay, K., Colby, A., & Sullivan, W. M. (2009). *Educating engineers*. San Francisco, CA: Jossey-Bass.

Sherin, B. (2006). Common sense clarified: The role of intuitive knowledge in physics problem solving. *Journal of Research in Science Teaching*, *43*(6), 535–555. doi:10.1002/tea.20136

Shih, J. L., Chu, H. C., Hwang, G. J., & Kinshuk, . (2011). An investigation of attitudes of students and teachers about participating in a context-aware ubiquitous learning activity. *British Journal of Educational Technology*, *42*(3), 373–394. doi:10.1111/j.1467-8535.2009.01020.x

Shroff, R. H., Vogel, D. R., & Coombes, J. (2008). Assessing individual-level factors supporting student intrinsic motivation in online discussions: A qualitative study. *Journal of Information Systems Education*, *19*(1), 111–125.

Shulman, L. S. (1986). Those who understand: Knowledge growth in teaching. *Educational Researcher*, *15*(2), 4–14. doi:10.3102/0013189X015002004

Shulman, L. S. (1986a). Paradigms and research programs in the study of teaching: A contemporary perspective. In M. C. Wirrtock (Ed.), *Handbook of Research on Teaching* (pp. 3–36). New York: Collier Macmillan.

Shulman, L. S. (1987). Knowledge and teaching: Foundations of the new reform. *Harvard Educational Review*, *57*(1), 1–22. doi:10.17763/haer.57.1.j463w79r56455411

Shulman, L. S. (2005). Signature pedagogies in the professions. *Daedalus*, *134*(3), 52–59. doi:10.1162/0011526054622015

Simondon, G. (1958). *Du mode d'existence des objets techniques*. Paris: Aubier.

Singer, S., Hilton, M., & Schweingruber, H. (Eds.). (2005). *America's Lab Report: Investigations in high school science. Committee on High School Science Laboratories: Role and Vision, Board on Science Education, Center for Education, Division of Behavioral and Social Sciences and Education, National Research Council*. Washington, DC: The National Academies Press.

Siraj-Blatchford, I., & Siraj-Blatchford, J. (2006). *A guide to developing the ICT curriculum for early childhood education*. Stoke-on-Trent, UK: Trentham Books.

Sitthiworachart, J., & Joy, M. (2003). Web-based peer assessment in learning computer programming. In *Advanced Learning Technologies, 2003. Proceedings. The 3rd IEEE International Conference on* (pp. 180–184). Retrieved from http://ieeexplore.ieee.org/xpls/abs_all.jsp?arnumber=1215052

Skaalvik, E. M., & Skaalvik, S. (2008). Teacher self-efficacy: conceptual analysis and relations with teacher burnout and perceived school context. In R. Craven, H. W. Marsh, & D. McInerney (Eds.), *Self-processes, learning, and enabling human potential* (pp. 223–247). Information Age Publishing.

Slavin, R. E. (1996). Research on cooperative learning and achievement: What we know, what we need to know. *Contemporary Educational Psychology*, *21*(1), 43–69. doi:10.1006/ceps.1996.0004

Smetana, L. K., & Bell, R. L. (2012). Computer simulations to support science instruction and learning: A critical review of the literature. *International Journal of Science Education*, *34*(9), 1337–1370. doi:10.1080/09500693.2011.605182

Sokolowski, J. A., & Banks, C. M. (Eds.). (2010). *Modeling and Simulation Fundamentals. Theoretical Underpinnings and Practical Domains*. Hoboken, NJ: John Wiley & Sons. doi:10.1002/9780470590621

Songer, N., Lee, H.-S., & Kam, R. (2002). Technology-rich inquiry science in urban classrooms: What are the barriers to inquiry pedagogy? *Journal of Research in Science Teaching, 39*(2), 128–150. doi:10.1002/tea.10013

Song, Y. (2014). Bring Your Own Device (BYOD) for seamless science inquiry in a primary school. *Computers & Education, 74*, 50–60. doi:10.1016/j.compedu.2014.01.005

So, W.-M. W., & Ching, N.-Y. F. (2011). Pupil science learning in resource-based e-learning environments. *Journal of Computers in Mathematics and Science Teaching, 30*(2), 203–223.

Sparrow, B., Liu, J., & Wegner, D. M. (2011). Google effects on memory: Cognitive consequences of having information at our fingertips. *Science, 333*(6043), 776–778. doi:10.1126/science.1207745 PMID:21764755

Spektor-Levy, O. (2012). Chashiva al machshev bagan [Thoughts about ICT in preschool classes]. *Alon Da-Ga, 5*, 56–69.

Steinberg, R. (2000). Computers in teaching science: To simulate or not to simulate? *American Journal of Physics, 68*(S1S37), S37–S41. doi:10.1119/1.19517

Stein, M. K., Grover, B. W., & Henningsen, M. (1996). Building student capacity for mathematical thinking and reasoning: An analysis of mathematical tasks used in reform classrooms. *American Educational Research Journal, 33*(2), 455–488. doi:10.3102/00028312033002455

Stephen, C. (2006). *Early Years Education: Perspectives from a Review of the International Literature*. Academic Press.

Stephen, C., & Plowman, L. (2008). Enhancing learning with information and communication technologies in pre-school. *Early Child Development and Care, 178*(6), 637–654. doi:10.1080/03004430600869571

Stieff, M., & Wilensky, U. (2003). Connected Chemistry - incorporating interactive simulations into the chemistry classroom. *Journal of Science Education and Technology, 12*(3), 285–302. doi:10.1023/A:1025085023936

Strasburger, V. (2015). Should babies be watching and using screens? The answer is surprisingly complicated. *Acta Paediatrica (Oslo, Norway), 104*(10), 967–968. doi:10.1111/apa.13105 PMID:26378634

Strauss, A., & Corbin, J. (1990). *Basics of Qualitative Research: Grounded Theory Procedures and Techniques*. Newbury Park, CA: Sage Publications.

Swackhamer, L., Koellner, K., Basile, C., & Kimbrough, D. (2009, Spring). Increasing the self-efficacy of in-service teachers through content knowledge. *Teacher Education Quarterly*, 63-78.

Swaminathan, S., & Wright, J. L. (2003). Education technology in the early and primary years. In J. P. Isenberg & M. R. Jalongo (Eds.), *Major trends and issues in early childhood education: Challenges, controversies, and insights* (2nd ed.; pp. 136–149). New York: Teachers College Press.

Sweller, J. (2010). Element interactivity and intrinsic, extraneous and germane cognitive load. *Educational Psychology Review*, *22*(2), 123–138. doi:10.1007/s10648-010-9128-5

Sweller, J., & van Merriënboer, J. (2005). Cognitive load theory and complex learning: Recent developments and future direction. *Educational Psychology Review*, *17*(2), 147–177. doi:10.1007/s10648-005-3951-0

Sweller, J., van Merriënboer, J., & Paas, F. (1998). Cognitive architecture and instructional design. *Educational Psychology Review*, *10*(3), 251–296. doi:10.1023/A:1022193728205

Szpunar, K. K., Spreng, R. N., & Schacter, D. L. (2014). A taxonomy of prospection: Introducing an organizational framework for future-oriented cognition. *Proceedings of the National Academy of Sciences of the United States of America*, *111*(52), 18414–18421. doi:10.1073/pnas.1417144111 PMID:25416592

Taber, K. S. (2008). Conceptual Resources for Learning Science: Issues of Transience and Grain-Size in Cognition and Cognitive Structure. *International Journal of Science Education*, *30*(8), 1027–1053. doi:10.1080/09500690701485082

Tamim, R. M., Bernard, R. M., Borokhovski, E., Abrami, P. C., & Schmid, R. F. (2011). What forty years of research says about the impact of technology on learning a second-order meta-analysis and validation study. *Review of Educational Research*, *81*(1), 4–28. doi:10.3102/0034654310393361

Tang, K. S., Delgado, C., & Moje, E. B. (2014). An integrative framework for the analysis of multiple and multimodal representations for meaning-making in science education. *Science Education*, *98*(2), 305–326. doi:10.1002/sce.21099

Tang, K. Y., Li, M. C., Hsin, C. T., & Tsai, C. C. (2016). A Co-Citation Network of Young Children's Learning with Technology. *Journal of Educational Technology & Society*, *19*(3), 294–305.

Tan, Y. S. M., & Caleon, I. S. (2016). Dealing with the ambiguities of science inquiry. *The Physics Teacher*, *54*(March), 168–172. doi:10.1119/1.4942141

The Scale of the Universe. (n.d.). Retrieved January 10, 2017, from http://htwins.net/scale/

Thomas, J. W. (2000). *A review of research on project-based learning executive summary*. San Rafael, CA: The Autodesk Foundation. Retrieved April 13, 2014 from http://www.bobpearlman.org/BestPractices/PBL_Research.pdf

Thomas, D. R. (2006). A General Inductive Approach for Analyzing Qualitative Evaluation Data. *The American Journal of Evaluation*, *27*(2), 237–246. doi:10.1177/1098214005283748

Thomas, K. (2013). Changing mathematics teaching practices and improving student outcomes through collaborative evaluation. *Teacher Education and Practice*, *26*(4), 778–795.

Thomas, R. A., & Hooper, E. (1991). Simulations: An opportunity we are missing. *Journal of Research on Computing in Education, 23*(4), 497–513. doi:10.1080/08886504.1991.10781978

Topi, H., Valacich, J. S., Wright, R. T., Kaiser, K., Nunamaker, J. F. Jr, Sipior, J. C., & De Vreeda, G.-J. (2010). IS 2010: Curriculum Guidelines for Undergraduate Degree Programs in Information Systems. *Communications of the Association for Information Systems, 26*, 359–428.

Toulmin, S. (1958, 2003). The uses of argument. Cambridge, UK: Cambridge University Press.

Toulmin, C. N., & Meghan, G. (2007). *Building a science, technology, engineering and math agenda*. Washington, DC: National Governor's Association.

Toulmin, S., Rieke, R., & Janik, A. (1984). *An introduction to reasoning*. Upper Saddle Ridge, NJ: Prentice Hall.

Trauth, E. M., Farwell, D. W., & Lee, D. (1993). The IS expectation gap: Industry expectations versus academic preparation. *Management Information Systems Quarterly, 17*(3), 293–307. doi:10.2307/249773

Tremblay, M. S., LeBlanc, A. G., Carson, V., Choquette, L., Connor Gorber, S., Dillman, C., & Kho, M. E. et al. (2012). Canadian sedentary behaviour guidelines for the early years (aged 0–4 years). *Applied Physiology, Nutrition, and Metabolism, 37*(2), 370–380. doi:10.1139/h2012-019 PMID:22448609

Trickett, S. B., & Trafton, J. G. (2007). What if...: The use of conceptual simulations in scientific reasoning. *Cognitive Science, 31*(5), 843–875. doi:10.1080/03640210701530771 PMID:21635319

Tricot, A. (2007). Apprentissages et documents numériques. Paris, France: Belin.

Trumper, R. (2006). Factors affecting junior high school students interest in physics. *Journal of Science Education and Technology, 15*(1), 47–58. doi:10.1007/s10956-006-0355-6

Tsai, C.-C., Lin, S. S., & Yuan, S.-M. (2002). Developing science activities through a networked peer assessment system. *Computers & Education, 38*(1), 241–252. doi:10.1016/S0360-1315(01)00069-0

Tsai, C., Jack, B., Huang, T., & Yang, J. (2012). Using the cognitive apprenticeship web-based argumentation system to improve argumentation instruction. *Journal of Science Education and Technology, 21*(4), 476–486. doi:10.1007/s10956-011-9339-7

Tseng, S.-C., & Tsai, C.-C. (2007a). On-line peer assessment and the role of the peer feedback: A study of high school computer course. *Computers & Education, 49*(4), 1161–1174. doi:10.1016/j.compedu.2006.01.007

Turkle, S. (Ed.). (2008). *Falling for science: Objects in mind*. Cambridge, MA: MIT Press.

Turner, F. (2006). From counterculture to cyberculture. Stewart Brand, the whole earth network, and the rise of digital utopianism. Chicago: University of Chicago Press.

Twitter. (2016). *Twitter usage: Company facts*. Retrieved October 14, 2016, from https://about.twitter.com/company

U.S. Department of Education. (2016). *National Center for Education Statistics, Schools and Staffing Survey (SASS), "Public school teacher data file," 1987-88 through 2011-12*. Retrieved from: http://nces.ed.gov/programs/digest/d14/tables/dt14_209.10.asp?referrer=report

Uluyol, Ç., & Sahin, S. (2016). Elementary school teachers ICT use in the classroom and their motivators for using ICT. *British Journal of Educational Technology*, *47*(1), 65–75. doi:10.1111/bjet.12220

UNESCO. (2008). ICT Competency Standarts for Teachers: Policy Framework. Paris: UNESCO.

UNESCO. (2011). *ICT competency framework for teachers*. Paris: UNESCO. Retrieved from: http://iite.unesco.org/pics/publications/en/files/3214694.pdf

UNESCO. (2014). *L'éducation pour tous en Afrique subsaharienne en 2010. Rapport pour la réunion du Groupe de Haut Niveau*. Retrieved from: http://www.unesco.org/new/fr/cfit

Valanides, N., & Angeli, C. (2008). Learning and teaching about scientific models with a computer modeling tool. *Computers in Human Behavior*, *24*(2), 220–233. doi:10.1016/j.chb.2007.01.005

Van Eperen, L., & Marincola, F. M. (2011). How scientists use social media to communicate their research. *Journal of Translational Medicine*, *9*(199), 1–3. PMID:22085450

Van Joolingen, W. R., De Jong, T., & Dimitrakopoulou, A. (2007). Issues in computer supported inquiry learning in science. *Journal of Computer Assisted Learning*, *23*(2), 111–119. doi:10.1111/j.1365-2729.2006.00216.x

van Zundert, M., Sluijsmans, D., & van Merriënboer, J. (2010). Effective peer assessment processes: Research findings and future directions. *Learning and Instruction*, *20*(4), 270–279. doi:10.1016/j.learninstruc.2009.08.004

Van't Hooft, M., Swan, K., Cook, D., & Lin, Y. (2007). What is ubiquitous computing? In M. Van't Hooft & K. Swan (Eds.), *Ubiquitous computing in education*. Mahwah, NJ: Lawrence Erlbaum Associates.

Vernadakis, N., Avgerinos, A., Tsitskari, E., & Zachopoulou, E. (2005). The use of computer assisted instruction in preschool education: Making teaching meaningful. *Early Childhood Education Journal*, *33*(2), 99–104. doi:10.1007/s10643-005-0026-2

Vernier-Technology. (2016). *Logger Pro* (Version 3.6.1). Portland, OR: Vernier Technology. Retrieved from www.vernier.com

Vigliocco, G., Meteyard, L., Andrews, M., & Kousta, S. (2009). Toward a theory of semantic representation. *Language and Cognition*, *1*(2), 219–247. doi:10.1515/LANGCOG.2009.011

Vroom, R. W., & Horvath, I. (2014). *Cyber-physical augmentation: An exploration. In* Tools and methods of competitive engineering. *Digital Proceedings of the 10th International Symposium on Tools and Methods of Competitive Engineering TMCE 2014.*

Vygotsky, L. (1978). *Mind in Society: The Development of Higher Psychological Processes*. Cambridge, MA: Harvard University Press.

Vygotsky, L. S. (1978). *Mind in society*. Cambridge, MA: MIT Press.

Vygotsky, L. S. (1978). *Mind in Society: Interaction between learning and development*. Cambridge, MA: Harvard University Press.

Vygotsky, L. S. (1978). *Mind in society: The development of higher psychological processes*. Cambridge, MA: Harvard University Press.

Walter-Herrmann, J., & Büching, C. (Eds.). (2013). *FabLab: Of machines, makers and inventors*. Bielefeld, Germany: Transcript Verlag. doi:10.14361/transcript.9783839423820

Wang, Y., & Baker, R. (2015). Content or platform: Why do students complete MOOCs? *MERLOT Journal of Online Learning and Teaching*, *11*(1), 17–30.

Ward, L., & Parr, J. M. (2010). Revisiting and Reframing Use: Implications for the Integration of ICT. *Computers & Education*, *54*(1), 113–122. doi:10.1016/j.compedu.2009.07.011

Warsh, D. (2007). *Knowledge and the Wealth of Nations: A story of economic discovery*. W.W. Norton & Company.

Warwick, P., Mercer, N., Kershner, R., & Staarman, J. K. (2010). In the mind and in the technology: The vicarious presence of the teacher in pupils learning of science in collaborative group activity at the interactive whiteboard. *Computers & Education*, *55*(1), 350–362. doi:10.1016/j.compedu.2010.02.001

Wasserman, S., & Faust, K. (1994). *Social Network Analysis: Methods and Applications* (1st ed.). Cambridge University Press. doi:10.1017/CBO9780511815478

Webb, M. (2005). Affordances of ICT in science learning: Implications for an integrated pedagogy. *International Journal of Science Education*, *27*(6), 705–735. doi:10.1080/09500690500038520

Webb, N. M., Nemer, K. M., & Ing, M. (2006). Small-group reflections: Parallels between teacher discourse and student behavior in peer-directed groups. *Journal of the Learning Sciences*, *15*(1), 63–119. doi:10.1207/s15327809jls1501_8

Wegener, M., McIntyre, T. J., McGrath, D., Savage, C. M., & Williamson, W. (2012). Developing a virtual physics world. *Australasian Journal of Educational Technology*, *28*(3), 504–521. doi:10.14742/ajet.847

Weiss, I. R., & Pasley, J. D. (2004). What is high-quality instruction? *Educational Leadership*, *61*(5), 24.

Weiss, I. R., Pasley, J. D., Smith, P. S., Banilower, E. R., & Heck, D. J. (2003). *Looking inside the classroom: A study of K-12 mathematics and science education in the United States*. Chapel Hill, NC: Horizon Research.

Weiss, I., Kramarski, B., & Tallis, S. (2006). Effects of multimedia environments on kindergarten childrens mathematical achievements and style of learning. *Educational Media International*, *43*(1), 3–17. doi:10.1080/09523980500490513

Wells, S. H., Warelow, P. J., & Jackson, K. L. (2009). Problem based learning (PBL): A conundrum. *Contemporary Nurse*, *33*(2), 191–201. doi:10.5172/conu.2009.33.2.191 PMID:19929163

Wertsch, J. (1991). *Voices of the mind: A sociocultural approach to mediated action*. Cambridge, MA: Harvard University Press.

West, M. A. (1996). Reflexivity and work group effectiveness: A conceptual integration. In M. A. West (Ed.), *Handbook of work group psychology* (pp. 555–579). Chichester, UK: Wiley.

White, B. Y., & Fredericksen, J. R. (1998). Inquiry, Modeling and Metacognition: Making Science Accessible to All Students. *Cognition and Instruction*, *16*(1), 3–118. doi:10.1207/s1532690xci1601_2

Whitehead, A. N., Griffin, D. R., & Sherburne, D. W. (1979). *Process and reality: An essay in cosmology*. New York, NY: Free Press.

Wieman, C. E. (2012). Applying new research to improve science education. *Issues in Science and Technology*, (Fall): 1–9.

Wieman, C. E., Adams, W. K., Loeblein, P., & Perkins, K. K. (2010). Teaching physics using PhET simulations. *The Physics Teacher*, *48*(April), 225–227. doi:10.1119/1.3361987

Wieman, C. E., & Perkins, K. (2005). Transforming physics education. *Physics Today*, *58*(11), 36–42. doi:10.1063/1.2155756

Wierzbicka, A. (1992). *Semantics, Culture, and Cognition: Universal Human Concepts in Culture-Specific Configurations*. Oxford University Press.

Wijnen, W. (2000). Towards design-based learning. Eindhoven: Eindhoven University of Technology. Retrieved from http://w3.tue.nl/fileadmin/osc/OSC_oo/doc/OGO_brochure_1_EN.pdf

Wilensky, U., & Rand, W. (2015). *An Introduction to Agent-Based Modeling: Modeling Natural, Social, and Engineered Complex Systems with NetLogo*. Cambridge, MA: The MIT Press.

Wilensky, U., & Resnick, M. (1999). Thinking in levels: A dynamic systems perspective to making sense of the world. *Journal of Science Education and Technology*, *8*(1), 3–19. doi:10.1023/A:1009421303064

Williams, R. L. (1999). The behavioral perspective in contemporary education. *Teacher Educator*, *35*(2), 44–60. doi:10.1080/08878739909555225

Winkel, B. (2013). Computers have taken us to the brink in mathematics ... and we have balked. *Computers in the Schools*, *30*(1-2), 148–171. doi:10.1080/07380569.2013.768940

Winsberg, E. (2010). *Science in the Age of Computer Simulation*. Chicago, IL: The University of Chicago Press. doi:10.7208/chicago/9780226902050.001.0001

Wolf, F., Hobby, R., Lowry, S., Bauman, A., Franza, B. R., Lin, B., & Kolker, E. (2011). Education and data-intensive science in the beginning of the 21st century. *OMICS: A Journal of Integrative Biology*, *15*(4), 217–219. doi:10.1089/omi.2011.0009 PMID:21476844

Wong, L.-H., & Looi, C.-K. (2011). What seams do we remove in mobile-assisted seamless learning? A critical review of the literature. *Computers & Education*, *57*(4), 2364–2381. doi:10.1016/j.compedu.2011.06.007

Wright, W. (2007, March). *Will Wright: Spore, birth of a game* [Video file]. Retrieved from https://www.ted.com/talks/will_wright_makes_toys_that_make_worlds

Wright, D. (2005). Graphical Calculators: Tools for Mathematical Thinking. In S. Johnston-Wilder & D. Pimm (Eds.), *Teaching Secondary Mathematics with ICT* (pp. 145–158). New York: Open University Press.

Wu, J.-H., Chen, Y.-C., & Chang, J. (2007). Critical IS professional activities and skills/knowledge: A perspective of IS managers. *Computers in Human Behavior*, *23*(6), 2945–2965. doi:10.1016/j.chb.2006.08.008

Wu, J., & Zhang, Y. (2010). Examining potentialities of handheld technology in students academic attainments. *Educational Media International*, *47*(1), 57–67. doi:10.1080/09523981003654977

Wu, W. H., Jim Wu, Y. C., Chen, C. Y., Kao, H. Y., Lin, C. H., & Huang, S. H. (2012). Review of trends from mobile learning studies: A meta-analysis. *Computers & Education*, *59*(2), 817–827. doi:10.1016/j.compedu.2012.03.016

Xie, C. (2012a). Framing Mixed-Reality Labs. *@Concord*, *16*(1), 8-9.

Xie, C. (2012b). Interactive heat transfer simulations for everyone. *The Physics Teacher*, *50*(4), 237–240. doi:10.1119/1.3694080

Yackel, E., & Hanna, G. (2003). Reasoning and proof. In J. Kilpatrick, W. G. Martin, & D. Schifter (Eds.), *A research companion to principles and standards for school mathematics* (pp. 227–236). Reston, VA: NCTM.

Yair, Y., Mintz, R., & Litvak, S. (2001). 3D-Virtual reality in science education: An implication for astronomy teaching. *Journal of Computers in Mathematics and Science Teaching*, *20*(3), 293–305.

Yee, E., & Thompson-Schill, S. L. (2016). Putting concepts into context. *Psychonomic Bulletin & Review*. doi:10.3758/s13423-015-0948-7

Yuan, E. J. (2013). A culturalist critique of online community in new media studies. *New Media & Society*, *15*(5), 665–679. doi:10.1177/1461444812462847

Yu, F. Y., & Yu, H. J. J. (2002). Incorporating e-mail into the learning process: Its impact on student academic achievement and attitudes. *Computers & Education*, *38*(1), 117–126. doi:10.1016/S0360-1315(01)00085-9

Yu, J., Dong, K., & Zheng, Y. (2016). VLAB-C: Collaborative Virtual Laboratory in Cloud Computing and Its Applications. In *Big Data Applications and Use Cases* (pp. 145–174). Springer International Publishing. doi:10.1007/978-3-319-30146-4_7

Yun Dai, D., & Sternberg, R. J. (2004). Beyond constructivism: toward an integrated understanding of intellectual functioning and development. In D. Yun Dai & R. J. Sternberg (Eds.), *Integrative perspectives on intellectual functioning and development* (pp. 3–38). London, UK: Lawrence Erlbaum Associates.

Yurdakul, I. K., Odabasi, H. F., Kilicer, K., Coklar, A. N., Birinci, G., & Kurt, A. A. (2012). The development, validity and reliability of TPACK-deep: A technological pedagogical content knowledge scale. *Computers & Education*, *58*(3), 964–977. doi:10.1016/j.compedu.2011.10.012

Zhang, J., & Sun, Y. (2011). Quantified measures of online discourse as knowledge building indicators. In *International conference on computer-supported collaborative learning (CSCL 2011). Proceedings, Long papers*, (p. 72). Hong Kong, China: Academic Press.

Zhang, B. H., Looi, C.-K., Seow, P., Chia, G., Wong, L.-H., Chen, W., & Norris, C. et al. (2010). Deconstructing and reconstructing: Transforming primary science via a mobilized curriculum. *Computers & Education*, *55*(4), 1504–1520. doi:10.1016/j.compedu.2010.06.016

Zhang, Z., & Martinovic, D. (2008). ICT in teacher education: Examining needs, expectations and attitudes. *The Canadian Journal of Learning and Technology*, *34*(2), 149–166.

Zhao, Y. (2012). *World class learners: Educating creative and entrepreneurial students*. Corwin Press.

Zimlich, S. L. (2015). Using technology in gifted and talented education classrooms: The teachers' perspective. *Journal of Information Technology Education: Innovations in Practice*, *14*, 101–124.

Zimmer, C. (2013). Bringing them back to life. *National Geographic*. Retrieved from http://ngm.nationalgeographic.com/2013/04/125-species-revival/zimmer-text

Zion, M., Slezak, M., Shapira, D., Link, E., Bashan, N., Brumer, M., & Valanides, N. et al. (2004). Dynamic, open inquiry in biology learning. *Science Education*, *88*(5), 728–753. doi:10.1002/sce.10145

Zollman, A. (2012). Learning for STEM literacy: STEM literacy for learning. *School Science and Mathematics*, *112*(1), 12–19. doi:10.1111/j.1949-8594.2012.00101.x

Zuiker, S. J., & Wright, K. (2015). Learning in and beyond school gardens with cyber-physical systems. *Interactive Learning Environments*, *23*(5), 556–577. doi:10.1080/10494820.2015.1063512

Zydney, J. M., & Warner, Z. (2016). Mobile apps for science learning: Review of research. *Computers & Education*, *94*, 1–17. doi:10.1016/j.compedu.2015.11.001

About the Contributors

Ilya Levin received his Ph.D. degree in Computer Engineering in 1987 from the Institute of Computer Technologies, Latvian Academy of Science. From 1987 he was the Head of the Computer Science Department in the Leningrad Institute of New Technologies (in the former USSR). He moved to Israel 1990. Between 1993 and 1997, Ilya Levin was the Head of the Computer Systems Department in Holon Institute of Technology. In 1997, he worked as a Research fellow in the Computer Science Department of University of Massachusetts. During four years between 2003-2006, Dr. Ilya Levin was an Associate Professor of the School of Engineering in the Bar Ilan University. Presently, Prof. Ilya Levin is a Full Professor in the School of Education of Tel Aviv University. His recent research interests include: Computer Design, Cultural Studies of Information Society. Ilya Levin is the author of around 160 research papers both in Computer Engineering and in Humanities.

Dina Tsybulsky received her PhD degree in Science Education in 2014 from the Hebrew University, Jerusalem. Her research interests include: inquiry-based learning biology and chemistry, nature of science and science education in digital age. Currently, Dina is a lecturer and a research fellow of the School of Education of Tel Aviv University. Her postdoctoral research deals with transformations of science teachers' worldviews and their implementations in teachers' "digitalized" educational practices.

* * *

Sergei Abramovich, PhD (mathematics), is Professor in the School of Education and Professional Studies at State University of New York at Potsdam. He is the author/editor of seven books and numerous articles on the use of computers in mathematics education. His service to educational community includes membership on editorial boards of nine professional journals published in the United States and elsewhere.

Marilyn Ault, PhD, is a Senior Research Associate and Director of ALTEC at the University of Kansas Center for Research on Learning. Her research interests focus on the instructional use of existing and emerging technologies, including the design and use of educational games, to affect teaching and learning at all levels of education.

Miri Barak is an Asst. professor in the Technion IIT Faculty of Education in Science and Technology. She is the Head of the Science and Learning Technologies (SLT) group, and her research involves evaluation of 21st century skills and the use of online environments for science and engineering learning. She is a PI and co-PI of a number of national and international projects on cloud pedagogy and MOOCs - massive online open courses. Her studies examine sociocultural aspects of online collaborative learning in large and small groups, self-regulated learning in distance education, and the promotion of innovative and flexible thinking. Her academic work is presented in peer-reviewed papers and book chapters in leading international and national publications.

Michael L. Connell, Ph.D., is Professor of Mathematics Education at University of Houston-Downtown. Dr. Connell has over 30 years of mathematics education experience at both graduate and undergraduate levels working with students in field based teacher certification programs. His research interests lie at the intersection between educational technology, learning theory, and mathematics education. He has written, presented and published extensively in these areas.

Jana Craig-Hare, PhD, is an Assistant Research Professor and Associate Director of ALTEC at the University of Kansas Center for Research on Learning. Her research focuses on the increased need to develop digitally-enhanced, active learning environments through teaching and learning strategies, instructional coaching, educational games, and program evaluation.

Isha DeCoito is an Assistant Professor in Curriculum Studies in the Faculty of Education at Western University. She is the Coordinator of the Science, Technology, Engineering, and Mathematics (STEM) Specialty Focus in the Teacher Education program and is the Lead Program and Curriculum Developer for the Certificate in Teaching and Learning in Higher Education in STEM Disciplines. Her research focuses on enhancing scientific and technological literacy in different contexts for all students, and investigates approaches and tools for improving teaching and learning in and about STEM education. She has a keen interest in initiatives promoting STEM education and is currently conducting a 7-year longitudinal multi-stakeholder partnership study exploring students' attitudes and interest in STEM, their future course selections and STEM career choices.

James D. Ellis is an associate professor in the Department of Curriculum and Teaching at the University of Kansas. His research interests focus on reform in science education, science teacher education, and use of technology to enhance teaching and learning in science.

Jacques Ginestié is professor at Aix-Marseille Université (France). He is currently dean of the school of education, director of the research unit EA 4671 ADEF, director of the research federation FED 4238 SFERE-Provence and chair-holder of the Unesco chair "Sciences & Technology Education and Teacher Training". He develops researches about the processes of teaching-learning and their efficiency, and about the interactions between student, teacher and knowledge. He is involved in many international programs and has a special focus on education for all, specifically in sciences and technology education for all. He is author or co-author of 45 papers in international journals and 52 books or book's chapters, and reviewer for several international journals.

Jean-François Hérold is senior lecturer at Aix Marseille University (France) since 2014 after teaching industrial computer science at the College for nearly twenty years. His research interests focus on learning processes, in the fields of technological education and science education, in order to help teachers develop effective teaching strategies. He studies the cognitive processes that occur when students solve problems, when students learn complex knowledge (like scientific and/or technical concepts); the role of emotions on cognition (especially stress effect). He focuses his research studies mainly on teaching-learning situations with ICT tools by taking a learner-centered approach through a cognitive analysis of students' activity.

Naama Israeli is a master student at the school of education, Bar Ilan University. Her expertise is on mobile learning and her study focuses on mobile learning in preschool frameworks and scientific literacy among preschoolers.

Franco Landriscina, as an expert in the fields of Instructional design and Technology-Enhanced Learning, has been working for strengthening the relation between learning and innovation. His competencies have been pivotal in the success of projects aimed at offering organizations a competitive advantage. His academic research focuses on the role of simulation in learning, and in applying the latest results from cognitive science to education and training.

Marina Milner-Bolotin is an Associate Professor at the Department of Curriculum and Pedagogy at the University of British Columbia (UBC), Vancouver, Canada, where she teaches undergraduate (B.Ed.) and graduate STEM education courses. She also teaches in the Master of Educational Technology Program at UBC Faculty of Education – a fully online graduate program. Her areas of research include STEM education, physics education, educational technology in STEM education, and STEM teacher education. She has been teaching mathematics and science in K-12 and at the undergraduate level for more than 20 years in Ukraine, Israel, USA and Canada. She is actively involved in provincial, national and international organisations focused on improving science and mathematics education, such as American Association of Physics Teachers, Canadian Association of Physicists, Canadian Association for the Studies in Education, BC Physics Teachers' Association, etc. Her research publications include more than 40 peer-reviewed papers, five book chapters, a collection of mathematics problems for gifted students, and an undergraduate physics textbook used in Canada and internationally (Hawkes, Iqbal, Mansour, Milner-Bolotin, & Williams, 2014). She received a number of local, national, and international research grants to study the implementation of technology in mathematics and science education, including the NSTA Vernier Award (2006), HP Educational Innovation Grant (2008), UBC Teaching and Learning Enhancement Fund grant. In 2010 she received a Canadian Association of Physicists Teaching Medal for Excellence in Undergraduate Physics Teaching and in 2014 she received a UBC Faculty of Education Killam Teaching Prize. In 2015 she has been elected a Fellow of the American Association of Physics Teachers. Since 2008 she serves as an Associate Editor of the Canadian Journal of Physics (in charge of the Physics Education section). Nowadays she studies how active engagement of physics teacher-candidates with technologies promotes the development of their Technological Pedagogical Content Knowledge and their interest in student-centered science education.

Yevgeny D. Patarakin received his MS in Molecular Biology and Microbiology from the State University of Nizhny Novgorod in 1984. He received PhD from Moscow Open Pedagogical University in 1993. He was the head of Laboratory at Program Systems Institute, Russian Academy of Sciences in 1993 - 2000, Associate professor in Nizhny Novgorod State Pedagogical University 2005 - 2014. Yevgeny D. Patarakin is Co-Founder and Chief Community Officer at WikiVote! Company, and is the author of Network communities and education (2006, Per Se) and Social interaction and learning 2.0 (2009).

Netta Perry is a PhD student at the School of Education, Bar Ilan University. Her expertise is in architecture and designing learning environments. Her study focus on preschooler's views and preferences regarding the preschool learning environments and the ways the design of the learning environment influence preschoolers exploration behavior and inquiry.

Inna Plutov is a PhD student at the School of Education, Bar Ilan University. Her study and her expertise focus on the integration of technology into preschool curriculum and the development of information literacy among preschool teachers and their young students.

Tasha Richardson holds an undergraduate degree in Biophysics, and a Masters of Arts degree in Education. She is currently a PhD Candidate in the Curriculum Studies and Teacher Development program at the Ontario Institute for Studies in Education at the University of Toronto. Her research interests include: informal learning spaces, digital video games and identity development, and physics education.

Amber Rowland, PhD, is an Assistant Research Professor at the University of Kansas Center for Research on Learning and Director of Professional Learning at ALTEC. She specializes in collaborative, practical and engaging adult professional learning with an emphasis on instructional practices and the powerful integration of technology. Currently, her research and development efforts center around supporting all students in the development of writing skills and the use of social media.

Ornit Spektor-Levy is a faculty member at the Science Education Center, School of Education, Bar Ilan University. Her educational research focuses on: development of scientific curiosity and literacy; engineering thinking; information literacy; professional development of science teachers; information and communication technologies. She is the director of the Israeli National Teacher Center for STEM in Preschool.

Joycelyn Streator teaches Information Technology Management at Georgia Institute of Technology. Her research interests include information and communication technologies, digital media, and technology entrepreneurship.

Abeer Watted is a Ph.D. student at the Faculty of Education in Science and Technology at the Technion IIT. Her research focuses on multicultural collaborative learning in massive open online courses (MOOCs). She received her B.Sc. and M.Sc. in Transportation and Highways Engineering, and a second master degree in Education in Science and Technology, from the Technion IIT. Currently, Abeer works part time as a lecturer at Al-Qasemi Academic College of Education, where she serves also as the head of Civil Engineering Department.

Index

E

F

G

I

K

L

M

N

P

www.ingramcontent.com/pod-product-compliance
Ingram Content Group UK Ltd.
Pitfield, Milton Keynes, MK11 3LW, UK
UKHW050948290926
15470UKWH00005B/1706

9 781522 525257